Foundations of Engineering Mechanics

Series Editors: V. I. Babitsky, Jens Wittenburg

Leonid I. Manevitch · Oleg V. Gendelman

Tractable Models of Solid Mechanics

Formulation, Analysis and Interpretation

With 136 Figures

Springer

Prof. Leonid I. Manevitch
Russian Academy of Sciences
Inst. Chemical Physics
Kosygina St. 4
119991 Moscow
Russia

Prof. Oleg V. Gendelman
Technion Israel Institute of Technology
Fac. Mechanical Engineering
32000 Haifa
Israel
ovgend@tx.technion.ac.il

Series Editors
V.I. Babitsky
Department of Mechanical Engineering
Loughborough University
LE11 3TU Loughborough
Leicestershire
Great Britain

J. Wittenburg
Universitat Karlsruhe (TH)
Institute für Technishe
Mechanik
Kaiserstr. 12
76128 Karlsruhe
Germany

ISSN 1612-1384
e-ISSN 1860-6237
ISBN 978-3-642-26662-1
ISBN 978-3-642-15372-3 (eBook)
DOI 10.1007/978-3-642-15372-3
Springer Heidelberg Dordrecht London New York

Cover design: deblik, Berlin

Printed on acid-free paper

Springer is part of Springer Science+Business Media (www.springer.com)

Preface

All what is necessary is simple and all what is too complex is not necessary
Nikolay Timofeev – Resovsky

To clarify the main objective of this book, we would like to compare the viewpoints of two famous Russian scientists.

A.M. Lyapunov wrote that after its formulation, every mechanical problem has to be solved analytically as a problem of pure Mathematics in all its rigor, no matter how complex this problem will be. The idea of N.E. Zhukovsky was that mechanicians have to derive the equations reasonably simple to be integrated and analyzed.

It is now clear that the first viewpoint in its initial meaning turned out to not be realistic, since many important mechanical problems do not allow any rigorous mathematical treatment. Besides, all mathematical formulations of real-world problems are based on approximate physical models. Should one strive for complete mathematical rigor if the accuracy of conclusions anyway cannot exceed that of the initial model?

As for the viewpoint of N.E. Zhukovsky, its direct application to the contemporary situation may lead to somewhat paradoxical conclusions. Indeed, it sometimes seems that there is no need in Mechanics at all because almost any reasonable mathematical formulation of any mechanical problem can be explored numerically to large extent. Therefore, almost any model with any initial and boundary conditions can eventually be integrated.

As we can see, the two viewpoints have some point of convergence – almost any problem can to some extent be examined numerically as a purely mathematical one and very often almost nothing is available beyond these numerical results. This convergence of both viewpoints led to the absolute domination of numerical simulation in Mechanics.

There is no need to explain the strength of numerical methods in mechanical problems, but some concerns should be mentioned. Sometimes one reads papers which try to solve a certain mechanical problem numerically while taking as many factors as possible into account without checking their relative significance.

Consequently, the investigator obtains a huge amount of numerical information but fails to interpret the results – despite all efforts, no understanding of the mechanical situation is gained. In fact, a lot of papers are published with rather sophisticated calculations that add nothing to our knowledge.

We suppose that the main goal of Mechanics is to develop the models providing an understanding. The latter concept is rather vague; therefore we try to refine it by using the term "tractable model". The model is considered to be tractable if it is based on clear physical assumptions, which allow the selection of significant effects and a relatively simple mathematical formulation. These models may be obtained either by phenomenological consideration (based on some physical hypotheses) or by asymptotic reduction of a more general and often non-tractable model. The mathematical formulation should be simple enough to provide direct relationship between the results (no matter if analytic or numeric) and the initial assumptions. The principal point is that the mathematical analysis of a tractable model has to deliver a clear sense of the mechanical phenomenon described by it, providing genuine understanding. As this takes place, computer simulation is a powerful tool for examination, confirmation and sometimes for refutation of the hypotheses used to formulate the model.

Successful tractable models are important milestones in any field of exact science. Besides a wide field of application for the model itself, one often uses it for comparison with the results provided by more refined models.

Very often, the tractable models have a clear asymptotic nature. Historically, some of them appeared as a result of a mere guessing of appropriate asymptotics; others appeared with the help of regularized construction of the asymptotics. The development of tractable models, including their formulation, analysis and interpretation often leads to surprises and paradoxes.

This book tries to describe some of the significant tractable models widely used in modern solid mechanics as well as some new ones. The models are selected in order to illustrate main ideas which allowed scientists to describe complicated effects in a rather simple manner and to clarify basic notions of Solid Mechanics. Of course, the choice of the models is sole responsibility of the authors and no attempt is made to cover the whole variety of mechanical models.

We restrict ourselves to problems related to mechanics of solids. The book is divided into four chapters. The first chapter is introductory and reviews the historical development of basic models in Solid Mechanics in general. The second chapter is devoted to a more or less systematic review of the models with finite number of the degrees of freedom. The third chapter deals with some infinite discrete systems, such as chains and systems of these chains. The fourth chapter treats some continuous models.

Some results and models presented in this book were obtained and formulated by the authors, in cooperation with a number of other scientists. At this opportunity, we would like to reveal our deep gratitude to our numerous co-authors, whose names can be found in the references.

We are especially grateful to Prof. A.I. Manevich, Prof. I.V. Andrianov and Dr. Y. Starosvetsky, who read the manuscript and made a lot of useful comments and recommendations. Needless to say, the authors are solely responsible for all possible mistakes.

Last but not the least: the authors are deeply indebted and grateful to members of their families: Anechka, Miriam, Sheina, Hava, Yelena, Elina and Oleg. Without their support and inspiration the book would never have been conceived and written.

Abbreviations

CR	chain of rotators
CV	Cattaneo–Vernotte (equation)
nD	n-dimensional (n=1,2,3)
DB	discrete breather
DNA	deoxyribonucleic acid
DOF	degree of freedom
FK	Frenkel–Kontorova (model)
FPU	Fermi–Pasta–Ulam (model)
ILM	intrinsic localized modes
LO	linear oscillator
LPT	limiting phase trajectory
KdV	Korteveg–de Vries (equation)
KG	Klein–Gordon
NES	nonlinear energy sink
NNM	nonlinear normal mode
PDE	partial differential equation
PE	polyethylene
PTFE	polythetrafluoroethylene
SIM	slow invariant manifold
SMR	strongly modulated response
TET	targeted energy transfer

Abbreviations

Contents

Chapter 1
Introduction: Historical Development of Tractable Models

The necessary precondition for the formulation of tractable models of mechanics is an adequate understanding of the main notions and concepts involved in the theoretical considerations. The lack of such understanding jeopardizes efficient scientific development. This is clearly seen in the historical context – the way to formation of mechanics was rather tortuous.

As it is well-known, the problem of two bodies in celestial mechanics was the first dynamic model of theoretical physics. And fortunately, it was a tractable model which allowed Newton to explain Kepler laws, derived by fitting experimental observations, and to provide a solid basis for a universal gravitation theory.

The problem of three bodies turns out to be unsolvable in a general case and as it is seen clearly now, there are very profound reasons for that. Therefore the only possible way to calculate the deviations of the motion of the planets from trajectories predicted by Kepler laws was to develop and apply perturbation techniques. Such techniques were primarily based on power expansions by a small parameter characterizing the ratio of the masses of the third body and the smaller one of two others. Quite obviously, these techniques are based on the closeness to the tractable model of two bodies.

After impressive achievements using the approach based on this model, G. Green noted in 1828 "that a time when astronomy, from the state of perfection to which it has attained, leaves little room for further applications of their (i.e., of Laplace, Poisson, Cauchy and Fourier) art, the rest of the physical sciences should show themselves daily more and more willing to submit to it..." (cited by Truesdell, 1968). The key point in the complexity of celestial mechanics from a mathematical viewpoint is its essential nonlinearity that implies absence of common analytical methods of its treatment. It is not surprising that the development of a linear theory of vibrations (alongside with that – a theory of linear differential equations) as one of the great achievements of theoretical science in the eighteenth century was not strongly connected to celestial mechanics. This theory was initiated by dynamical problems for massless deformable bodies with lumped masses and deformable bodies with distributed mass, admitting the construction of tractable models in terms of ordinary or partial differential equations.

L.I. Manevitch, O.V. Gendelman, *Tractable Models of Solid Mechanics*, Foundations of Engineering Mechanics, DOI 10.1007/978-3-642-15372-3_1,

We understand today that it was necessary to develop low-dimensional, first of all one-dimensional models of a deformable continuum before the general concepts of stress and strain and of the relationship between them were properly understood and elaborated. The formulation of such models also was rather tortuous and intricate. The reason is clearly seen from Truesdell's analysis of the *Principia*: "While we may regard Newton's laws as equivalent to the differential equations called 'Newton's equations' in modern textbooks, there is no evidence that Newton himself thought of or ever used his principles in any general mathematical form. In Book 1 the problem of two bodies is skillfully reduced to an equivalent problem of one body attracted to a fixed center. Newton could indeed examine problems of this kind by means of differential equations, expressed in his usual style in terms of components tangent and normal to the trajectory, and then solve. But the three-body problem cannot be reduced this way. Not only does Newton give no solution or approximate solution for it in the modern sense, but he also shows no sign of any attempt even to set up equations of motion. The year in which the 'Newtonian equations' for celestial mechanics were first published is not 1687 but 1749 (by Euler)... Moreover, there was no direct way to derive equations of equilibrium or motion in other areas of Mechanics... Even the basic necessary notions were not clarified. Is the Second Law a mere definition of force? If so, does it bring us any nearer to the laws of nature? If not, then what is force, and how do we measure and know it?" (Truesdell, 1968).

One can say the same about the concepts of *body*, *inertia*, and *momentum*. Therefore the followers of Newton were frequently guided by the old ideas of his predecessors. Not accidentally, the first clear enough tractable model in Solid State Mechanics was a string subjected to arbitrary static loading along its length. In 1704, Jacob Bernoulli obtained differential equations and was able to determine both the shape and the tension of the string due to the applied load. Two crucial points lead to the success of this model: the possibility to find the restoring forces in the string acting on the masses without necessity to involve stresses, strains and relations between them, and the application of a static device.

The case of a transversally loaded bar (beam) is more complicated than the string under transversal loading. It is easy to construct a one-dimensional (tractable) model for the case of the loaded bar, assuming uniform distribution of stresses in the cross-section. Such an assumption was accepted by Galileo for bending of a beam and led him to an erroneous estimation of the breaking load. In spite of Hooke's observation that in the case of transversal loading of a beam (a piece of dry wood) the fibers on the convex side are stretched while the fibers on the concave side are compressed (Truesdell, 1968; Timoshenko, 1983), the correct description was not achieved even by J. Bernoulli and Euler, led by wrong "similarities" with levers and pendulums. The full understanding of the problem has been achieved only by Coulomb (1773). This was almost a 100 years after Leibnitz' statement that the bending moment is proportional to the moment of inertia of the cross section (1684); the latter in fact implied the introduction of bending stiffness for an effective one-dimensional continuum. The other obstacles were a lack of clear understanding of the stress distribution in the transversal section (J. Bernoulli, Euler) and of linear stress-strain relations as a material property of the deformable medium (J. Bernoulli).

Actually, the adequate approach (before Coulomb's one) had been elaborated by Parent (1713). He was the first to mention that the areas under the curves representing the tensions and compressions have to be equal. It leads to the conclusion that the neutral plane must be the central (middle) one. But such a viewpoint has not been taken into account before Coulomb; he has written down all conditions of equilibrium for the forces acting upon the cross-section of the loaded beam and has proven that the shear stresses in this section are not only possible (as Parent has concluded) but are also necessary for adequate description.

Complete understanding of the reduction problem has allowed Coulomb (1773) to outline the basis for a tractable one-dimensional model of torsion of a circular bar in which, contrary to the long transversally loaded beam, the shear stresses play a dominant role. Coulomb has successfully calculated the torsional stiffness of the circular shaft. Still, only Saint-Venant succeeded to derive these results on the basis of a common theory of elasticity (1853).

Before the transition to dynamical models it is necessary to mention the stability problem for a longitudinally loaded bar for which a first tractable model was developed by Euler (1744). He has formulated a well-known linear differential equation valid in the case of small transversal deflections. Lagrange (1766) removed the latter restriction and integrated the nonlinear equation of the problem in terms of series. In these studies, from the very beginning the authors dealt with one-dimensional models. These models, similarly to other Euler's studies, were based on J. Bernoulli's assumption that the curvature of an elastic beam at any point is proportional to the bending moment at that point (without any consideration of strains and stresses). These classical results by Euler and Lagrange are unique in the sense that they have laid the foundation of the theory of stability in physics and of spectral theory in mathematics.

The first tractable dynamical model was taut massless string loaded by n equal and equidistant masses suggested by Daniel Bernoulli (1733) (Newton in his "Principia", proclaiming the basic foundations of mechanics, as it was mentioned above, did not really deal with the systems having more than one essential degree of freedom – the only tractable dynamic model available at that time). The point was that the derivation of equations of motion for specific problems turned out to be rather complicated till Euler's "Discovery of a new principle of mechanics" (1747) in which "Newton's equations of motion" $Ma_x = F_x$; $Ma_y = F_y$; $Ma_z = F_z$ (mass M being either finite or infinitesimal) were formulated for the first time.

This was very important, because, in spite of Newton's formulation of his Second Law, there existed very different opinions at that time about the priority of different mechanical relations (e.g., equations of motion were not regarded as primary with respect to general theorems of dynamics and conservation laws).

D'Alembert even tried to expel the notion of force from dynamics. Nevertheless he "tacitly adopted the Newton approach deriving first famous linear wave equation as governing the small vibration of a string" (1746) (Truesdell, 1968). This equation

$$\mu \frac{\partial^2 w}{\partial t^2} - N \frac{\partial^2 w}{\partial x^2} = 0 \tag{1.1}$$

where μ is the linear density, w is the transversal displacement and N is the tensile force, had later become one of three fundamental equations of mathematical physics. It is a basis for many tractable dynamical models in different fields of mechanics and physics. (As for the vibration problem for a discrete system of linearly coupled masses, its adequate understanding has been attained by Lagrange in the theory of small oscillations, a part of his Analytical Mechanics. One can compute normal modes; then, the superposition principle is valid and therefore, the linear theory is a tractable one).

But even Euler circumvented the solution of partial differential equations and used the quasi-static representation when determining the natural frequencies of a cantilever beam. At the first stage, he derived a static equation introducing (by the similarity to the restoring force for a pendulum) the force providing the equilibrium of the beam for a given distribution of the displacements. Then this ordinary differential equation was solved and both the profile of displacement, corresponding to normal mode, and the length of an "equivalent pendulum" were calculated. Finally, the frequency of normal vibration was calculated based on the well-known solution for the pendulum.

The formulation of the first tractable models of solid state mechanics became possible when a differential equation of motion was applied to a finite or an infinitesimal part of a one-dimensional body, with explicit or implicit hypotheses about strains and stresses. One of the key points was reducing the three-dimensional beam to a one-dimensional continuum. This procedure is directly related to the extension of the notion of pressure to consider non-uniform distribution of stresses. Such extension leads to the general concept of normal stresses and also forces one to treat shear stresses, absent in ideal liquid and gas, as well as in the string and the 1D longitudinally loaded bar. Besides, it was necessary to formulate Hooke's law for stresses as a fundamental property of material but not of specific structure (otherwise the parameters of specimens are also present in the formulation of Hooke's law).

It is significant that earlier experimental studies of natural frequencies for stressed strings had no effect on the theoretical studies. Advancement in the solution of one-dimensional problems allowed Euler to consider a closely related two-dimensional problem – vibrations of a membrane (1759), because there was no need to introduce any new concepts. It was the first tractable two-dimensional continuum model in mechanics of solids. Still, its extension to the case of a plate turned out to be a more complicated problem, which was left to scientists of the nineteenth century.

Beginning from Galileo, the scientists of the seventeenth and eighteenth centuries were guided by pre-Hookean perception of a solid as a predominantly absolutely rigid object; they did not take into account its elastic properties. As for the practical problems, the ultimate load was the focus of interest. The elastic deformations seemed to be so small that their influence on this load could not be of any significance. Navier (1826) was probably the first who understood that it is very important to know the upper limit up to which structures behave perfectly elastically.

This new understanding allowed him to synthesize the achievements and to overcome the delusions of his great predecessors. Contrary to Euler and his followers, which considered several particular cases of loaded beams (cantilevers and symmetrically loaded simply supported beams) on the basis of the equation

$$M = EI\frac{d^2w}{dx^2} \tag{1.2}$$

where w, E, I, M are the transversal displacement, elastic (Young) modulus, moment of inertia and bending moment respectively, Navier has actually considered Eq. (1.2) as a universal relation (in contemporary terms this is Hooke's law in the case of bending). He applied it in a general case, including statically indeterminate problems. In this latter case the role of even very small elastic deformations becomes absolutely decisive.

One can say that in 1831, when Poisson wrote the equation of the equilibrium of the beam

$$EI\frac{d^4w}{dx^4} = q \tag{1.3}$$

where q is an arbitrary distributed transversal load, the creation of one of the key tractable models was completed. The other key tractable models were D. Bernoulli's model of a taut string

$$N\frac{d^2w}{dx^2} = q \tag{1.4}$$

or the mathematically equivalent model of a longitudinally stressed bar

$$ES\frac{d^2u}{dx^2} = p \tag{1.5}$$

S, N, u, p being the area of cross-section, tensile force, longitudinal displacement and distributed longitudinal load respectively. A further key model was a combination of (1.3) and (1.5), yielding the equation of the bar stability

$$EI\frac{d^4w}{dx^4} + N\frac{d^2w}{dx^2} = 0 \tag{1.6}$$

These models constitute the basis of that we call now strength of materials.

The phenomenological model of the beam having a plane of symmetry and loaded in this plane is based on three main hypotheses:

(1) The cross sections of the beam remain planar during bending;
(2) There are no normal stresses in longitudinal sections parallel to the middle plane;
(3) Relations between normal stresses and strains are determined by Hooke's law.

This approach is valid if the length of the beam is much greater than the dimensions of the cross section and deformations are small enough, so that the elasticity remains linear. Otherwise, the first two, the third or all three hypotheses are not valid.

It is significant that almost simultaneously with the formulation of the first tractable models of solid state mechanics the first one-dimensional hydrodynamics model that was introduced by D. Bernoulli in his "Hydraulics" (1744). The main important point here was a clear distinction between the concepts of normal stress and of pressure on the walls of a vessel or tube; the concepts were commonly mixed before. Accounting for a possible nonuniformity of pressure (normal stress) together with the clear formulation of the condition of mass conservation paved the way to a description of the general 3D hydrodynamics of ideal (non-viscous) incompressible liquid and gas (Euler, J. Bernoulli, 1746). This event was a great achievement of mechanics because 3D hydrodynamics was the first three-dimensional mechanical theory in which an infinitesimally small mass of liquid was considered as Newton's particle governed by Newton's Second Law. That, in turn, yielded Euler's famous partial differential equations of motion. "By creating a simple field model for fluids, defined by a set of partial differential equations, Euler opened to us a new range of vision in physical science. It is the range we all work in today. In this great insight, looking within the interior moving fluid, where neither eye nor experiment may reach, he called upon the 'imagination, fancy and invention' which Swift could find neither in music nor in mathematics" (Truesdell, 1968).

The general character of this ideologically very simple theory makes it too complicated for an analytical solution of specific hydrodynamic problems. Therefore, its appearance has opened a new stage in the construction of tractable models of mechanics: they could be obtained as particular cases of the general theory or as its asymptotic approximations.

The first possibility was realized in the wide application of the hydrodynamic 2D model in the nineteenth and twentieth centuries. Such a model after its linearization has become tractable because it turned out to be an ideal field for the application of a potential theory, complex variables and conformal mapping which were elaborated in close connection to 2D hydrodynamics. Linearization, which in the particular case of Euler's hydrodynamics is based on an assumption about the relative smallness of the velocities, has become a universal procedure which led to the formation of linear mathematical physics in the nineteenth century.

In the three-dimensional case, analytical solutions can only be obtained in highly symmetric cases. Complete, nonlinear hydrodynamics of liquids with an a priori unknown free boundary (surface) does not allow, as a rule, the application of any conventional analytical technique even in the two-dimensional case. This circumstance was a reason why the very important observation and description of solitary waves in the shallow channel by Scott Russel was not accepted by his outstanding contemporaries and has not been fully understood until Korteveg and de Vries had found a tractable model for shallow liquid (KdV-equation). On the basis of this model, the solitary waves were obtained as particular solutions. Derivation of the KdV-equation from two-dimensional hydrodynamic equations is a striking example of how the second possibility opened by Euler's hydrodynamics may be realized.

As for a compressible non-viscous liquid (gas), the simplest tractable model available was one-dimensional acoustic approximation based on the linearization of the equation of motion near the stationary state of gas or liquid with given constant pressure, density and speed. This approximation goes back to Newton's estimation of sound velocity. In the nineteenth century it was developed into linear acoustics described by 1D, 2D or 3D linear wave equations. The tractable models of linear acoustics were comprehensively reflected in the "Theory of Sound" by Rayleigh and in our time in the "Linear and Nonlinear Waves" by Whitham (1973). The new element that acoustic models involve is that it is necessary to consider a relationship between deflection of the pressure and density from their stationary values. In classical acoustics this relation is a linear one similarly to the relation between stresses and deformations in the modified Hooke's law for solids.

As for the extension of hydrodynamics to the general case of non-viscous compressible liquid (gas), the first tractable model in both 1D and 2D cases ("gas dynamics") has been suggested by Riemann. This model requires determination of the nonlinear relation between pressure and density, which leads, in particular, to the existence of triple-valued solutions as shown in the Riemann equation

$$\frac{\partial \rho}{\partial t} + c(\rho)\frac{\partial \rho}{\partial x} = 0 \tag{1.7}$$

One way to avoid such a problem is the introduction of discontinuity (shock wave) with different densities at two sides of the break. In gas dynamics, the existence of rapidly changing solutions forces one to invoke the effects of viscosity and also of non-mechanical factors like temperature and entropy. Such consideration yields smooth solutions.

The viscosity of liquid has been taken into account in the famous Navier-Stokes equations, where the linear relation between the shear stresses and the strain rates was accepted, following Newton. The presence of viscosity complicates the problem in comparison to that for an ideal, non-viscous medium. This complication has become a reason for the elaboration of one of the most efficient tractable models of theoretical physics that is the theory of boundary layer. The point is that even if the "viscous terms" in the equations of motion are relatively small (if they are formed by higher order derivatives of unknown functions with relatively small coefficients) they can be essential, e. g., near the surface of a streamlined body because of the rapid changes of velocities in the vicinity of the surface region. This circumstance has allowed L. Prandtl (1904) to formulate a tractable model of the boundary layer for the cases where the liquid mentioned above is an ideal, non-viscous one. Prandtl's theory of the boundary layer has become the basis for the construction of similar tractable models in different areas of Physics and of the Theory of Singular Perturbations in Mathematics.

The first version of the theory of elasticity (Navier, 1821) was essentially based on Boscovich's remarks about molecular forces in elastic substances. In further attempts to construct 3D theories of elasticity, A. Cauchy (1822) from the very beginning considered a 3D linearly elastic continuum. He introduced both stress and strain tensors as well as partial differential equations of motion and relations between the six stress components and the six components of strain for an isotropic body. This theory turned out to be more complex than hydrodynamics of a non-viscous incompressible liquid since the shear stresses had to be taken into account. In a certain sense it is similar to the linearized version of Navier-Stocks equations.

Full clarification of the basic concepts of the theory of elasticity was achieved when Green (1828) introduced a potential function and explicitly rejected any molecular models of elastic solids. The problem was that the idea of molecular forces proportional to the changes in the distances between the molecules and acting in the directions of lines joining them (accepted by Navier, Poisson and supported even by Cauchy, Lame and Clapeyron), led to "one-constant" model in the isotropic case. Green's approach has shown clearly that two independent elastic constants exist in an isotropic case as opposed to 21 constants (instead of 15 predicted by molecular theory with "central forces") for generic material without symmetries. This approach was the next very important step after Euler and Cauchy in the development of field representations in physics, today absolutely dominant (one should mention that for several decades after Green's famous paper his approach was considered by outstanding physicists as a purely mathematical exercise contrary to the "truly physical" approach of Navier and his followers). With its acceptance, the clarification of the main concepts of linear elasticity was completed. From this point onwards, the 3D theory of elasticity became a foundation for tractable models of a linear elastic body.

Naturally, first of all, similarly to 3D hydrodynamics, the highly symmetric problems were solved, such as the propagation of waves in an infinite elastic continuum (Lame, 1852), an infinite body bounded by a plane upon which given normal forces are distributed (high symmetry allowed the use of the Fourier integral), a similar problem for a layer bounded by two parallel planes, a circular cylinder of infinite length (Lame, Clapeyron,1833), where the symmetry has for the first time become explicit due to introduction of cylindrical coordinates. In order to deal with more complicated problems, the conditions for the application of 2D models were gradually formulated in terms of planar deformation and planar stress state; this led, in turn, to the creation of the theories of plates and shells. If the former supposes elimination of one of dimensions in the cases of the planar deformation and planar stress state, the latter represents the remote extension of earlier phenomenological theories of one-dimensional beam models. This extension was realized by using several additional hypotheses (Kirchhoff, Love), generalizing the hypotheses of Bernoulli mentioned above. Justification of the phenomenological beam theory in the framework of the theory of elasticity became possible due to the proposition of the semi-inverse method by Saint-Venant (1855). The main achievement of this technique was a reduction of the problem of bar torsion to the 2D Laplace equation which is tractable for the same reasons as is the 2D linearized hydrodynamics of an ideal incompressible liquid-potential theory: complex variables and conformal mapping can be applied. It was the first result of the theory of elasticity that had real significance for engineers. It finally broke their skepticism with respect to mathematical constructions of this theory.

In the course of this development another important question arose: the total torque twisting the bar may be a given one, but the detailed distribution of the load on the ends of the bar "is determined by the method and cannot be assigned" (Truesdell, 1968). Different distributions of load may produce different effects near the end itself but far enough from it the difference is not essential if the total torque is the same.

Boussinesq (1885) expressed this Saint Venant's principle as follows: the difference between the effects produced by two different equivalent loads applied in a given part of the body becomes very small at great distances from that part, the equivalent loads being those which would have the same effect on the body if it was rigid.

The justification of tractable phenomenological one-dimensional models together with the construction of two-dimensional models of membrane, plate, shell, planar deformation and planar stress states with the help of the Saint Venant principle was a triumph of solid state mechanics of the nineteenth century. In all these models not only deformations, but also the internal rotations are assumed to be small enough to linearize the relations between the strains and displacements. Therefore, it is possible to consider the body as rigid while dealing with the equations of motion. Such an approach does not allow the consideration of stability of bars or plates. This is a reason why Euler considered the equilibrium of a deformed elastic line while dealing with compressed bar. As a result, the ordinary linear differential equation (1.6) generalizing this approach for the bent beam was derived. Bryan (1891) and R. Lorenz (1908) obtained its extension for the cases of compressed plates and shells, respectively.

In fact, Navier (1820) was their predecessor but the tractable differential equation of stability for the bi-axially compressed (by the same distributed forces N) rectangular plate

$$D\left(\frac{\partial^4 w}{\partial x^4}+2\frac{\partial^4 w}{\partial x^2 \partial y^2}+\frac{\partial^4 w}{\partial y^4}\right)+N\left(\frac{\partial^2 w}{\partial x^2}+\frac{\partial^2 w}{\partial y^2}\right)=0 \tag{1.8}$$

and the equation of the plate bending were obtained on the basis of molecular representations by Boscovich. Therefore, the expression for the bending rigidity contained only one elastic constant. As for the dynamic tractable equation for transversal vibrations of a plate, this was finally obtained by Lagrange (1812) within the framework of earlier ideas of D. Bernoulli and L. Euler about the bent elastic line.

Unfortunately, 2D models of an isotropic elastic body, for both planar deformation and planar stress state, cannot be formulated in terms of potential theory

(2D Laplace equation) with its extraordinary efficient tool-conformal mapping. They are described by a biharmonic (not harmonic!) equation; on this basis only the simplest 2D problems were solved in the nineteenth century. The situation changed drastically at the beginning of the twentieth century after reducing the complete 2D planar problem to the computation of two harmonic functions by Kolosov (1908) and further development of corresponding techniques based on conformal mapping by N. I. Muskhelishvili. This development turned 2D theory of elasticity into a tractable model comparable at this point with 2D hydrodynamics.

The development of the mathematical tools of mechanics which was strongly connected with the construction of its tractable models seems in retrospect to have been very tortuous and intricate.

With notable exceptions of celestial mechanics and gas dynamics, the success of tractable models was based on the linearity of their governing equations. Linearization turned out to be a rather universal procedure, and this fact has both physical and mathematical substantiation. From a physical viewpoint it, is a consequence of a general principle: the reaction of the system on weak enough deflection from an initial state in the lowest order of approximation is proportional to this deflection itself. From a mathematical point of view, the linearization allows to find a complete solution of the problem due to an appropriate number of internal symmetries, additional to space-time symmetries. It is often possible to find the change of variables which splits the system of equations of motion into a number of uncoupled equations. In a dynamic case, a normal mode satisfies all equations of motion and preserves its parameters during the evolution of the system.

As for complex behavior, this is caused by superposition of a number of the elementary excitations (normal modes). The validity of the superposition principle constitutes one of the most important consequences of linearization.

The change of the variables mentioned above means a transition to "collective coordinates". This approach turns out to be more appropriate in the case of strong interparticle interaction because the interaction between the modes is absent in the linear problem.

In linear static problems, the normal modes are substituted by elementary equilibrium forms, corresponding, e.g., to external actions which are harmonic with respect to the space coordinates. Then, the superposition allows us to find a reaction of the system to complex actions (with respect to their space distribution).

In problems of heat and mass transfer energy dissipation should be taken into account. Thus, relaxation and damping will modify the notions of vibrations and waves. Still, due to linearity, collective modes can be revealed, and their superposition gives a possibility to describe a complex relaxation of the system. In essence, the elementary solutions in statics and the relaxation modes play a role similar to normal modes in dynamics.

The area of efficient application of the linear theory is powerful rather wide. Still, it does not take into account the important effects caused by even weak nonlinearity and interaction of normal modes. One can mention, e.g., the amplitude dependence of natural frequencies, modulational instability of quasi-harmonic waves, abrupt change of amplitude frequency under minor change of the frequency of excitation, self-sustained vibrations assuming a specific energy exchange with surrounding media. It turned out that in the mentioned cases it is possible to consider the nonlinear effects as weak, though they qualitatively determine the type of the process. As a result, a powerful quasi-linear approach to these processes and phenomena was developed. It is also based on mutually independent unperturbed modes. In spite of its magnificence and deepness, this approach also turns out to be insufficient for understanding of certain very important phenomena. They are strongly connected with the discovery of essentially nonlinear localized excitations. In this book, we discuss the tractable models of quasi-linear and essentially nonlinear solid state mechanics together with particularly complex problems of the linear theory (e.g., due to complicated boundary conditions).

References

Timoshenko S.P.: History of Strength of Materials. Dover, New York (1983)
Truesdell C.: Essays in the History of Mechanics. Springer, Berlin (1968)
Whitham G.B.: Linear and Nonlinear Waves. Wiley, New York (1973)

Chapter 2
Discrete Finite Systems

An asymptotic approach is rather efficient when dealing with the theory of oscillations, since one often can figure out a number of relatively simple limit cases which can be efficiently treated and completely understood. Such an approach allows the deepest possible simplification but preserves the most significant features of dynamical behavior. At the same time, one can use the expansions by small parameters characterizing the deviation of the system from the tractable limit case.

The mathematician supposes usually that the small parameters are already known and that construction and substantiation of the asymptotic expansions is the only problem. Contrary to this, for physicists and engineers the choice of the appropriate limit cases and corresponding small parameters is the most important step.

Systems with a relatively small number of degrees of freedom offer convenient frameworks for the explanation of the basic concepts and ideas related to the simplification and formulation of the tractable models. Thus, in this part of the book we are going to discuss the discrete models of a dynamical system – from one to few degrees of freedom. The step-by-step approach will reveal essential new features arising from each stage of complication of the discussed models.

2.1 Linear Oscillators

Beyond any doubt, a linear oscillator and a system of coupled linear oscillators are the most popular and the best understood models of discrete dynamical systems. If parameters of the oscillators are time-independent, then a general analytic solution can be provided in the form of exponents (combined with polynomials in resonant cases). This theory is widely known and can be found in standard textbooks (see, e.g. Den Hartog, 1956; Meirovitch, 2000), so there is no reason to repeat it here. Instead, we will concentrate on asymptotic aspects of behavior of individual and coupled linear oscillators, keeping in mind possible generalizations for less common systems.

L.I. Manevitch, O.V. Gendelman, *Tractable Models of Solid Mechanics*, Foundations of Engineering Mechanics, DOI 10.1007/978-3-642-15372-3_2,

2.1.1 Linear Conservative Oscillator

The system consists of a point-like mass m and a linear spring (with stiffness c) connecting the mass with an immobile point (Fig. 2.1) is one of the most simple and important models of mechanics and physics. A pendulum with small amplitudes (without friction) and oscillatory contour in radiophysics (without resistance) are well known realizations of this model. This mathematical model is described by the well known differential equation

$$m\frac{d^2U}{dt^2} + cU = 0 \tag{2.1}$$

(U is the displacement of point-like mass with respect to its equilibrium state, m is the mass and c – the rigidity of the elastic spring). The initial conditions in certain initial instant have to be supplied as $U^0 = U(0)$ and $V^0 = \frac{dU}{dt}(0)$.

In dimensionless variables $\tau = \omega_0 t, \quad u = U/U^0, \omega_0 = \sqrt{c/m}$ one obtains

$$\frac{d^2u}{d\tau^2} + u = 0 \tag{2.2}$$

where $u^0 = u(0) = 1$ and $v^0 = v(0) = \frac{du}{d\tau}(0) = {}^{V^0}\!/_{\omega_0 U^0}$ for $\tau = 0$. Due to linearity of Eq. (2.1), its solution has to be easily found as

$$u = a\cos(\tau - \tau_0) \tag{2.3}$$

(harmonic vibration with period $T = 2\pi$); the amplitude a and phase shift τ_0 are determined by the initial conditions. One easily obtains $\omega = 1$ substituting (2.3) to (2.2).

Equation (2.2) has no small parameters; still, it is not the simplest possible description of the oscillator. After a reduction to two coupled first order differential equations we can write

$$\frac{dV}{d\tau} = -u; \quad \frac{du}{d\tau} = v$$

Then it is possible to introduce conjugate complex variables

$$\psi = v + iu; \quad \psi^* = v - iu$$

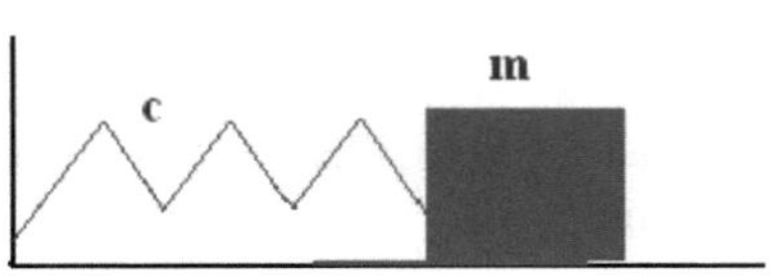

Fig. 2.1 Sketch of the linear oscillator

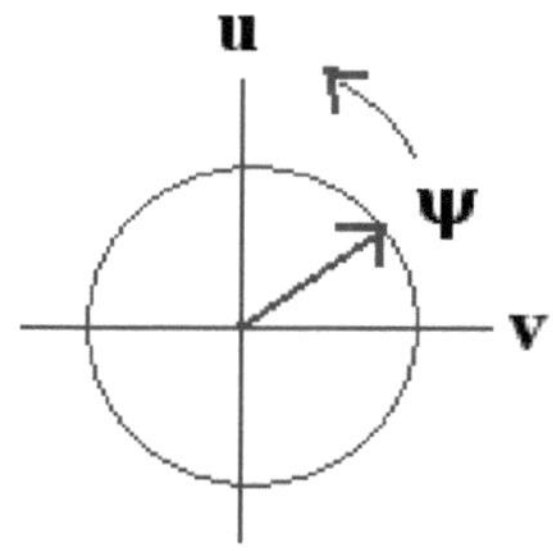

Fig. 2.2 Rotation of the complex vector

Complex representation was initially considered as a useful phenomenological model for analyzing nonlinear effects in different fields of physics (Scott et al., 1985; Kosevitch and Kovalyov, 1989). Recently it has been shown (Manevitch, 1999, 2001; Manevich and Manevitch, 2005) that complexification is instrumental for the analysis of enharmonic oscillators and nonlinear oscillatory chains. The approach of (Manevitch, 1999, 2001) employs a complex combination of displacements and velocities (similar to those applied in the linear problem of secondary quantization in quantum mechanics) and replaces the initial second order equations of motion by an array of first-order equations; the subsequent separation of fast and slow time scales allows obtaining leading-order approximations of a nonlinear process. We will often use this method below.

The equations of motion can be expressed as

$$\frac{d\psi}{d\tau} - i\psi = 0, \quad \frac{d\psi^*}{d\tau} + i\psi^* = 0 \tag{2.4}$$

Only one of these equations is independent. The solution of the Eq. (2.4) is $\psi = ce^{i\tau}$ and $\psi^* = c^* e^{-i\tau}$ ($c = c_1 + ic_2$ is a complex number). These complex functions have a very simple geometric sense (Fig. 2.2). The first of them corresponds to counterclockwise rotation in the complex plane of the vector $\psi(v, u)$ ("positive" rotation) and the second one to clockwise rotation of the vector ψ^* ("negative" rotation). Each one of these two rotations conveys complete information about the process.

One can speculate that each of the Eq. (2.4) is simpler than (2.2) since it is of the first order. In this respect, we consider its solution as more "fundamental" than (2.3), although of course they are mathematically equivalent. We replace the oscillations by rotations. We will see below that this transition significantly simplifies the solution of much more complicated problems.

In the considered case, the function ψ^* is absent in the equation for ψ and vice versa. However, account of the viscous friction changes this situation.

2.1.2 Linear Oscillator with Viscous Damping

In this case the equation of motion can be written as follows

$$m\frac{d^2U}{dt^2} + 2\eta\frac{dU}{dt} + cU = 0 \tag{2.5}$$

under initial conditions $U = U^0$, $V = \dfrac{dU}{dt} = V^0$ for $t = 0$.

After introduction of the dimensionless variables and parameters one obtains

$$\frac{d^2u}{d\tau^2} + 2\varepsilon\frac{du}{d\tau} + u = 0 \tag{2.6}$$

where $\tau = \omega_0 t$, $\varepsilon = {}^{\eta}/_{m\omega_0}$, $\omega_0 = \sqrt{{}^{c}/_{m}}$. Due to linearity of (2.6) one can find again the exact analytical solution. Standard anzats $u = ce^{\lambda\tau}$ yields the algebraic equation for λ:

$$\lambda^2 + 2\varepsilon\lambda + 1 = 0$$

Depending on the magnitude of ε, these quadratic equations has two real or two complex roots

(a) $\lambda_{1,2} = -\varepsilon \pm \sqrt{\varepsilon^2 - 1}$ for $\varepsilon^2 > 1$
(b) $\lambda_{1,2} = \varepsilon \pm i\sqrt{1 - \varepsilon^2}$ for $\varepsilon^2 < 1$

so that in the cases a and b

(a) $u = c_1 e^{\lambda_1\tau} + c_2 e^{\lambda_2\tau}$
(b) $u = ce^{-\varepsilon\tau}\cos\left(\sqrt{1 - \varepsilon^2}\,\tau - \varphi\right)$

In the special resonant case of $\varepsilon = 1$ the solution is

(c) $u = (c_1 + c_2\tau)e^{-\tau}$

Certainly, it is possible to find the power expansions of the solution both for small and large ε. But we would like to show that even in linear case further simplification is possible before beginning to solve the differential equations. This simple example allows introducing the ideas and tools which become necessary in more general and complicated cases.

2.1.2.1 Strong Energy Dissipation ($\varepsilon >> 1$)

This case physically corresponds to monotonous approach to equilibrium (relaxation). After dividing of the Eq. (2.6) by ε one obtains

$$\varepsilon^{-1}\left(\frac{d^2u}{d\tau^2} + u\right) + 2\frac{du}{d\tau} = 0 \tag{2.7}$$

The main problem is choice of appropriate limit cases for $\varepsilon^{-1} << 1$. How to choose the limiting systems? One has no a priori knowledge about the characteristic decrease rate of function u(t) with respect to parameter ε due to time differentiation (the order of the variable itself is irrelevant due to linearity). To determine this characteristic rate, let us perform a new change of the independent variable $\tau_1 = \varepsilon^{\alpha}\tau$, where the exponent α characterizes change of time scale due to damping. Then the equation of motion is written as follows:

$$\varepsilon^{-1}\left(\varepsilon^{2\alpha}\frac{d^2u}{d\tau_1^2}+u\right)+2\varepsilon^{\alpha}\frac{du}{d\tau_1}=0.$$

Consistent asymptotic expansion may be obtained if in this equation two terms have the same order and the third one has smaller order of magnitude (with respect to ε^{-1}). Only two such cases are possible:

1. $2\alpha - 1 = \alpha,\ \alpha = 1;$
2. $\alpha = -1.$

These exponents correspond to fast and slow change of solution respectively. The main approximation in the first case has the following form (after one integration):

$$\frac{du}{d\tau_1}+2u=0 \tag{2.8}$$

taking into the account the decay condition (for $\tau \to \infty$) that results in the slowly decaying solution:

$$u = c_1 e^{-2\tau_1} = ae^{-2\varepsilon\tau}.$$

In the second case the main approximation is written as follows

$$2\frac{du}{d\tau_1}+u=0 \tag{2.9}$$

Solution of (2.9) describes a rapid relaxation:

$$u = c_2 e^{-\frac{1}{2}\tau_1} = c_2 e^{-\frac{1}{2}\varepsilon^{-1}\tau}.$$

Equations (2.8) and (2.9) provide a tractable model of the strongly damped oscillator. They describe (in main approximation) equilibrating the damping and elastic forces (Eq. 2.8) or that of inertial and damping forces (Eq. 2.9). In other terms, for every particular case the tractable model corresponds to the balance of two certain forces, whereas the other terms in the equations of dynamics yield only small corrections. As one can see, limit Eq. (2.9) describes early stage of a fast relaxation, whereas Eq. (2.8) describes the late stage of relatively slow relaxation. However, it is not enough in order to solve the initial value problem (IVP) for Eq. (2.5) – one should also know how to combine these expansions together and to satisfy the initial conditions. Due to the linearity of the initial equation, one should be tempted to simply add together these two asymptotic expressions. Unfortunately, this idea is first of all wrong from a mathematical viewpoint, since expressions (2.8) and (2.9) yield asymptotic formulas with respect to different orders of ε not with the same accuracy. Besides, it is instructive to see a more systematic way of getting the asymptotic expansions including terms with different time scales, since the luxury of linear superposition is often unavailable.

Let us rescale time in Eq. (2.6) as $\xi = \tau\varepsilon$ and introduce the new parameter $\delta = \varepsilon^{-2}$. The resulting equation obtains the following form:

$$u_{\xi\xi} + 2u_{\xi} + \delta u = 0 \tag{2.10}$$

Let us introduce the sequence of time scales

$$\xi_k = \delta^k \xi, k = 0, 1, \ldots$$

Then, the dependent function is considered as a function of these multiple time scales:

$$u = u(\xi_0, \xi_1, \ldots)$$

The differentiation with respect to time is performed by a common chain rule. For the case of the multiple scales expansion, this rule can be conveniently written in the form of expansion of the differentiation operator:

$$d/d\xi = \partial/\partial\xi_0 + \delta\partial/\partial\xi_1 + \ldots$$

Substituting this expansion in Eq. (2.10), we get in the zero order of approximation the following equation:

$$u_{\xi_o\xi_o} + 2u_{\xi_o} = 0 \Rightarrow u = C_1(\xi_1, \ldots)\exp(-2\xi_0) + C_2(\xi_1, \ldots) \tag{2.11}$$

where C_1 and C_2 are yet undefined functions of higher time scales – "constants" with respect to the scale ξ_0. In order to determine these functions, we have to take into account the terms of order δ:

$$2\partial^2 u/\partial\xi_0\partial\xi_1 + 2\partial u/\partial\xi_1 + u = 0 \Rightarrow \begin{cases} -2\partial C_1/\partial\xi_1 + C_1 = 0 \\ 2\partial C_2/\partial\xi_1 + C_2 = 0 \end{cases} \tag{2.12}$$

Finally, coming back to initial variables of Eq. (2.6), we obtain the following asymptotic expansion of the solution:

$$\begin{aligned} u(\tau) =& C(\tau O(1/\varepsilon^3))\exp(-\tau(2\varepsilon - 1/2\varepsilon)) + \\ &+ D(\tau O(1/\varepsilon^3))\exp(-\tau/2\varepsilon) \end{aligned} \tag{2.13}$$

where C and D are arbitrary functions. In order to establish their exact shape, further orders of approximation should be analyzed. One can see that the accuracy of expansion (2.13) is the same in both terms (the first one is a boundary layer with

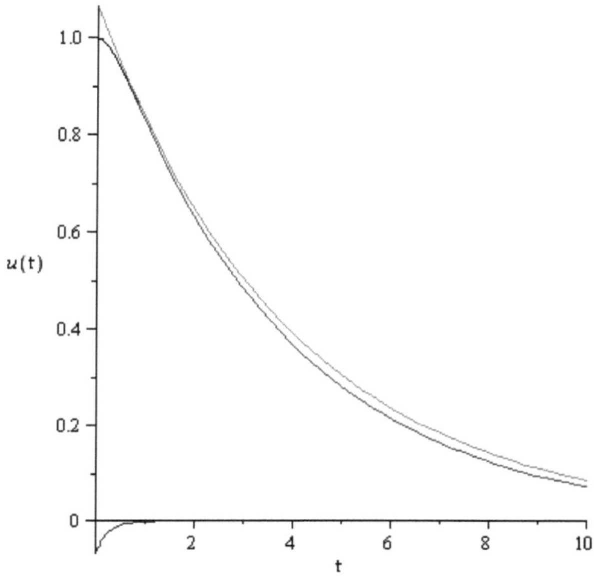

Fig. 2.3 Response of the linear oscillator – the case of the strong damping, ε = 2, u(0)=1, du/dt(0)=0. *Red line* – exact solution, *black* – approximate solution in accordance with (2.13) (they almost coincide), *green* (upper line in the figure) – slow dissipation, *blue* (below t axis) – boundary layer (fast initial dissipation)

a fast decay and the second one corresponds to a slow dissipation). For this sake, two orders should be kept in the first term although only the fast one is required to understand the qualitative behavior. Needless to say, expansion (2.13) can be easily obtained from the exact solution. The time history of the typical solution for this case is illustrated at Fig. 2.3.

2.1.2.2 Weak Energy Dissipation (ε<<1)

If $\varepsilon \ll 1$, the dissipative force manifests itself in (1) decrease of the oscillations amplitude, (2) shift of the frequency. The former effect is more significant since it reveals itself in the first order of ε.

To obtain the tractable model, let us use the complex representation of the equations of motion (as well as in the case of a conservative oscillator). Application of the complex representation to Eq. (2.6) yields

Then one can write

$$\begin{aligned} &\frac{d\psi}{d\tau} - i\psi + \varepsilon\left(\psi + \psi^*\right) = 0 \\ &\frac{d\psi^*}{d\tau} + i\psi^* + \varepsilon\left(\psi + \psi^*\right) = 0 \end{aligned} \tag{2.14}$$

The equations are now coupled due to the friction. To obtain the tractable model we use a change of variables

$$\psi = \varphi\, e^{i\tau},\ \psi^* = \varphi^*\, e^{-i\tau}$$

Such a transformation means that we describe the process in a rotating reference system. After substitution (2.15) into (2.14) one obtains

$$\frac{d\varphi}{d\tau} + \varepsilon\left(\varphi + \varphi^*\, e^{-2i\tau}\right) = 0 \tag{2.15}$$

and the conjugate equation.

In accordance with multiple scale procedure, we introduce a "slow" time $\tau_1 = \varepsilon\tau$ alongside with "fast" time $\tau_0 = \tau$, so that

$$\frac{d}{d\tau} = \frac{\partial}{\partial\tau_0} + \varepsilon\frac{\partial}{\partial\tau_1} \tag{2.16}$$

and look for the solution as power expansion

$$\varphi = \varphi_0 + \varepsilon\varphi_1 + \cdots \tag{2.17}$$

Substituting (2.17) into (2.15) with taking into account (2.16), selecting the terms of similar order by parameter ε and equating their sum to zero, one obtains

$$\varepsilon^0 : \frac{\partial\varphi_0}{\partial\tau_0} = 0,\ \text{so}\ \varphi_0 = \varphi_0\,(\tau_1) \tag{2.18}$$

$$\varepsilon^1 : \frac{\partial\varphi_1}{\partial\tau_0} + \frac{\partial\varphi_0}{\partial\tau_1} + \left(\varphi_0 + \varphi_0^*\, e^{-2i\tau_0}\right) = 0$$

Integrating the second equation by the fast time, one obtains:

$$\varphi_1 = -\left(\frac{\partial\varphi_0}{\partial\tau_1} + \varphi_0\right)\tau_0 - \frac{i}{2}\varphi_0^*\, e^{-2i\tau_0} + C(\tau_1)$$

The first terms in the right-hand side of this equation imply secular growth of the solution with respect to the fast time scale. In order to avoid it, one must require that this coefficient disappears:

$$\frac{\partial\varphi_0}{\partial\tau_1} + \varphi_0 = 0 \tag{2.19}$$

describing in the main asymptotic approximation the effect of the damping – exponential decrease of the length of complex vector φ_0.

$$\varphi_0 = c\, e^{-\tau_1}$$

The Eq. (2.19) provides the tractable model of the oscillator if the damping is weak.

2.1.2.3 Exercise

Find the asymptotic solution (similar to (2.13)) for the strongly damped oscillator after action of initial impact (zero initial coordinate, non-zero initial velocity).

2.1.3 Linear Oscillator with Viscous Damping and Periodic (Harmonic) Forcing

The equation of motion may be written as follows

$$m\frac{d^2U}{dt^2} + 2\eta\frac{dU}{dt} + cU = F\cos\omega t \tag{2.20}$$

General system (2.20) can be solved exactly (Den Hartog, 1956; Meirovitch, 2000), but we are primarily interested in various simplified limit cases due to different possible small parameters (Fig. 2.4)

For the beginning, let us consider $\dfrac{\omega}{\omega_0} = \varepsilon$ as a characteristic parameter and let us also introduce the following rescaling:

$$F = \varepsilon^{\beta} f, \quad u = {}^{Uc}\!/_{f}$$

Then the equation of motion may be written as follows:

$$\varepsilon^2\frac{d^2u}{d\tau^2} + 2\varepsilon\eta_1\frac{du}{d\tau} + u = \varepsilon^{\beta}\cos\tau \tag{2.21}$$

where $\eta_1 = \dfrac{\eta}{\sqrt{cm}}$.

Two limit cases yield consistent asymptotic expansions:

1. $\varepsilon \to 0,\ \beta = 0$
2. $\varepsilon \to \infty,\ \beta = 2.$

In the first case, the elastic and external forces are the most important (they have 0th order with respect to small parameter ε). Such a situation is typical for static loading; therefore one can speak about a quasi-static approach.

In the second case, the inertial and external forces dominate and we deal with a "quasi-dynamic" approach. Corresponding limiting equations look like this:

$$u = \cos\tau$$

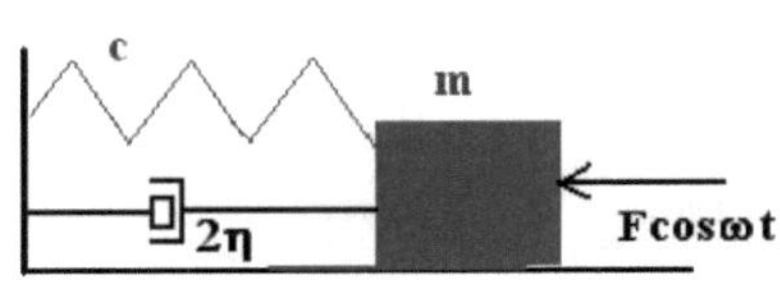

Fig. 2.4 Forced and damped linear oscillator

$$\frac{d^2u}{d\tau^2} = \cos\tau, \text{ so } u = -\cos\tau.$$

These are the tractable models because they take the most significant effects into account and provide a clear physical understanding of the problem.

Besides, if $\varepsilon = \dfrac{\omega}{\omega_0} \sim 1$, it is possible to introduce the small parameter

$$\varepsilon_1 = \frac{\omega_0 - \omega}{\omega_0} = 1 - \frac{\omega}{\omega_0}$$

because the case $\varepsilon_1 << 1$ is of special interest. The conditions $\varepsilon \sim 1$ or $\varepsilon_1 << 1$ mean that inertial and elastic forces are very close by their magnitude so the process is similar to that of the free vibrations. To provide this, the force parameter also has to be small. As this takes place, the small dissipative term already in the main approximation may be important. Let us perform the transition to the complex variables and present external force and dissipative coefficients as

$$F_1 = \varepsilon_1^{\beta}\frac{f}{2}\left(e^{i\omega t} + e^{-i\omega t}\right),\ \eta = \varepsilon_1^{\gamma}\upsilon, \tag{2.22}$$

and introduce the change of variables

$$\tau = \omega_0 t, \quad u = {}^{Uc}\!/_{f}$$

After transition to complex variables and with account of (2.22) one obtains

$$\frac{d\psi}{d\tau} - i\psi + \varepsilon_1^{\gamma}\upsilon\left(\psi + \psi^*\right) = \frac{1}{2}\varepsilon_1^{\beta}\left(e^{i\varepsilon t} + e^{-i\varepsilon t}\right)$$

$$\frac{d\psi^*}{d\tau} + i\psi^* + \varepsilon_1^{\gamma}\upsilon\left(\psi + \psi^*\right) = \frac{1}{2}\varepsilon_1^{\beta}\left(e^{i\varepsilon t} + e^{-i\varepsilon t}\right)$$

Using the multiple scale procedure with

$$\beta = 1, \gamma = 1, \tau_n = \varepsilon_1\tau,\ \frac{d}{d\tau} = \frac{\partial}{\partial\tau_0} + \varepsilon_1\frac{\partial}{\partial\tau_1} + \varepsilon_1^2\frac{\partial}{\partial\tau_2} + \cdots$$

$$\psi = \psi_0 + \varepsilon_1\psi_1 + \varepsilon_1^2\psi_2 + \cdots,$$

we obtain the system

$$\begin{aligned} &\frac{\partial\psi_0}{\partial\tau_0} - i\psi_0 = 0 \\ &\frac{\partial\psi_1}{\partial\tau_0} - i\psi_1 = -\frac{\partial\psi_0}{\partial\tau_1} - \upsilon\left(\psi_0 + \psi_0^*\right) + \frac{f}{2}\left(e^{i(1-\varepsilon_1)\tau_0} + e^{-i(1-\varepsilon_1)\tau_0}\right) = 0 \end{aligned} \tag{2.23}$$

Change of dependent variable

$$\psi_0 = \varphi_0 \, e^{i\tau_0}$$

leads to relation $\varphi_0 = \varphi_0 \left(\tau_1, \tau_2, \ldots\right)$ and the tractable model of the main asymptotic approach may be written as follows (from the second equation of (2.23)):

$$\frac{\partial \varphi_0}{\partial \tau_1} + \upsilon \varphi_0 = \frac{f}{2} e^{i\tau_1}$$

so that

$$\phi_0 = \Phi\left(\tau_1, \ldots\right) e^{-i\upsilon\tau_1} + \frac{f}{2\eta_1} e^{i\tau_1}$$

The first term describes the damped free oscillations and the second one corresponds to stationary forced vibrations describing their amplitude and phase shift due to damping (Fig. 2.5).

2.1.3.1 Exercise

Calculate the frequency correction caused by the linear damping (use the second equation of (2.23)).

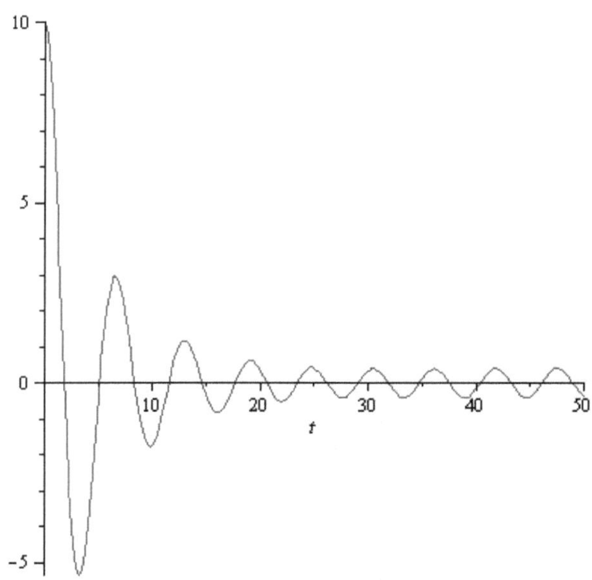

Fig. 2.5 Response of the forced oscillator close to the resonance. One can clearly see the exponentially decaying initial transient and further stationary response

2.1.4 Two Coupled Oscillators

The system of two oscillators coupled by an elastic spring is the next step after the isolated oscillator by degree of complexity. This is also the fundamental model of the theory of vibrations (Fig. 2.6).

This system is described by the following equations:

$$\begin{cases} m_1\ddot{u}_1 + c_{11}u_1 + c_{12}(u_1 - u_2) = 0 \\ m_2\ddot{u}_2 + c_{22}u_2 + c_{12}(u_2 - u_1) = 0 \end{cases} \tag{2.24}$$

This system is linear and therefore can be solved by an exact analytic procedure. Still, the results are not easy to interpret and we would like to develop the asymptotic approach instead in order to find the simplest description of the dynamics which reflects the main features of the system. It turns out that these qualitative features are determined both by coupling between the oscillators and by the relationship between their natural frequencies.

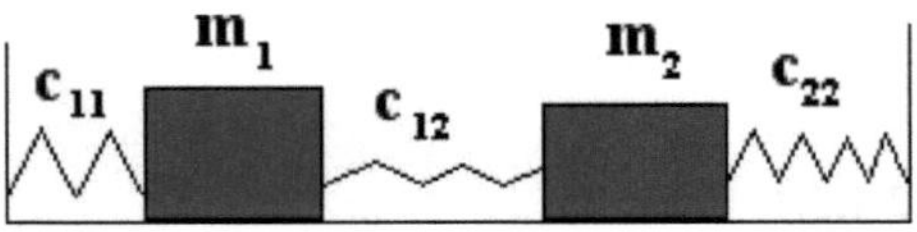

Fig. 2.6 System of two coupled linear oscillators

2.1.4.1 Weakly Coupled Oscillators with Strongly Different Frequencies

Let us consider the case when two oscillators with the frequencies $\omega_1 = \sqrt{\dfrac{c_{11}}{m_1}}$ and $\omega_2 = \sqrt{\dfrac{c_{22}}{m_2}}$ respectively are connected by the linear spring with stiffness c_{12}, so that $\dfrac{\omega_1^2}{\omega_2^2} = \varepsilon << 1$ and $\dfrac{c_{12}}{c_{11}} = \mu\varepsilon << 1$ (parameters of coupling and ratio of the squared frequencies are supposed to be of the same order with respect to the small positive dimensionless parameter ε).

Smallness of the parameter ε can be caused by strong difference of masses or stiffness; therefore it is reasonable to consider two different limit cases:

(a) $c_{11} = c_{22}$, but $\dfrac{m_2}{m_1} = \varepsilon << 1$

(b) $m_1 = m_2$, but $\dfrac{c_{11}}{c_{22}} = \varepsilon << 1.$

The exact equality of the masses or the stiffness's is of course not compulsory, but it does not change the main idea and somewhat simplifies the treatment. Using dimensionless variable $\tau_1 = \omega_1 t$, one can see that the equations of motion for cases (a) and (b) are different:

For the case (a):

$$\frac{d^2u_1}{d\tau_1^2} + u_1 + \mu\varepsilon\,(u_1 - u_2) = 0$$
$$\varepsilon\frac{d^2u_2}{d\tau_1^2} + u_2 + \mu\varepsilon\ (u_2 - u_1) = 0 \tag{2.25}$$

For the case (b):

$$\frac{d^2u_1}{d\tau_1^2} + u_1 + \mu\varepsilon\,(u_1 - u_2) = 0$$
$$\varepsilon\frac{d^2u_2}{d\tau_1^2} + u_2 + \mu\varepsilon^2\,(u_2 - u_1) = 0 \tag{2.26}$$

Due to the introduction of dimensionless time we have implicitly introduced the small parameter into the equation of motion. Looking for a solution in the form of harmonic vibration $u_i = A_i \cos\omega\tau\ (i = 1{,}2)$ leads to the system of two coupled linear algebraic equations with respect to amplitudes A_i with coefficients dependent on a yet unknown frequency ω. For an existence of a nontrivial solution of the system the determinant has to be equal to zero; that leads to a biquadratic equation with respect to frequency (instead of a quadratic one in the case of an isolated oscillator). Substitution of the ω values satisfying to the biquadratic equation gives us two independent particular solutions, corresponding to natural frequencies (eigenfrequences). Relations between amplitudes of the particles can be found after substitution of the normal frequencies into the linear algebraic equations. In this manner, the normal modes corresponding to each of two eigenfrequencies are determined. Arbitrary linear combination of normal modes (due to linear superposition principle) gives a general solution of the problem with the amplitudes and phases being calculated from the initial conditions.

We will examine the limit systems of differential equations (2.26). For this, it is necessary to determine the yet unknown relations (by small parameter ε) between two dependent variables as well as between these variables and their derivatives. Therefore we will use the new dependent variables having the same order by ε and new independent variable τ, differentiation by which does not change the order of smallness:

$$u_1 = w_1,\ u_2 = w_2\varepsilon^{\beta},\ \tau = \varepsilon^{-\alpha}\tau_1$$

(parameter β characterizes the relationship between displacements of two particles). After such a change the equation of motion can be written in the case (a) as

$$\varepsilon^{-2\alpha}\frac{d^2w_1}{d\tau^2} + w_1 + \mu\varepsilon\left(w_1 - \varepsilon^{\beta}w_2\right) = 0$$
$$\varepsilon^{-2\alpha+\beta+1}\frac{d^2w_2}{d\tau^2} + \varepsilon^{\beta}w_2 + \mu\varepsilon\left(\varepsilon^{\beta}w_2 - w_1\right) = 0. \tag{2.27}$$

Taking into account that $\varepsilon << 1$, it is possible to determine the consistent values of the yet unknown exponents α and β. It means that corresponding limit systems have to contain no parameter ε but to allow subsequent determination of unknown functions. It is easy to see that there are only two different limiting cases:

1. $2\alpha = 0, \beta = 1$;
2. $2\alpha = 1, \beta = -2$

In the first case the limiting system is:

$$\begin{aligned} &\frac{d^2 w_1}{d\tau^2} + w_1 = 0, \\ &w_2 - \mu\, w_1 = 0. \end{aligned} \tag{2.28}$$

The first equation describes free vibrations of the low frequency oscillator; the second oscillator has small displacements (of order ε) and feels a coupling with first oscillator as action of external force $\mu\, w_1(\tau)$. As this takes place, the equation of motion for a weakly excited oscillator corresponds to a quasi-static regime (the inertial term plays a secondary role). From the mathematical viewpoint, we obtain an essential simplification and replace the system of two differential equations by one simple algebraic relation. The solution of the limiting system is extraordinarily simple:

$$\begin{aligned} w_1 &= c \cos(\tau - \tau_0), \\ w_2 &= \mu c \cos(\tau - \tau_0), \end{aligned} \tag{2.29}$$

where the arbitrary constants c and τ_0 have to be determined from the initial conditions for a low-frequency oscillator. Returning to the initial variables, one can write

$$\begin{aligned} u_1 &= c \cos(\omega_1 t - \tau_0), \\ u_2 &= \varepsilon \mu c \cos(\omega_1 t - \tau_0). \end{aligned}$$

In the second case, the limiting system can be written as follows

$$\frac{d^2 w_2}{d\tau^2} + w_2 = 0, \quad \frac{d^2 w_1}{d\tau^2} - \mu\, w_2 = 0 \tag{2.30}$$

Here, the first equation describes the free vibrations of the oscillator with a higher frequency. Its displacements can be found using the corresponding initial conditions. Then the other oscillator perceives the action of the first one as an external force $\mu\, w_2(\tau)$. In this case, the second oscillator moves in quasi-dynamic regime, with weak effect of the elastic force. The solution of this limiting system is:

$$w_2 = c \cos(\tau - \tau_0), \quad w_1 = -\mu c \cos(\tau - \tau_0) \tag{2.31}$$

and in the initial variables:

$$u_1 = c\cos(\omega_2 t - \tau_0), \quad u_2 = -\varepsilon^{-2} c\cos(\omega_2 t - \tau_0).$$

In the case (b), when the stiffnesses of the oscillators are strongly different, the limiting systems are the same, although the relationship between the unknown functions for the first combinations of α and β is different. Because there is no qualitative difference here, we restrict ourselves by this notion.

Let us summarize several points. We see that in the system of weakly coupled oscillators with essentially different frequencies, the motions of two types occur. In both cases, it is possible to say that the "individuality" of the oscillators is preserved. As this takes place, their interaction manifests itself in such a manner that the first oscillator in the case 1 and the second in the case 2 produces periodic force acting on the other oscillator. Corresponding forced motion occurs in weakly excited subsystem, periodic force being generated by a strongly excited subsystem. In principle, arbitrary motion of the system, can due to its linearity be presented as a linear combination of two distinguished types of motion, corresponding to normal (one-frequency) modes of vibration. Let us underline once more that every one of weakly coupled oscillators with strongly different frequencies nearly preserves its individuality. Weak interaction between two oscillators can be understood in terms of the combination of free and forced one-particle models.

2.1.4.2 Exercise

Describe the vibrations of weakly coupled oscillators with strongly different frequencies after action of given initial pulse applied to one of the oscillators.

2.1.4.3 Weakly Coupled Oscillators with Close Frequencies

Let us suppose that both oscillators considered have close frequencies. As in the previous case, small parameter ε, characterizing a weak coupling between the oscillators, can be introduced to the system. However, instead of the frequencies ratio, now their relative difference

$$\frac{\Delta\omega}{\omega_1} = \frac{\omega_2 - \omega_1}{\omega_1} << 1$$

turns out to be small.

For the sake of simplicity, we here consider the case of weakly coupled oscillators with equal frequencies. After transition to dimensionless variables in the system (2.24) one can write

$$\begin{aligned} &\frac{d^2u_1}{d\tau^2} + u_1 + \varepsilon(u_1 - u_2) = 0 \\ &\frac{d^2u_2}{d\tau^2} + u_2 + \nu\varepsilon(u_2 - u_1) = 0 \end{aligned} \qquad (2.32)$$

$$\tau = \omega_0 t, \omega_0 = \sqrt{\frac{c_{11}}{m_1}} = \sqrt{\frac{c_{22}}{m_2}}, \ \frac{k}{c_{11}} = \varepsilon << 1, \ \nu = \frac{c_{11}}{c_{22}} = \frac{m_1}{m_2},$$

($\nu = 1$ in the particular case of the identical oscillators). We admit exact equality of the frequencies of both oscillators; this restriction does not affect qualitative characteristics of the response. The direct asymptotic approach to the equations of motion distinguishes the system of independent oscillators as a limiting one. However, both intuition and experience tell us that interaction between the oscillators is important even for very weak coupling (phenomenon of beating). Does it mean that we can not use a smallness of coupling and have to consider the starting system as a tractable model? Similar to the case of isolated damped oscillator, the direct expansion of the solution by small parameter leads to "resonance" (appearance of secular non-periodic terms). To avoid this obstacle, we reduce the system to the form:

$$(1) \ \frac{dv_1}{d\tau} = -u_1 - \varepsilon\,(u_1 - u_2) = 0,$$

$$(2) \ \frac{du_1}{d\tau} = v_1,$$

$$(3) \ \frac{dv_2}{d\tau} = -u_2 - \nu\varepsilon\ (u_2 - u_1) = 0,$$

$$(4) \ \frac{du_2}{d\tau} = v_2$$

Multiplying the Eqs. (2) and (4) from the above system by imaginary unity i and summing them up with (1) and (3) respectively, one can obtain a system of complex equations

$$\frac{d\psi_n}{d\tau} - i\psi_n - i\frac{\nu_n}{2}\varepsilon\left[\left(\psi_n - \psi_n^*\right) - \left(\psi_{3-n} - \psi_{3-n}^*\right)\right] = 0$$
$$n = 1, 2; \ \ \nu_1 = 1; \ \ \nu_2 = \nu \tag{2.33}$$

Using the change of variables $\psi_n = \varphi_n e^{i\tau}$, we transform the system (2.33) to the equations

$$\frac{d\varphi_n}{d\tau} - i\frac{\nu_n}{2}\varepsilon\left[\left(\varphi_n - \varphi_{3-n}\right) - \left(\varphi_n^* - \varphi_{3-n}^*\right) e^{-2i\tau}\right] = 0 \tag{2.34}$$

and conjugate equation. Exactly as in the case of an isolated oscillator, one can now introduce a "slow" time $\tau_1 = \varepsilon\tau$ together with a "fast" time τ and look for a solution in the form of power series by parameter ε

$$\varphi_n = \sum_j \varphi_{n,j}\varepsilon^j \tag{2.35}$$

Substituting (2.35) into the equations of motion, selecting the terms of the same order of the small parameter and equating them to zero, one obtains (for $\nu_n = 1$):

$$\frac{\partial \varphi_{n,0}}{\partial \tau} = 0, \quad \varphi_{n,0} = \varphi_{n,0}(\tau_1) \tag{2.36}$$

$$\frac{d\varphi_n}{d\tau} = -\frac{d\varphi_{n,0}}{d\tau_1} - i\left[\left(\varphi_{n,0} - \varphi_{3-n,0}\right) - \left(\varphi^*_{n,0} - \varphi^*_{3-n,0}\right) e^{-2i\tau}\right] = 0 \tag{2.37}$$

Integration of (2.37) by τ yields secular growth. To avoid it, we should put:

$$\frac{d\varphi_{n,0}}{d\tau_1} + i\left(\varphi_{n,0} - \varphi_{3-n,0}\right) = 0 \tag{2.38}$$

and it is the main asymptotic approximation and tractable model of the problem. The conventional manner of its solution uses linearity and superposition of in-phase ($\varphi_{n,0} = \varphi_{3-n,0}$) and out-of phase ($\varphi_{n,0} = -\varphi_{3-n,0}$) normal modes. Such an approach is not applicable to nonlinear problems. Therefore we will use an alternative way, based on integrals of motion. Such integrals will appear also in the nonlinear problems. Let us make an additional change of variables

$$\varphi_{n,0} = e^{-i\tau} f_{n,0} \tag{2.39}$$

Then the equations of motion are written as follows:

$$\frac{df_{n,0}}{d\tau_1} - if_{3-n,0} = 0 \tag{2.40}$$

and the integrals of motion have the form:

$$\begin{aligned} H &= i\left(f_{1,0} f^*_{2,0} + f_{1,0}{}^* f_{2,0}\right) = H_0 \\ &\left|f_{1,0}\right|^2 + \left|f_{2,0}\right|^2 = N. \end{aligned} \tag{2.41}$$

The second of these integrals motivates the following change of variables:

$$f_{1,0} = \sqrt{N}\cos\theta\,(\tau_1)\,e^{-i\delta_1\tau_1},\ f_{2,0} = \sqrt{N}\sin\theta\,(\tau_1)\,e^{-i\delta_2\tau_1} \tag{2.42}$$

Substituting (2.42) into (2.40) and separating real and imaginary parts, one obtains the following equations for dependent variables θ and $\Delta = \delta_1 - \delta_2$:

$$\frac{d\theta}{d\tau_1} = \sin\Delta; \quad \frac{d\Delta}{d\tau_1} = ctg\theta\cos\Delta \tag{2.43}$$

This system has two different fixed points (1) $\theta = \frac{\pi}{2}$, $\Delta = 0$; (2) $\theta = \frac{\pi}{2}$, $\Delta = \pi$, corresponding to the in-phase and out-in-phase linear normal modes respectively. The phase trajectories close to the stationary points describe periodic motions with weak energy exchange between the oscillators (the normal modes are cooperative motions in which an identity of every oscillator is not seen). But for phase trajectories lying far from the stationary points the situation is very different. Let us consider the limiting phase trajectory (LPT) (Manevitch, 2005), which is maximally distant from the stationary points. This trajectory corresponds to a complete periodic energy exchange between the oscillators. The period of such an energy exchange equals $2\pi/\varepsilon$ (if to return to starting time variable). If, in the initial instant, all energy is concentrated on one of the oscillators (e.g., $\theta = 0$), the energy exchange with other oscillators requires a very large time if $\varepsilon << 1$. So, over a rather long time we can distinguish the oscillators as, to a certain degree, independent ones. However, consideration of the tractable model in slow time demonstrates *beating* due to the internal resonance. This is a transition to a true two-particle model without domination of any oscillator.

In spite of the linearity of the starting equations of motion, the Eq. (2.43) in the variables θ, Δ are nonlinear. They may be linearized in the vicinities of both stationary points:

$$\begin{cases} \ddot{\theta}_1 + \theta_1 = 0; & \theta_1 = \theta - \dfrac{\pi}{4} \\ \ddot{\Delta} + \Delta = 0 \end{cases} \tag{2.44}$$

$$\begin{cases} \ddot{\theta}_1 + \theta_1 = 0; & \Delta_1 = \Delta - \pi \\ \ddot{\Delta}_1 + \Delta_1 = 0 \end{cases} \tag{2.45}$$

The dots denote the derivatives with respect to τ_1. So, these variables characterizing the relationship between the amplitudes and phase shift behave as a linear oscillator for $\theta_1 << 1$, $\Delta_1 << 1$.

The solution corresponding to the LPT can be also written in a rather simple form:

$$\theta = \tilde{\tau}, \ \ \Delta = \frac{\pi}{2} e \tag{2.46}$$

where $\tilde{\tau}(\tau_1)$ and $e(\tau_1)$ are non-smooth functions, presented in Fig. 2.7 (Pilipchuk, 1999b).

The solution which is close to LPT may be presented as

$$\begin{aligned} \theta &= X_1(\tau_1) + Y_1(\tau_1)\ e(\tau_1), \\ \Delta &= X_2(\tau_1) + Y_2(\tau_1)\ e(\tau_1). \end{aligned} \tag{2.47}$$

In turn, the smooth functions X_1, Y_1 can be determined as a power series with respect to τ_1:

$$X_i = \sum_j X_{i,j} {\tau_1}^j; \quad Y_i = \sum_j Y_{i,j} {\tau_1}^j \tag{2.48}$$

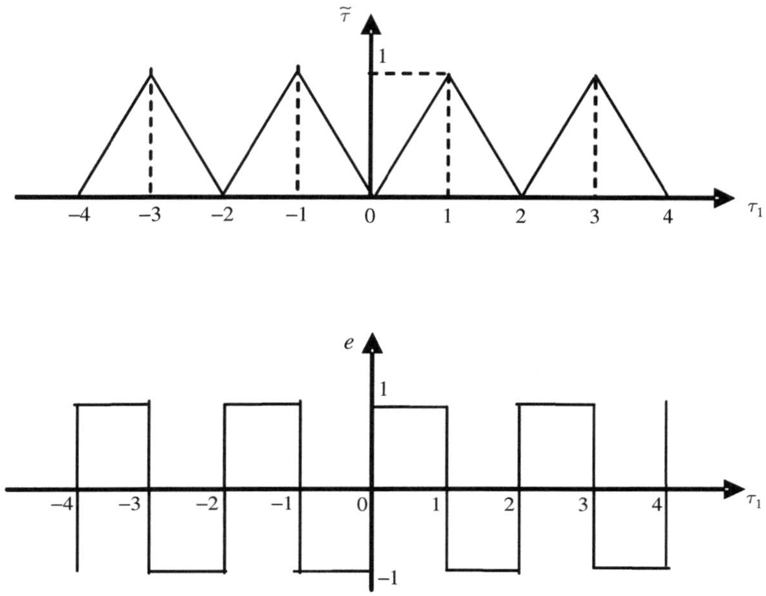

Fig. 2.7 Saw – tooth function $\tilde{\tau}(\tau_1)$ and its derivative $e(\tau_1)$

2.1.4.4 Exercise

Find the first correction to the functions $X_i(\tau)$, $Y_i(\tau)$ using series (2.48)

2.1.4.5 Strongly Coupled Oscillators with Essentially Different Frequencies

Going over to the system of strongly coupled oscillators, let us begin from the case of a strong difference between the frequencies due to a large magnitude of the masses ratio or of the stiffnesses ratio. In the first case (the stiffnesses are of the same order), the equations of motion are reduced to the form

$$\begin{aligned} &\frac{d^2u_1}{d\tau_1^2} + u_1 + \mu\,(u_1 - u_2) = 0 \\ &\varepsilon\frac{d^2u_2}{d\tau_1^2} + \frac{\mu}{\nu}u_2 + \mu\,(u_2 - u_1) = 0 \end{aligned} \tag{2.49}$$

where $\tau_1 = \sqrt{\frac{c_{11}}{m_1}}t$, $\mu = \frac{k}{c_{11}}$, $\nu = \frac{k}{c_{22}}$, $\varepsilon = \frac{m_2}{m_1}$, and u_i are normalized, as previously, by U_i^0

In the second case (the masses m_i are adopted to be identical),

$$\begin{aligned}&\frac{d^2u_1}{d\tau_1^2}+u_1+\mu\left(u_1-u_2\right)=0\\&\delta\frac{d^2u_2}{d\tau_1^2}+u_2+\mu\delta\left(u_2-u_1\right)=0\end{aligned}\tag{2.50}$$

where $\delta=\dfrac{c_{11}}{c_{22}}<<1.$

Let us begin with the first case. We use the new variables having the same order again, with respect to the small parameter ε and new time variable, differentiation by which does not change the orders of unknown functions

$$u_1=w_1,\ u_2=w_2\varepsilon^{\beta},\ \tau=\varepsilon^{\alpha}\tau_1\tag{2.51}$$

Substitution of (2.51) into (2.49) gives the equations

$$\begin{aligned}&\varepsilon^{2\alpha}\frac{d^2w_1}{d\tau^2}+w_1+\mu\left(w_1-\varepsilon^{\beta}w_2\right)=0\\&\varepsilon^{2\alpha+\beta+1}\frac{d^2w_2}{d\tau^2}+\varepsilon^{\beta}\frac{\mu}{\nu}w_2+\mu\left(\varepsilon^{\beta}w_2-w_1\right)=0\end{aligned}$$

A consistent asymptotic approximation is obtained for the following values of the parameters:

(1) $\alpha=0,\beta=0;$
(2) $2\alpha=-1,\beta=-1$

In the first case one obtains the following system:

$$\begin{aligned}&\frac{d^2w_1}{d\tau^2}+w_1+\mu\left(w_1-w_2\right)=0\\&w_2+\mu\left(w_2-w_1\right)=0\end{aligned}\tag{2.52}$$

If, for simplicity, one supposes that $c_{11}=c_{22},\ \nu=\mu$, the following simplified equations are obtained:

$$\frac{d^2w_1}{d\tau^2}+\frac{1+2\mu}{1+\mu}w_1=0,\ w_2=\frac{\mu}{1+\mu}w_1,\tag{2.53}$$

This type of motion corresponds to cooperative mode because the displacements (and the energies) of both oscillators have the same order with respect to parameter ε.

In case (2) we have the limiting system

$$\frac{d^2w_2}{d\tau^2}+(1+\mu)\ w_2=0,\ \frac{d^2w_1}{d\tau^2}-\mu\,w_2=0.$$

We come to one of the important notions of the theory of vibrations – "partial system". The main equation with respect to the dominating displacement u_2 now describes the vibration of the system which may be obtained from the initial one if one "fixes" the other oscillator – $u_1 = 0$. This regime is rather different from the vibration of the isolated oscillator. The partial system lies in the "middle" between the isolated oscillator and cooperative normal modes; one of the oscillators is excited more intensively and its interaction with the weakly excited oscillator is taken into account in the main approximation. The additional equation describes, similarly to the case of weak coupling, a quasi-dynamical excitation of the other oscillator. One of the tractable models with one degree of freedom is again obtained asymptotically from a more complicated model.

In the case described by Eq. (2.50), after substitution (2.51) with small parameter δ one obtains:

$$\delta^{2\alpha}\frac{d^2w_1}{d\tau^2} + w_1 + \mu\left(w_1 - \delta^{\beta}w_2\right) = 0$$

$$\delta^{2\alpha+\beta+1}\frac{d^2w_1}{d\tau^2} + \delta^{\beta}w_2 + \mu\delta\left(\delta^{\beta}w_2 - w_1\right) = 0$$

The consistent pairs of the exponents α, β are somewhat different from the previous ones:

(1) $\alpha = 0, \beta = 0$;
(2) $\alpha = -1, \beta = -1$

and corresponding limiting systems are as follows:

For the case (1):

$$\frac{d^2w_1}{d\tau^2} + (1+\mu)\,w_1 = 0, w_2 - \mu\,w_1 = 0.$$

For the case (2):

$$\frac{d^2w_2}{d\tau^2} + w_2 = 0, \frac{d^2w_1}{d\tau^2} - \mu\,w_2 = 0.$$

Here, in both limiting cases, the main approximations correspond to the vibrations of the partial systems. So, if the essential difference of the frequencies of the oscillators is due to the strong mass asymmetry, then the strong coupling brings about the cooperative normal mode (in only one of two limiting systems).

2.1.4.6 Exercise

Describe the vibrations of the strongly coupled oscillators with essentially different frequencies after action of impact applied to one of oscillators.

2.1.4.7 Strongly Coupled Oscillators with Close Frequencies

It is in this case only that we completely lose a connection, not only with the notion of the isolated oscillator, but also with the concepts of the partial system and of the beats (which are in fact a weakened manifestation of the cooperative effects). As well as in the case of weak coupling between the oscillators with identical frequencies, the equations of motion have the shape (2.32), but the parameter ε is not small any more and the initial system does not contain any small coefficients. Therefore we change ε by γ ($\varepsilon = \gamma$). However, for the system with strong coupling between the oscillators having identical or close frequencies, a small parameter can be related to the deflection from the system of two identical oscillators. On particular, supposing that $\nu = 1 + \varepsilon$, one obtains in the limit $\varepsilon \to 0$ the following pair of equations

$$\begin{aligned} \frac{d^2u_1}{d\tau_1^2} + u_1 + \gamma\,(u_1 - u_2) = 0 \\ \frac{d^2u_2}{d\tau_1^2} + u_2 + \gamma\,(u_2 - u_1) = 0 \end{aligned} \tag{2.54}$$

This system has rather high symmetry (it is invariant under change $u_1 \leftrightarrow u_2$). As a consequence, one can easily derive an independent equations for the normal modes for the functions

$$W_1 = \frac{u_1 + u_2}{2}, \quad W_2 = \frac{u_2 - u_1}{2} \tag{2.55}$$

(e.g. after summing the equations and subtracting the first equation from the second one):

$$\frac{d^2W_1}{d\tau_1^2} + W_1 = 0, \quad \frac{d^2W_2}{d\tau_1^2} + (1 + 2\gamma)\,W_2 = 0$$

Both these equations describe the cooperative normal modes (in-phase and out-in-phase respectively). Linear combination of W_1 and W_2 gives the general solution for arbitrary initial conditions. There are also the motions of the oscillators, but these are "cooperative" ones.

2.1.4.8 Exercise

Describe the vibrations of the strongly coupled oscillators with close frequencies after action of initial pulse.

2.2 Single-DOF Nonlinear Oscillator

The next level of complexity corresponds to the model of a 1DOF nonlinear autonomous oscillator. Such a model is, generally speaking, described by the following equation:

$$\ddot{u} + f(u, \dot{u}) = 0 \tag{2.56}$$

Of course, no general analytic procedure exists for the description of Eq. (2.56). It seems that the only positive statement possible here is that the system can be completely represented in two-dimensional state space $(u, \dot{u})$. This fact has some significant consequences for the global dynamics – for instance, absence of chaotic regimes and a very restricted set of possible limit trajectories (Guckenheimer and Holmes, 2002). Still, these facts are not enough to understand the dynamics.

If one agrees that the function f in (2.56) depends only on the displacement, but not on the velocity, the situation becomes simpler and the solution is readily obtained in quadratures:

$$t - t_o = \int \frac{du}{\sqrt{2(E - F(u))}}, \quad F'(u) = f(u) \tag{2.57}$$

where E is the constant energy of the oscillator. The shape of the function $F(u)$ determines whether the trajectories of the system are bounded or unbounded for different values of E (Landau and Lifshits, 1976). From a certain viewpoint, this model is already tractable. Let us consider the oscillator with the potential function $F(u)$ depicted in Fig. 2.8.

According to (2.57), the motion is possible if the expression under the square root has a nonnegative value. Then, for different values of E the following motions are possible:

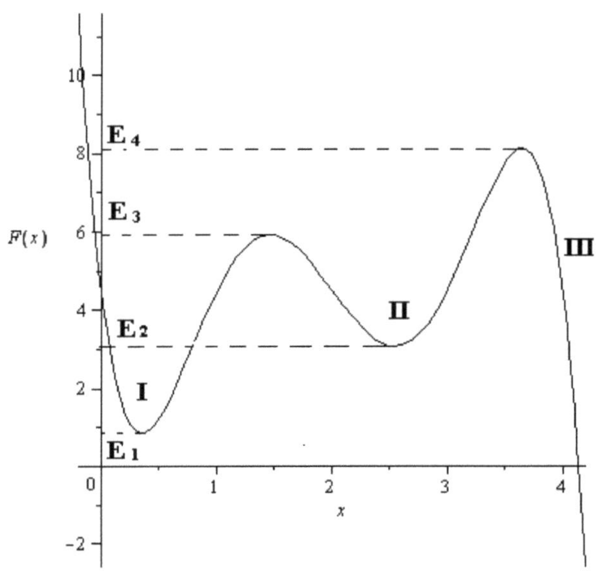

Fig. 2.8 Model potential function

- Unbounded escape into zone III – all values of energy (for $E < E_4$ the motion occurs only in zone III, for $E \geq E_4$ it will pass also over zones I and II).
- Oscillations in zone I – $E_1 \leq E \leq E_3$.
- Oscillations in zone II – $E_2 \leq E \leq E_3$.
- Oscillations over zones I and II – $E_3 \leq E \leq E_4$

A particular example of such qualitative analysis of the motion will be presented in Sect. 2.2.3.

If one is interested in more detailed information about the dynamics, it is necessary to compute and invert the integral (2.57). It may be a rather awkward task since no general algorithm exists for it. So, in order to assess main effects caused by the nonlinearity, we are going to resort to even simpler models, which allow deeper treatment.

2.2.1 Quasilinear Oscillator

The first and the most popular model to be considered here is that of the quasilinear oscillator, often referred to as Duffing model (Nayfeh and Mook, 1995). We restrict ourselves to the lowest order symmetric nonlinearity and linear viscous damping. The system may be written down as follows:

$$m\ddot{u} + \gamma\dot{u} + ku + au^3 = 0 \tag{2.58}$$

where m is the mass if the oscillator, k and γ-linear stiffness and damping coefficients respectively, a – the coefficient of the nonlinear term. Rescaling $u = Uw$, $t = \tau/\omega_0$ yields the following equation:

$$w_{\tau\tau} + \lambda w_\tau + w + \alpha w^3 = 0 \tag{2.59}$$

where $\omega_0 = \sqrt{k/m}$, $\lambda = \gamma/\omega_o m$, $\alpha = aU^2/m\omega_0^2$.

Equation (2.59) can be integrated exactly for the case $\lambda = 0$ in terms of Jacobi elliptic functions (Kosevitch and Kovalyov, 1989), but here we are interested in the *quasilinear* case, which corresponds to small deviations from the regime of linear oscillations and, therefore, to small values of α. It should be mentioned, that such a regime of motion can be realized for *any* values for the parameters in the initial Eq. (2.58), provided that the characteristic amplitude U is small enough. Hence, we can adopt $\alpha = 4\varepsilon/3$, $\varepsilon << 1$. The damping is also considered to be small and therefore is also adopted to be of order ε: $\lambda = \varepsilon\sigma$. Finally, Eq. (2.59) is considered in the form

$$w_{\tau\tau} + \varepsilon\sigma w_\tau + w + \frac{4\varepsilon}{3}w^3 = 0 \tag{2.60}$$

Transition to complex variables and change of the dependent variable yield:

$$\begin{aligned} &\varphi = e^{-i\tau}(w_\tau + iw) \\ &\varphi_\tau + \frac{\varepsilon\sigma}{2}(\varphi + \varphi^* e^{-2i\tau}) + \frac{i\varepsilon}{6}e^{-i\tau}(\varphi e^{i\tau} - \varphi^* e^{-i\tau})^3 = 0 \end{aligned} \tag{2.61}$$

Equation (2.61) can be approximately solved with the help of multiple-scales expansion:

$$\begin{aligned} &\tau_k = \varepsilon^k \tau, k = 0, 1, \ldots \\ &\varphi = \varphi_0 + \varepsilon\varphi_1 + \ldots, \; \varphi_k = \varphi_k(\tau_0, \tau_1, \ldots) \end{aligned} \tag{2.62}$$

Expansion of Eq. (2.61) according to (2.62) yields in orders ε^0 and ε^1:

$$\begin{aligned} &\frac{\partial \varphi_0}{\partial \tau_0} = 0 \Rightarrow \varphi_0 = \varphi_0(\tau_1, \ldots) \\ &\frac{\partial \varphi_1}{\partial \tau_0} + \left[\frac{\partial \varphi_0}{\partial \tau_1} + \frac{\sigma}{2}\varphi_0 - \frac{i}{2}|\varphi_0|^2 \varphi_0\right] + \varphi_0^* e^{-2i\tau_0} + \\ &+ \frac{i}{6}\left(\varphi_0^3 e^{2i\tau_0} + 3|\varphi_0|^2 \varphi_0^* e^{-2i\tau_0} - \varphi_0^{*3} e^{-4i\tau_0}\right) = 0 \end{aligned} \tag{2.63}$$

The expression in the square brackets does not depend on τ_0, therefore it will cause secular linear growth of φ_1, inconsistent with bounded oscillations. Therefore, for the sake of consistency, this secular term should disappear:

$$\frac{\partial \varphi_0}{\partial \tau_1} + \frac{\sigma}{2}\varphi_0 - \frac{i}{2}|\varphi_0|^2 \varphi_0 = 0 \tag{2.64}$$

Equation (2.64) can be easily solved as follows:

$$\begin{aligned} &\varphi_0 = N\exp(i\delta) \\ &N(\tau_1) = N_0 \exp\left(-\sigma\tau_1/2\right), \; \delta(\tau_1) = \frac{1}{2}N_0^2 \exp(-\sigma\tau_1)\tau_1 + \zeta_0 \end{aligned} \tag{2.65}$$

The dependence on higher time scales is not analyzed here and is omitted for the sake of brevity. Finally, the approximate "tractable" solution of Eq. (2.60) is written as follows:

$$\begin{aligned} &w(\tau) = N_0 \exp\left(-\frac{\varepsilon\sigma\tau}{2}\right) \times \\ &\times \sin\left(\tau(1 + \frac{\varepsilon}{2}N_0^2 \exp(-\varepsilon\sigma\tau) + O(\varepsilon^2)) + \xi_0\right) + O(\varepsilon) \end{aligned} \tag{2.66}$$

where N_0 and ξ_0 are determined by the initial conditions of the problem.

Expression (2.66) requires some further comments. First of all, from a formal viewpoint it is asymptotically not completely consistent – some corrections of order ε are "inside" the main approximation and some are "outside". Such distinction is justified since the remaining ε-order terms describe the principal effects of the damping (decrease of the amplitude) and the nonlinearity (dependence of the frequency on the amplitude). At Fig. 2.9a, b, the accuracy of approximate solution (2.66) is cross-checked with the result of the numeric solution of initial Eq. (2.60).

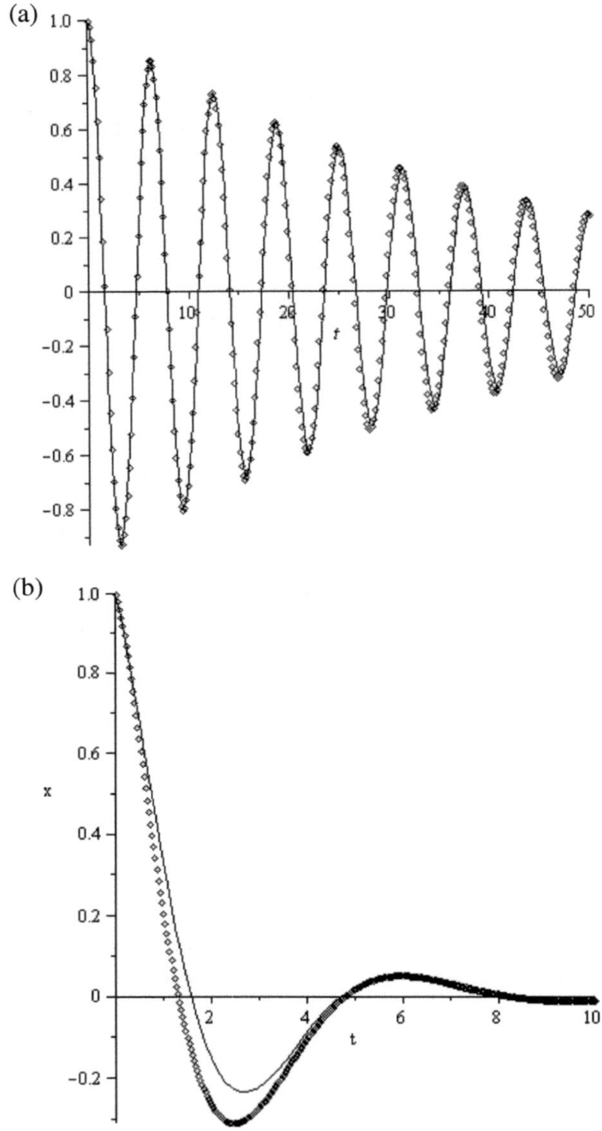

Fig. 2.9 Comparison between exact numeric and approximate analytic solution for the damped nonlinear oscillator. *Dotted line* corresponds to the numeric solution, *solid* to the approximate analytic solution, (**a**) $\varepsilon = 0.05$, $\sigma = 1$, x(0)=1, dx/dt(0)= –0.025; (**b**) $\varepsilon = 1$, $\sigma = 1$, x(0)=1, dx/dt(0)= –0.5

As one can see, for weak nonlinearity the coincidence is very good, although the discrepancy due to growing phase shift can be observed for larger times. Even for relatively strong nonlinearity (Fig. 2.9b) the approximation still works, but the inaccuracy is noticeable.

2.2.2 Strongly Nonlinear Oscillator and Vibro-Impact Approximation

As it was demonstrated in the previous section, the weakly nonlinear (or quasilinear) oscillator can be efficiently treated on the basis of the linear model; the effects of nonlinearity (such as frequency dependence on the amplitude) can be taken into account by asymptotic procedure. If the nonlinearity is strong, such an approach will become irrelevant and one should look for alternative ways to simplify the system.

An impact is arguably the strongest possible manifestation of the nonlinearity in mechanics. As for an appropriate potential function, the impact interaction corresponds to a vertical wall, which restricts the motion of the particle. Simple description of this process, dating back to Newton, interprets the impact as an abrupt change of the particle velocity

$$v(t_0 + 0) = -kv(t_0 - 0) \tag{2.67}$$

if the impact occurs at time t_0. The coefficient k lies in the interval $0 \leq k \leq 1$ and is referred to as the *restitution coefficient.* It characterizes the loss of mechanical energy in the impact.

Empiric relationship (2.67) allows efficient treatment of mechanical systems with impacts. It can be viewed as a boundary condition for the matching of trajectories before and after the impact. In addition, it allows one to use the tools of theory of generalized functions for investigation of the vibro-impact motions, as demonstrated in the next section.

2.2.2.1 Special Solutions for Vibro-Impact Motions

Let us consider the case of an absolutely elastic interaction between the particle and rigid barriers (the restitution coefficient is equal to the unity). The particle's motion in this case is described by the Newton equation:

$$m\frac{d^2u}{dt^2} + P\left(u, \frac{du}{dt}\right) = 0 \tag{2.68}$$

$$P\left(u, \frac{du}{dt}\right) = 2\frac{du}{dt}[\delta(u + \Delta) - \delta(u - \Delta)], \ \ |u \leq \Delta| \tag{2.69}$$

Coefficient 2 on the right-hand side of (2.69) appears since every elastic impact changes the velocity by $2du/dt$ at the point of impact. There are three natural approaches to the description of vibro-impact processes. The first one was proposed in (Zhuravlev and Klimov, 1988) and uses the evident solution of Eqs. (2.68 and 2.69) which is saw-tooth sine of period 2π and amplitude 1:

$$u(t) = \Delta\tau(\varphi) = \frac{2\Delta}{\pi}\arcsin[\sin(\varphi)], \ \frac{d\varphi}{dt} = (1/2)\pi v/\Delta \tag{2.70}$$

where ϕ is the phase and v the constant velocity of the particle. A non-smooth change of the dependent variable transforms Eq. (2.68) to the simplest form:

$$\frac{d^2\phi}{dt^2} = 0 \tag{2.71}$$

Another possibility proposed in papers (Manevitch et al., 1998; Pilipchuk, 1999a, b, 2001) is based on the change of the right-hand part of expression (2.69) by a periodic function reflecting a sequence of periodic pulses. A Delta-function describes impact interaction of the particle with a rigid barrier with the intensity depending on impulse $2p$. The generalized function can also be presented in a more compact form as the second derivative (in the sense of generalized functions, see e.g. Courant and Hilbert, 1962) of saw-tooth periodic function:

$$P\left(u, \frac{du}{dt}\right) = 2p\left\{\begin{array}{l}\sum\limits_{j=-\infty}^{\infty} \delta[t-(T/4+jt-\Phi)]- \\ -\delta[t-(-T/4+jt-\Phi)]\end{array}\right\} = p\frac{d^2\tau}{dt^2} \tag{2.72}$$

where T is the period of the oscillations, Φ is the phase. The period on the right-hand part of expression (2.72) is determined by the impulse intensity.

It is important to note that, contrary to initial system, when using the presentation (2.72), the system becomes formally linear, although with a non-smooth right-hand part.

This approach is closely connected with a non-smooth change of the independent (temporal) variable that may be naturally made by introducing of non-smooth sine-like and cosine-like basic functions $\tau(t), e(t)$ (see Fig. 2.7). The latter is considered as a generalized derivative of the former one. As it was mentioned above, these basic functions were first introduced by V. Pilipchuk (Pilipchuk, 1988, 1999b) and the corresponding procedure was further developed later (Pilipchuk, 2001); see also (Azeez et al., 1999).

Looking for the solution as

$$u = U(\tau) \tag{2.73}$$

we find, quite similarly to (2.71):

$$\frac{d^2U}{d\tau^2} = 0 \tag{2.74}$$

and

$$U(\tau) = \Delta\tau \tag{2.75}$$

So, the solution of the problem in non-smooth basis is even simpler than that for classical linear oscillator and is presented as the straight line: in real time it corresponds to periodic motion with periodically repeating impacts. Both Eqs. (2.71 and

2.74) are tractable models of the vibro-impact system and they can be extended to the cases of many degrees of freedom and forced vibrations.

The third approach is based on a canonical transformation of displacement and velocity to the action and angle variables. If, following Lin and Reichl (1986), one considers a particle of mass $1/2$ oscillating in an infinite square well potential, its Hamiltonian is written as follows:

$$H = p^2 + b[\eta(w-1) + \eta(-w-1)] \tag{2.76}$$

Where, again, w is the position of the particle, p is the momentum, η is the unit step function, b is the height of the square well.

New variables can be introduced by transformation:

$$w = -1 + (2\vartheta/\pi)sign[\sin(\vartheta)],\ p = (\pi I/2)sign[\sin\vartheta] \tag{2.77}$$

$\vartheta \in [-\pi, \pi], w \in [-1, 1]$ and p are periodic functions of ϑ with period 2π. This transformation can be easily predicted because it reproduces the uniform periodic motion of the particle between the rigid barriers with period $T = 2/|p|$ over the time period $[-\tau/2, \tau/2]$, if $I = (2\sqrt{E_0})/\pi$, $\vartheta = \pi^2 It/2$. This solution is related to a transformed Hamiltonian

$$H = \pi^2 I^2/4. \tag{2.78}$$

and equations of motion

$$\frac{dI}{dt} = \frac{dH}{d\vartheta}, \frac{d\vartheta}{dt} = -\frac{dH}{dI} \tag{2.79}$$

with initial conditions $t = 0 : I = (2/\pi)\sqrt{E_0}, \vartheta = 0$.

Let us now use the generalized functions for the description of a linear oscillator with natural frequency ω_0 and impact clearance Δ. If a linear oscillator vibrates between two rigid barriers, after introduction of non-smooth basic functions and presentation of the solution in the form

$$u = U\left(\frac{4}{T}\tau\right),\ u \le \Delta \tag{2.80}$$

we obtain the tractable model (Pilipchuk, 2001) described by this equation:

$$\left(\frac{4}{T}\right)^2 \frac{d^2}{d\tau^2}U + \omega_0^2 U = 0, \frac{dU}{d\tau}[|\tau = \pm 1] = \left(\frac{T}{4}\right)^2 p \tag{2.81}$$

Taking into account (together with (2.80)) the condition $U=\pm\Delta$ when $\tau=\pm 1$, one can find the solution of the boundary value problem in the form:

$$u = \frac{\sin[(T/4)\omega_0\tau]}{\sin[(T/4)\omega_0]} \tag{2.82}$$

If $u(0) = 0$ it means that $\phi = 0$ and total energy is expressed as

$$E = 1/2\left(\frac{du}{dt}\right)^2 \tag{2.83}$$

and

$$\omega_0(T/4) = \pm\frac{1}{2}\arccos\left(1 - \frac{\omega_0^2\Delta^2}{E}\right) + k\pi, k = 0, 1, 2\ldots \tag{2.84}$$

There is a critical value of energy

$$E = E_{cr} = \frac{\omega_0^2\Delta^2}{2} \tag{2.85}$$

of the extent that the oscillator can reach the constraints if $E \geq E_{cr.}$.

In the case of a unilateral barrier (e.g. if the left hand barrier is removed) the problem can be solved similarly (Pilipchuk, 2001).

There is a natural limit case that can be found from analysis of Eq. (2.79). It corresponds to condition $\omega_0 \rightarrow 0$ when

$$u \rightarrow \Delta\tau = \Delta\tau\left(\frac{t}{T/4} + \Phi\right) \tag{2.86}$$

corresponding to the solution for the free particle between two rigid barriers.

2.2.2.2 Vibro-Impact Systems Treatable by Analytic Functions

The vibro-impact model can be considered as a limiting case for the systems with strongly nonlinear but smooth potentials of interaction which may be of the power or exponential type. This situation is typical when considering solids on the molecular level, because atomic potentials of interaction in the compression region are commonly described by power or exponential functions (Lennard–Jones or Morse potentials respectively).

Let us briefly discuss how to formally change a strongly nonlinear but smooth elastic force and impact interaction. The former can be described, e.g. by power nonlinearity of high degree such as $w^{2n+1}, n. >> 1$. After introducing the dimensionless variable $W = w/\Delta$, where Δ is the amplitude, the latter corresponds to the limiting case $n \rightarrow \infty$. As it is shown by Andrianov (1993), one can formally change a smooth nonlinearity by impact interaction with the help of Laplace transformation of the power elastic force

$$\phi(p) = p^{-n-1}\gamma(n+1, p). \tag{2.87}$$

where $\gamma(n+1, p)$ is an incomplete Gamma-function. After its expansion by a small parameter $1/n$ and returning to originals, the elastic force is presented as follows (in terms of generalized functions):

$$f(W) = (1/n+1)[\delta(W-1) - \delta(W+1)]- \\ -[1/(n+1)(n+2)]\left[\frac{d\delta}{dt}(W-1) - \frac{d\delta}{dt}(W+1)\right] + \ldots \tag{2.88}$$

Thus we replace the strongly nonlinear power force by impact interaction and the equation of motion in principal approximation is presented as follows:

$$\frac{d^2 W}{dt^2} + \frac{\Delta^{n-1}}{n+1}[\delta(W-1) - \delta(W+1)] = 0, \tag{2.89}$$

where the multiplier before squared brackets plays the role of the impulse change which is the result of "collision".

So, the system is transformed to a view typical for the particle vibrating between two rigid barriers, in striking similarity to the procedure described in the previous paragraph. Construction of the corrections to the main asymptotic approximation is also discussed by Andrianov (1993).

The argument can also be used in the opposite direction – instead of imposing exact impact conditions, it is possible to simulate the impact-like motion with the help of smooth potential functions (Gendelman, 2006; Gendelman and Meimukhin, 2007; Babitsky and Veprik, 1998; Sokolov et al., 2007). Two common models of this sort are the models of two-sided impact, where the particle is allowed to move between two restraints, and one-sided impact – with only one restraint. In the former case, the smooth-function approximation takes advantage of the potential of the shape x^{2n} with n-positive integer and the motion is considered for x in the interval $(-\infty,+\infty)$. Still, if n is large enough, the motion of the particle is localized in the vicinity of the interval (–1,1). Therefore, impact restraints at points ± 1 are simulated. In the case of one-sided impact the potential $x^{-\alpha}$, $\alpha > 0$ is used, simulating the impact restraint at $x = 0$. In this case, the motion is restricted to the interval $(0, \infty)$.

The smooth-function models of impact interactions described above were designed for the case of purely elastic impacts (with unity restitution coefficient). One should mention, that there exist a number of models which simulate the inelastic impact with the help of viscoelastic elements (Babitsky and Veprik, 1998; Sokolov et al., 2007), but this approach is completely linear.

Let us generalizations of the strongly nonlinear smooth models for the case of the inelastic impact for both the cases of two-sided and one-sided impacts. A common one-dimensional model for simulating the elastic impacts by means of smooth functions is formulated with the help of the following equation:

$$\ddot{u} + (n+1)u^{2n+1} = 0 \tag{2.90}$$

where u denotes the displacement of the particle, n is the positive integer. Limit $n \rightarrow \infty$ corresponds to the motion of a free particle between impact restraints at $u = \pm 1$.

Our goal is to generalize the model described by Eq. (2.90) in order to describe the impacts with non-unity restitution coefficient. In other terms, we are looking for the model described by the equation

$$\ddot{u}+f(\dot{u},u)+(n+1)u^{2n+1}=0 \tag{2.91}$$

which will satisfy the following conditions:

$$\begin{aligned}&As\ \mathrm{n}\rightarrow\infty,\ f(\dot{u},u)\rightarrow 0,|u|<1,\\ &f(\dot{u},u)\rightarrow\infty,|u|=1\end{aligned} \tag{2.92}$$

$$\begin{aligned}&For\ \mathrm{u}=0:\\ &\kappa\ |\dot{u}(0)|_{before\ impact}=|\dot{u}(0)|_{after\ impact}\end{aligned} \tag{2.93}$$

where κ is the restitution coefficient. The value of κ should not depend on the value of the initial velocity of the particle.

Condition (2.93) for invariance of the restitution coefficient with respect to velocity imposes severe restrictions on possible shapes of the function f. Scaling with respect to time should preserve the symmetry of the equation; therefore the function f must have a shape

$$f(\dot{u},u)=\dot{u}^{2m+1}g(u) \tag{2.94}$$

with m – nonnegative integer. Further simplifications are based on the fact that velocity independence of the "restitution coefficient" simulated by Eq. (2.93) manifests additional internal symmetry of the system, i.e. its invariance with respect to a certain nontrivial group. Infinitesimal Lie generators of such a group for Eq. (2.91) may be written as (Bluman and Kumei, 1989):

$$\begin{aligned}\mathrm{Z}&=\xi(u,t)\frac{\partial}{\partial t}+\eta(u,t)\frac{\partial}{\partial u}\\ \mathrm{Z}_1&=(\eta_t+p(\eta_u-\xi_t)-p^2\xi_u)\frac{\partial}{\partial p}\\ \mathrm{Z}_2&=(\eta_{tt}+p(2\eta_{ut}-\xi_{tt})+p^2(\eta_{uu}-2\xi_{ut})-\\ &\quad-p^3\xi_{uu}+r(\eta_u-2\xi_t)-3pr\xi_u)\frac{\partial}{\partial r}\end{aligned} \tag{2.95}$$

where Z_1 and Z_2 are the first and the second prolongations of the infinitesimal operator Z respectively, p ≡ du/dt, r ≡ $\mathrm{d}^2\mathrm{u}/\mathrm{dt}^2$. The symmetry condition for Eq. (2.91) with an account of (2.94) is:

$$(\mathrm{Z}+\mathrm{Z}_1+\mathrm{Z}_2)(r+p^{2m+1}g(u)+(n+1)u^{2n+1})=0 \tag{2.96}$$

Equation (2.96) may be easily solved by standard methods (Bluman and Kumei, 1989). The result is summarized below:

$$\eta = Cu, \xi = -Ctn \text{ provided that } m = 0, g(u) = \mu u^n$$

where C and μ are constants. It is convenient to rescale $n = 2k, \mu = \lambda(2k+1)$. Condition (2.92) imposes additional restrictions (the effective dissipation in the system should be positive) and finally the required system will be presented as follows:

$$\ddot{u} + \lambda(2k+1)\dot{u}u^{2k} + (2k+1)u^{4k+1} = 0 \tag{2.97}$$

with k – positive integer. Equation (2.97) is integrable, although non-Hamiltonian. Substituting $w(u) = \dot{u}, y(u) = u^{-2k-1}w(u)$, one finally gets the integral of Eq. (2.97):

$$(y^2 + \lambda y + 1)^{1/2} \exp\left(\frac{-\lambda}{\sqrt{1-\lambda^2/4}} \tan^{-1}\left(\frac{y}{\sqrt{1-\lambda^2/4}}\right)\right) u^{2k+1} = \text{const} \tag{2.98}$$

For the initial conditions $u(0) = 0, \dot{u}(0) = v_0$, one gets in the vicinity of the initial point: $y \to +\infty$, $w(u) \to v_0$. After one "impact" we have $y \to -\infty$, $w(u) \to -v_1$ as $u \to 0$. Thus, from Eq. (2.98) one obtains:

$$\kappa = \left|\frac{v_1}{v_0}\right| = \exp\left(-\frac{\pi\lambda}{2\sqrt{1-\lambda^2/4}}\right) \tag{2.99}$$

Velocity ratio at $u = 0$, expressed by Eq. (2.99), depends neither on the initial velocity nor on k. Consequently, it represents the genuine restitution coefficient in the limit $k \to \infty$. Therefore, Eq. (2.97) provides a suitable model for the description of an inelastic two-sided impact by means of smooth functions.

The standard model for description of the one-sided impact may be written as

$$\ddot{u} - \frac{\alpha - 1}{2u^\alpha} = 0 \tag{2.100}$$

$\alpha > 0$ may be not integer, since only the motion for $u > 0$ is considered in this problem. The velocity of the particle as $u \to \infty$ should be considered as velocity before and after impact. Equation (2.100) may also be modified in order to describe the inelastic one-sided impacts. The consideration is similar to the one presented in the previous section and it is not necessary to repeat it here. The modified model may be thus written as

$$\ddot{u} - \lambda\dot{u}\frac{\alpha - 1}{2u^{(\alpha+1)/2}} - \frac{\alpha - 1}{2u^\alpha} = 0 \tag{2.101}$$

The restitution coefficient will be also expressed by Eq. (2.99).

2.2.3 Oscillator with Multiple Equilibriums

In the previous sections we have considered the single-DOF oscillators with weakly nonlinear and extremely nonlinear (vibro-impact) potential functions. In all cases considered, the oscillator possessed only one state of equilibrium and all oscillations occurred around this point. However, in many applications it is necessary to consider the motion of the oscillator with multiple (or an infinite number of) states of equilibrium. Typical examples of this sort may be molecules with multiple conformational states (Grosberg and Khokhlov, 1989) or buckling structural elements (Manevitch et al., 1989).

The general qualitative treatment of such an oscillator has been described above in introduction to Sect. 2.2. For the generic shape of the potential function, each state of stable equilibrium is the bottom of a "potential well". Qualitatively, oscillations in every single potential well may be described by quasilinear approximation. If the motion is bounded and the potential function is "nonlinear enough", then for high energies the approximation of the vibro-impact type may be used.

There exists an additional interesting limit case, which corresponds to intermediate energy levels. If the states of the stable equilibrium are multiple and isolated, then they are separated by the unstable equilibrium points. Let us treat a motion in the vicinity of such a point for the particular example of the potential function with three equilibria:

$$F(u) = \frac{1}{2}(-u^2 + u^4) \tag{2.102}$$

Corresponding equation of motion

$$\ddot{u} - u + 2u^3 = 0 \tag{2.103}$$

has two stable fixed points for $u = \pm 1/\sqrt{2}$ and one unstable fixed point for $u = 0$. Phase portrait of the system is presented in Fig. 2.10

The general solution of Eq. (2.103) is written as:

$$t - t_0 = \int \frac{du}{\sqrt{2E - u^2 + u^4}} \tag{2.104}$$

The integral can be computed in terms of Jacobi elliptic functions. Even without such computation, it is easy to recognize that for $-1/8 < E < 0$ the trajectory will oscillate around one of the equilibriums and for $E > 0$ it will surround both equilibriums. The trajectory, which corresponds to $E = 0$, will separate these two qualitatively different types of motion. It is referred to as *separatrix*. Quite naturally, it will approach the saddle point (0, 0) at the phase portrait. The explicit equation for this trajectory will read

$$u_s(t) = \pm \frac{1}{\cosh(t - t_0)} \tag{2.105}$$

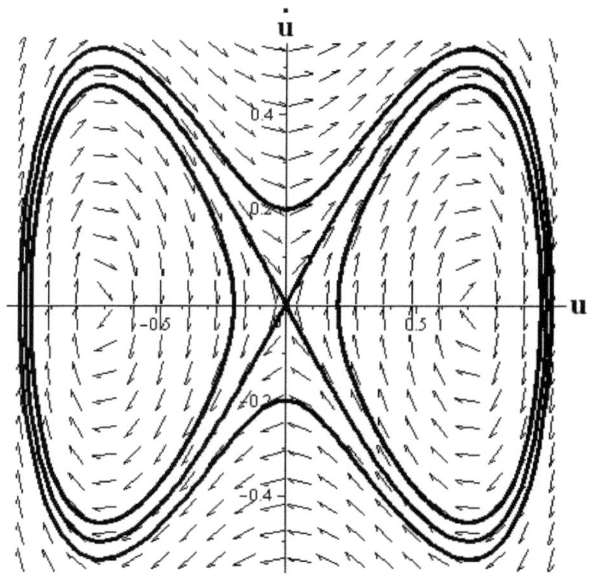

Fig. 2.10 Phase portrait of Eq. (2.103)

"Plus" and "minus" correspond to the two branches of the separatrix. It is easy to see that the velocity of the oscillator at the separatrix decreases exponentially with time as $t \to \pm\infty$; then, the whole loop of the separatrix will take an infinite time. It should be mentioned that the separatrix approaches the saddle point as $t \to \pm\infty$. Such phase trajectories are often referred to as *homoclinic connections* and are generic in single-DOF conservative systems with multiple states of equilibrium.

2.3 Forced Nonlinear Oscillator

2.3.1 General Remarks

The purpose of this section is to analyze the energy exchange between a source of harmonic excitation and the Duffing oscillator under the condition of a primary 1:1 resonance. It is well known that a nearly-resonance excitation may induce two stable solutions of small and large amplitudes. The solution corresponding to large amplitude is associated with the efficient energy exchange between the source and the oscillator.

As shown below, motion of the system can be divided into two stages. The first stage, associated with maximum energy pumping from the source to the oscillator, is characterized by fast large deviation from the initial position. At the second stage, motion of the system is approaching stationary resonance oscillations near the steady state. At the first stage, an approximate description of motion is based on the concept of the limiting phase trajectories (LPT) suggested in (Manevitch, 2007) and mentioned above (Sect. 2.1.4) when considering the linear weakly coupled oscillators with close natural frequencies.

2.3.2 Governing Equations

We investigate the transient response of the Duffing oscillator in the presence of resonance 1:1. The dimensionless equation of motion is written as

$$\frac{d^2u}{dt^2} + 2\varepsilon\gamma\frac{du}{dt} + u + 8\alpha\varepsilon u^3 = 2\varepsilon F sin[(1+\varepsilon s)t + \Theta] \tag{2.106}$$

where $\varepsilon > 0$ is a small parameter. We recall that maximum energy pumping from the source of excitation into the oscillator takes place if the oscillator is initially at rest; this corresponds to the initial conditions

$$t = 0, u = 0; \frac{du}{dt} = 0 \tag{2.107}$$

An orbit satisfying conditions (2.107) is in fact *the limiting phase trajectory* (LPT) – see above, Sect. 2.1.4.

Following (Manevitch, 1999, 2001), we introduce the complex variables as follows:

$$\varphi = e^{-it}\left(\frac{du}{dt} + iu\right), \varphi^* = e^{it}\left(\frac{du}{dt} - iu\right) \tag{2.108}$$

where $i = \sqrt{-1}$; the asterisk denotes a complex conjugate. Inserting φ, φ^* from (2.108) into (2.106), we obtain the following (still exact) equation of motion

$$\begin{aligned}&\frac{d\varphi}{d\tau} + \varepsilon\gamma(\varphi + \varphi^* e^{-2i\tau}) + \\ &+ \varepsilon i\alpha[\varphi^3 e^{2i\tau} - (\varphi^*)^3 e^{-4i\tau} - 3\,|\varphi|^2\,\varphi + 3\,|\varphi|^2\,\varphi^* e^{-2i\tau}] = \\ &= 2\varepsilon^2 e^{-i\tau} F \sin[(1+\varepsilon s)\tau + \Theta],\end{aligned} \tag{2.109}$$

The multiple time-scale approach (Kevorkian and Cole, 1996; Nayfeh, 2000) is applied to construct an approximate solution of (2.109). To this end, we introduce the following transformation

$$\begin{aligned}\varphi(t,\varepsilon) &= \varphi_0(\tau_0,\ \tau_1) + \varepsilon\varphi_1(\tau_0,\ \tau_1) + \ldots \\ \frac{d}{dt} &= \frac{\partial}{\partial\tau_0} + \varepsilon\frac{\partial}{\partial\tau_1}, \frac{d^2}{dt^2} = \frac{\partial^2}{\partial\tau_0^2} + 2\varepsilon\frac{\partial^2}{\partial\tau_0\partial\tau_1} + \ldots\end{aligned} \tag{2.110}$$

where $\tau_0 = t$ and $\tau_1 = \varepsilon t$ are the fast and slow time-scales, respectively. A similar representation is valid for the function φ^*. Then we substitute expressions (2.110) into (2.109) and equate the coefficients of equal powers of ε. In the leading-order (ε^0) approximation we obtain the equation

$$\frac{\partial \varphi_0}{\partial \tau_0} = 0, \tag{2.111}$$

which implies that a leading-order approximation for φ is a slowly-varying function to be found at the next level of approximation.

Equating the coefficients of order ε leads to

$$\begin{aligned} &\frac{\partial \varphi_1}{\partial \tau_0} + \frac{\partial \varphi_0}{\partial \tau_1} + \gamma(\varphi_0 + \varphi_0^* e^{-2i\tau_0}) - i\alpha[\varphi_0^{*^3} e^{-4i\tau_0} - 3|\varphi_0|^2 \varphi_0^* e^{-2i\tau_0} \\ &+ 3|\varphi_0|^2\varphi_0 - \varphi_0^3 e^{2i\tau_0}] = -iF\left(e^{i(s\tau_1+\Theta)} - e^{-i(2\tau_0+s\tau_1+\Theta)}\right). \end{aligned} \tag{2.112}$$

Equation (2.112) gives an $O(\varepsilon)$-approximation of the slow dynamics. In order to avoid the secular growth of $\varphi_1(\tau_0, \tau_1)$ with respect to the fast time τ_0, i.e., to avoid a response not uniformly valid with increasing time, we eliminate non-oscillating terms from (2.112) (Kevorkian and Cole, 1996). This yields the following equation for φ_0

$$\frac{\partial \varphi_0}{\partial \tau_1} + \gamma\varphi_0 - 3i\alpha|\varphi_0|^2\varphi_0 = -iFe^{i(s\tau_1+\Theta)}, \quad \varphi_0(0) = 0. \tag{2.113}$$

Then, as usual, we introduce the polar representation

$$\varphi_0 = ae^{i\delta} \tag{2.114}$$

where a and δ represent a real amplitude and a real phase of the process $\varphi_0(\tau_1)$. Inserting (2.114) into (2.113) and separating the real and imaginary parts, Equation (2.113) is transformed into the system

$$\begin{aligned} \frac{da}{d\tau_1} + \gamma\alpha &= -F\sin\Delta \\ a\frac{d\Delta}{d\tau_1} &= -sa + 3\alpha a^3 - F\cos\Delta \end{aligned} \tag{2.115}$$

where $a > 0$, $\Delta = \delta - (s\tau_1 + \Theta)$. It now follows from (2.108, 2.114) that

$$u(t,\varepsilon) = a(\tau_1)\sin[t + \delta(\tau_1)] + O(\varepsilon), \quad \tau_1 = \varepsilon t \tag{2.116}$$

This means that the amplitude $a(\tau_1)$ and the phase $\delta(\tau_1)$ completely determine the process $u(t,\varepsilon)$ (in the leading-order approximation). Note that $a = 0$ if the oscillator is not excited.

2.3.3 The Dynamics of the Oscillator without Dissipation and the LPT

In this section, we recall main definitions and results concerning the dynamics of the underlying non-dissipative system. In the absence of damping ($\gamma = 0$), system (2.115) is rewritten as

$$\begin{aligned} \frac{da}{d\tau_1} &= -F\sin\Delta \\ a\frac{d\Delta}{d\tau_1} &= -sa + 3\alpha a^3 - F\cos\Delta \end{aligned} \tag{2.117}$$

It is easy to prove that system (2.117) is integrable; it conserves the integral of motion

$$H = \frac{3}{4}\alpha a^4 - \frac{sa^2}{2} - Fa\cos\Delta = H_0, \tag{2.118}$$

where H_0 depends on initial conditions. In the phase plane, the LPT corresponds to the contour $H = 0$, as only in this case the LPT goes through the point $a = 0$. Taking $H_0 = 0$, we obtain the following expression

$$H = a\left(\frac{3\alpha a^3}{4} - \frac{sa}{2} - F\cos\Delta\right) = 0. \tag{2.119}$$

Excluding the solution $a \equiv 0$ from consideration, we obtain a as a solution of the cubic equation

$$\frac{3\alpha a^3}{4} - \frac{sa}{2} - F\cos\Delta = 0. \tag{2.120}$$

Equation (2.120) gives the second initial condition $a(0^+) = 0$, $\cos\Delta(0^+) = 0$. We suppose that $da/d\tau_1 > 0$ at $\tau_1 = 0^+$; under this assumption, $\Delta(0^+) = -\pi/2$. Hence the initial conditions for the LPT take the form

$$\tau_1 = 0^+, a(0^+) = 0, \Delta(0^+) = \frac{-\pi}{2}. \tag{2.121}$$

In the following, we will write 0 instead of 0^+, except as otherwise noted.

The steady state of system (2.117) can be found as the solution of the equations

$$\frac{da}{d\tau_1} = 0, \ \frac{d\Delta}{d\tau_1} = 0 \tag{2.122}$$

This yields

$$-sa + 3\alpha a^2 - F\cos\Delta = 0 \tag{2.123}$$

where $\cos\Delta = \pm 1$. Due to periodicity of the solution in Δ, it is sufficient to study only the cases $\Delta = 0$ or $\Delta = -\pi$. We analyze the properties of Eq. (2.123) through the properties of its discriminant D

$$D = \frac{1}{9\alpha^2}\left(\frac{F^2}{4} - \frac{s^3}{81\alpha}\right) \tag{2.124}$$

If $D < 0$, Eq. (2.123) has 3 different real roots; if $D > 0$, Eq. (2.123) has a single real and two complex conjugate roots; if $D = 0$, two real roots merge (Korn and Korn, 2000.). The latter condition gives a critical value of the parameter α

$$\alpha^* = \frac{4s^3}{81F^2} \tag{2.125}$$

A straightforward investigation proves that, if $\alpha < \alpha^*$ (weak nonlinearity), then there exist two stable centres C_-: $(-\pi, a_-)$, C_+: $(0, a_+)$, and an intermediate unstable hyperbolic point O: $(-\pi,\ a_0)$; if $\alpha > \alpha^*$ (strong nonlinearity), then there exists only a single stable centre C_+: $(0, a_+)$. Note that oscillations around the centre C_- is associated with the quasilinear dynamics of the system; oscillations around the centre C_+ correspond to strongly nonlinear motion.

As seen in Figs. 2.11 and 2.12, the shape of the LPT depends on the parameters of the system. We next find a critical value α_{cr} ensuring the transition from small quasilinear ($\alpha < \alpha^*$) to large-amplitude strongly nonlinear ($\alpha > \alpha^*$) oscillations. In the critical case $\alpha_{\text{cr}} = \alpha^*$, an unstable hyperbolic point coincides with the maximum of the left branch of the LPT at $\Delta = -\pi$ (Fig. 2.13).

The discriminant of Eq. (2.120) at $\Delta = -\pi$ equals

$$D = \frac{4}{\alpha^2}\left(F^2 - \frac{2s^3}{81\alpha}\right) \tag{2.126}$$

The condition $D = 0$ yields a critical parameter

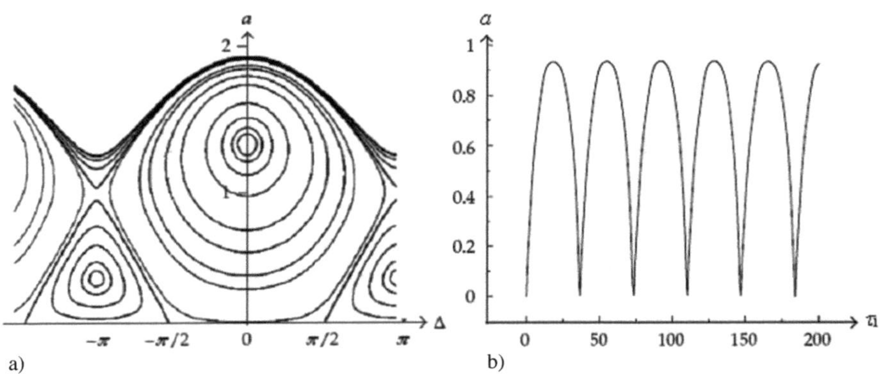

Fig. 2.11 Phase portrait (**a**) and plot of $a(\tau_1)$ (**b**) for quasi-linear oscillations: $s = 0.4$, $F = 0.13$, $\alpha = 0.093$

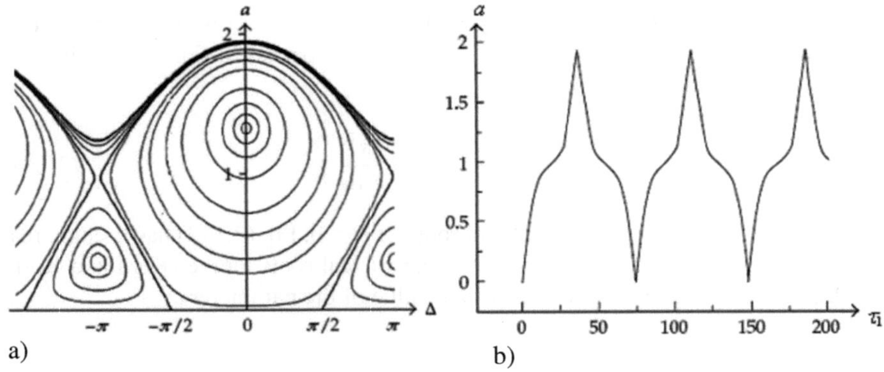

Fig. 2.12 Phase portrait (**a**) and plot of $a(\tau_1)$ (**b**) for strongly nonlinear oscillations: $s = 0.4$, $F = 0.13$, α=0.094

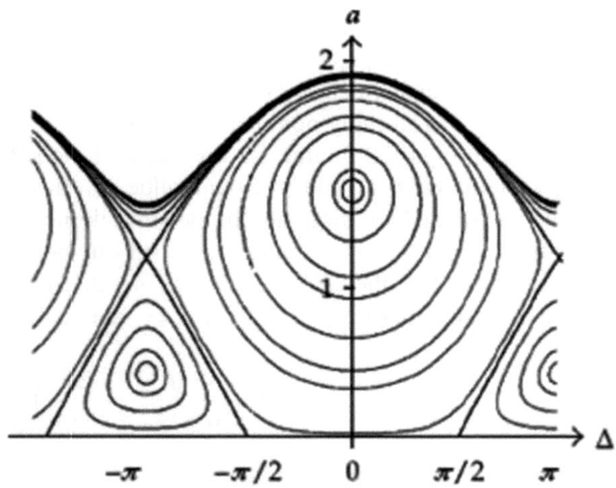

Fig. 2.13 Transition from small to large oscillations

$$\alpha_{cr} = \frac{2s^3}{81F^2} = \frac{\alpha^*}{2} \tag{2.127}$$

which defines a boundary between small quasi-linear ($\alpha < \alpha_{\mathrm{cr}}$) and large nonlinear ($\alpha > \alpha_{cr}$) oscillations.

In particular, for $s = 0.4$, $F = 0.13$ we obtain α_{cr}=0.0935. Figures 2.11 and 2.12 are plotted for $\alpha < \alpha_{cr}$ and $\alpha > \alpha_{cr}$, respectively. In Fig. 2.11a, one can see the LPT encircling the center C_- of relatively small oscillations; Fig. 2.12a shows the LPT encircling the centre C_+ of large oscillations; this case is associated with the maximum energy absorption. Figures 2.11b and 2.12b demonstrate the temporal behavior of the function $a(\tau_1)$ corresponding to the respective branch of the LPT. Note that both branches of the LPT begin at the same point (2.121).

If $\alpha = \alpha^* = 2\alpha_{cr}$, the above-mentioned coincidence of the stable and unstable points at $\Delta = -\pi$ results in the transformation of the phase portraits (Fig. 2.14) and

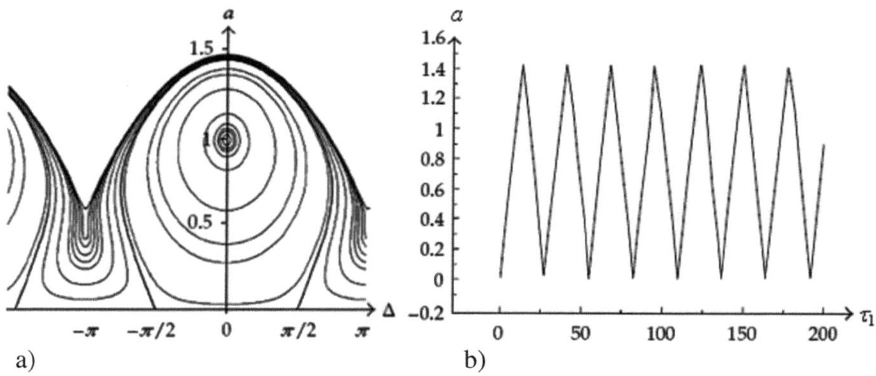

Fig. 2.14 Phase portrait (**a**) and plot of $a(\tau_1)$ (**b**) for $s = 0.4$, $F = 0.13$, $\alpha = 0.187$

disappearance of the stable centre C_{-}. Figure 2.14a demonstrates the existence of a single stable fixed point corresponding to $\Delta = 0$.

2.3.4 The Transient Dynamics of a Weakly Damped Oscillator

In this section we investigate the dynamics of a weakly damped oscillator with strong nonlinearity ($\alpha > \alpha_{cr}$). As mentioned previously, strongly nonlinear oscillations of large amplitude are associated with maximum energy absorption; it is often the case of particular interest.

As one can see in Fig. 2.15, the damped system exhibits strongly nonlinear behavior on the time interval $[0, \tau_1{}^*]$; an instant $\tau_1{}^*$ corresponds to the first

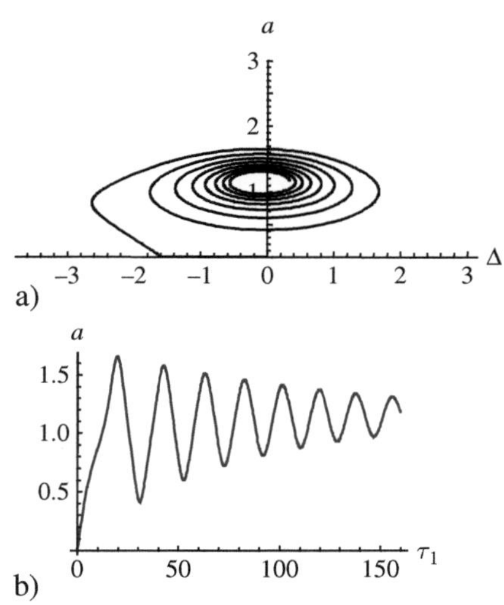

Fig. 2.15 Phase portrait (**a**) and plot of $a(\tau_1)$ (**b**) for strongly nonlinear oscillations: $s = 0.4$, $F = 0.13$, α=0.094, $\gamma = 0.01$

maximum of the function $a\ (\tau_1)$. After that, motion becomes similar to smooth oscillations around the stationary point. This allows separating the transient dynamics into two stages. While on the interval $0 \le \tau_1 \le \tau_1{}^*$ motion is close to the LPT of the undamped system, at the second stage, $\tau_1 \ge \tau_1{}^*$, motion is similar to quasi-linear oscillations.

In the remainder of this section we study nonlinear motion at the first stage of motion. Quasi-linear oscillations will be studied in Sect. 2.3.5.

2.3.4.1 Non-smooth Temporal Transformations

Figure 2.14b shows that strongly nonlinear oscillations are similar to motion of a particle moving with constant velocity between two symmetric motion-limiters. A connection between smooth and vibro-impact modes of motion in a smooth nonlinear oscillator has been revealed in (Pilipchuk, 1988). A detailed discussion of this effect can be found in (Vakakis et al., 1996).

Owing to similarity of the system dynamics to vibro-impact motion, the method of non-smooth transformations (Pilipchuk, 1999b, 2001) can be used to describe the first stage of motion. In order to extend this method to systems with damping, we introduce the new fast and slow time scales. The slow time scale is defined as $t_0 = \gamma\tau_1$; the fast time scale τ is represented as a saw-tooth sine (2.70) (see Fig. 2.7), and ${}^{d\varphi}\!/_{d\tau_1} = \omega(\tau)$.

We now construct the solution of (2.117) in the form

$$\begin{aligned} a(\tau_1) &= X_1(\tau,\ t_0) + e(\phi)Y_1(\tau,\ t_0) \\ \Delta(\tau_1) &= X_2(\tau,\ t_0) + e(\phi)Y_2(\tau,\ t_0). \end{aligned} \tag{2.128}$$

Using the slow and fast time scales, the derivatives with respect to τ_1 are

$$\frac{d}{d\tau_1} = \frac{\partial}{\partial\tau}\frac{\partial\tau}{\partial\phi}\frac{\partial\phi}{\partial\tau_1} + \frac{\partial}{\partial t_0}\frac{\partial t_0}{\partial\tau_1} + \frac{\partial}{\partial\phi}\frac{\partial\phi}{\partial\tau_1} = e\omega\frac{\partial}{\partial\tau} + \gamma\frac{\partial}{\partial t_0} + \omega\frac{\partial}{\partial\phi} \tag{2.129}$$

We now recall that

$$\frac{\partial e}{\partial\varphi} = \delta(\phi - n)$$

with Dirac's delta-function on the right-hand side, $n = 1, 2, \ldots$. The δ-singularity is excluded by requiring

$$Y_{1,2} = 0,\ \frac{\partial X_{1,2}}{\partial\tau} = 0 \text{ at } \tau = 1, 2, \ldots, \tag{2.130}$$

Using (2.128, 2.129), we obtain the following system

$$
\begin{aligned}
\frac{da}{d\tau_1} &= e\omega\frac{\partial X_1}{\partial \tau} + \gamma\frac{\partial X_1}{\partial t_0} + \omega\frac{\partial Y_1}{\partial \tau} + e\gamma\frac{\partial Y_1}{\partial t_0},\\
\frac{d\Delta}{d\tau_1} &= e\omega\frac{\partial X_2}{\partial \tau} + \gamma\frac{\partial X_2}{\partial t_0} + \omega\frac{\partial Y_2}{\partial \tau} + e\gamma\frac{\partial Y_2}{\partial t_0}
\end{aligned}
\tag{2.131}
$$

provided $Y_{1,2} = 0$ at $\tau = 1, 2, \ldots$. To obtain the equations for X_i, Y_i, $i = 1, 2$, we insert (2.131) into (2.117) and separate the terms with and without e. This yields the set of equations

$$
\begin{aligned}
&\omega\frac{\partial Y_1}{\partial \tau} + \gamma\left(\frac{\partial X_1}{\partial t^0} + X_1\right) = -F\sin X_2\cos Y_2\\
&\omega\frac{\partial X_1}{\partial \tau} + \gamma\left(\frac{\partial Y_1}{\partial t^0} + Y_1\right) = -F\sin Y_2\cos X_2\\
&\omega\left(X_1\frac{\partial Y_2}{\partial \tau} + Y_1\frac{\partial X_2}{\partial \tau}\right) + \gamma\left(X_1\frac{\partial X_2}{\partial t^0} + Y_1\frac{\partial Y_2}{\partial t^0}\right) +\\
&+sX_1 - 3\alpha X_1^3 - 9\alpha X_1Y_1^2 = -F\cos Y_2\cos X_2\\
&\omega\left(X_1\frac{\partial X_2}{\partial \tau} + Y_1\frac{\partial Y_2}{\partial \tau}\right) + \gamma\left(X_1\frac{\partial Y_2}{\partial t^0} + Y_1\frac{\partial X_2}{\partial t^0}\right) +\\
&+sY_1 - 3\alpha Y_1^3 - 9\alpha Y_1X_1^2 = F\sin Y_2\sin X_2
\end{aligned}
\tag{2.132}
$$

2.3.4.2 The Construction of a Generating Solution

Considering γ as a small parameter, we construct an approximate solution in the form of an expansion

$$
\begin{aligned}
&X_i(\tau, t_0) = X_{i0}(\tau, t_0) + \gamma X_{i1}(\tau, t_0)\ldots,\\
&Y_i(\tau, t_0) = Y_{i0}(\tau, t_0) + \gamma Y_{i1}(\tau, t_0) + \ldots, i = 1, 2;\\
&\omega(t_0) = \omega_0(t_0) + \gamma\omega_1(t_0) + \ldots
\end{aligned}
\tag{2.133}
$$

Inserting (2.133) into (2.132) and considering only the leading order terms, we find that $Y_{10} = 0$, $X_{20} = 0$; in this case the variables X_{10} and Y_{10} satisfy the equations

$$
\begin{aligned}
&\omega_0\frac{\partial X_{10}}{\partial \tau} + F\sin Y_{20} = 0\\
&\omega_0 X_{10}\frac{\partial Y_{20}}{\partial \tau} + sX_{10} - 3\alpha X_{10}^3 + F\cos Y_{20} = 0
\end{aligned}
\tag{2.134}
$$

The generating system (2.134) obviously corresponds to the undamped system ($\gamma = 0$). System (2.134) is integrable, yielding the following first integral of motion

$$
H = X_{10}\left(\frac{3}{4}\alpha X_{10}^3 - \frac{s}{2}X_{10} - F\cos Y_{20}\right) = 0. \tag{2.135}
$$

By analogy with (2.121), we obtain the initial conditions for system (2.134) in the form

$$X_{10} = 0, Y_{20} = -\pi/2, \omega_0 \frac{\partial X_{10}}{\partial \tau} = F \text{ at } \tau = 0, t_0 = 0. \tag{2.136}$$

For the further analysis, it is convenient to transfer (2.134) into the second-order form. Using (2.135) to exclude Y_{20}, the resulting equation is written as

$$\frac{\partial^2 X_{10}}{\partial \tau^2} + \lambda_0^2 f(X_{10}) = 0 \tag{2.137}$$

where $\lambda_0(t_0) = 1/\omega(t_0)$, and

$$f(X_{10}) = \frac{s^2}{4} X_{10} - \frac{3s\alpha}{2} X_{10}^3 + \frac{27\alpha^2}{16} X_{10}^5$$

In addition, we use the following approximations to conditions (2.130):

$$\frac{\partial X_{10}}{\partial \tau} = 0, Y_{20} = 0 \, at \, \tau = 1. \tag{2.138}$$

In contrast to (2.130), conditions in (2.138) are not independent; they are equivalent by virtue of the first Eq. (2.134). Using this equivalence, we involve only the first equality in the further analysis. It is worth noting that equalities (2.138) have a clear physical meaning: $X_{10}(\tau)$ attains its maximum at an instant τ such that $Y_{20}(\tau) = 0$.

In order to highlight the substantial features of the solution, we use an approach known as vibro-impact approximations (Vakakis et al., 1996). Following this approach, we represent the solution for problems (2.137, 2.138) as

$$X_{10} = x_0 + x_1 \tag{2.139}$$

where it is assumed that $|x_1(\tau,\ t_0)| << |x_0(\tau,\ t_0)|$. The correctness of the latter assumption will be shown below. The generating solution x_0 is chosen as the solution of the equation

$$\frac{\partial^2 x_0}{\partial \tau^2} = 0 \tag{2.140}$$

The initial conditions follow from (2.138)

$$x_0 = 0, \omega_0(0) \frac{\partial x_0}{\partial \tau} = F \, at \, \tau = 0, t_0 = 0. \tag{2.141}$$

The solution of Eq. (2.140) satisfying (2.141) is

$$x_0(\tau, t_0) = A_0\,(t_0)\,\tau;\ A_0\,(0)\,\omega_0\,(0) = F. \tag{2.142}$$

Formulas (2.142) show that if $\gamma = 0$, then the solution $x_0(\tau, 0) = A_0(0)\tau$ corresponds to the motion of a particle between two rigid stops.

The next approximation x_1 is governed by the following equation

$$\begin{aligned}\frac{\partial^2 x_1}{\partial \tau^2} &= -\lambda_0^2(t_0) f(x_0)\\ x_1(\tau, t_0) &= -\lambda_0^2(t_0)\int_0^\tau (\tau - \xi) f(A_0(t_0)\xi)\mathrm{d}\xi\end{aligned} \tag{2.143}$$

Combining solutions (2.142) and (2.143) and imposing constraints (2.140), we find

$$\frac{\partial X_{10}}{\partial \tau} = \frac{\partial x_0}{\partial \tau} + \frac{\partial x_1}{\partial \tau} = 0, \tau = 1 \tag{2.144}$$

and, therefore,

$$\begin{aligned}\omega_0^2 A_0 &= \int_0^1 f(A_0\xi)\mathrm{d}\xi,\ \omega_0^2 = A_0^{-2}\Psi(A_0)\\ \Psi(A_0) &= \int_0^{A_0} f(s)ds = \frac{A_0^2}{8}\left(\frac{3}{2}\alpha A_0^2 - s\right)^2\end{aligned} \tag{2.145}$$

Obvious transformations yield a connection between $\omega_0(t_0)$ and $A_0(t_0)$

$$\omega_0^2(t_0) = \frac{1}{8}\left[\frac{3}{2}\alpha A_0^2(t_0) - s\right]^2 \tag{2.146}$$

If we consider $A_0(t_0)$ and $\omega(t_0)$ as the amplitude and the phase of the saw-tooth sine function $x_0(\tau, t_0) = A_0(t_0)\tau(\phi(\tau_1))$, then (2.146) represents the slowly varying backbone curve of system (2.139) (Meirovitch, 2000).

It follows from (2.142, 2.146) that the initial values $\omega_0(0)$ and $A_0(0)$ can be found from the equations

$$\begin{aligned}&\Psi(A_0(0)) = [A_0(0)\omega_0(0)]^2 = F^2\\ &A_0^2(0)\left[\frac{3\alpha}{2}A_0^2(0) - s\right]^2 = 8F^2,\ \omega_0(0) = \frac{F}{A_0(0)}\end{aligned} \tag{2.147}$$

By definition, $t_0 = 0$ as $\gamma = 0$. This implies that the values $\omega_0(0)$ and $A_0(0)$ can be interpreted as the frequency and the amplitude of the undamped system.

Given $A_0(t_0)$ and $\omega_0(t_0)$, formula (2.143) yields

$$x_1(\tau, t_0) = -\frac{A_0(t_0)\tau^3}{4\omega^2(t_0)}\left\{\frac{9\alpha^2}{56}[A_0(t_0)\tau]^4 - \frac{3s\alpha}{5}[A_0(t_0)\tau]^2 + \frac{s^2}{6}\right\} \tag{2.148}$$

We now find the function $Y_{20}(t_0,\tau)$. Arguing as above, we construct a generating solution Y_{20} in the form

$$Y_{20} = y_0 + y_1 \tag{2.149}$$

where it is assumed that $|y_1(\tau,\ t_0)| << |y_0(\tau,\ t_0)|$ in an interval of interest. The leading order term y_0 can be found as a solution to the first Eq. (2.134), in which $X_{10} = x_0$. This yields

$$y_0\,(t_0) = -\arcsin\left(\frac{F}{A_0(t_0)\omega(t_0)}\right),\ y_0\,(0) = -\frac{\pi}{2} \tag{2.150}$$

The term $y_1(t_0,\tau)$ is defined by the second Eq. (2.134). As before, we take $X_{10} = x_0$ and exclude $\cos Y_{20}$ by (2.135). Then we obtain

$$\begin{aligned} \frac{\partial y_1}{\partial \tau} &= \frac{1}{\omega_0}\left(-\frac{s}{2} + \frac{9}{4}\alpha A_0^2\tau^2\right) \\ y_1(\tau,t_0) &= \frac{1}{\omega_0(t_0)}\left(-\frac{s\tau}{2} + \frac{3}{4}\alpha A_0^2(t_0)\tau^3\right) \end{aligned} \tag{2.151}$$

It will be shown below that the correction terms x_1, y_1 are negligibly small compared to x_0, y_0. This allows taking approximations $X_{10} = x_0$, $Y_{20} = y_0$ into the subsequent analysis.

2.3.4.3 Calculation of $A_0(t_0)$ and $\omega_0(t_0)$

We find $A_0(t_0)$ and $\omega_0(t_0)$ by proceeding to the next order of approximation. To this end, we use the procedure of iterative approximations. A detailed review of iterative approximations algorithms can be found e.g. in (Berinde, 2007).

The first iteration is constructed as:

$$X_1^{(1)} = X_{10} + \gamma X_{11}, Y_2^{(2)} = y_{20} + \gamma Y_{21}, X_2^{(2)} = \gamma X_{21}, Y_1^{(1)} = \gamma Y_{11}, \omega^{(1)} = \omega_0 + \gamma\omega_1 \tag{2.152}$$

Inserting $X_i^{(1)}$, $Y_i^{(1)}$ into (2.132), taking into account γ-order terms, and letting $X_{10} = x_0$, $Y_{20} = y_0$, we obtain two independent sets of equations

$$\begin{aligned} &\omega^{(1)}\frac{\partial Y_1^{(1)}}{\partial \tau} + F\sin\left(X_2^{(1)}\right)\cos(Y_2^{(1)}) = -\gamma\left(\frac{\partial x_0}{\partial t_0} + x_0\right) \\ &\omega^{(1)}X_1^{(1)}\frac{\partial X_2^{(1)}}{\partial \tau} + sY_1^{(1)} - 9\alpha Y_1^{(1)}\left(X_1^{(1)}\right)^2 - \\ &-F\sin(Y_2^{(1)})\sin(X_2^{(1)}) = -\gamma\ x_0\frac{\partial y_0}{\partial t_0}, \end{aligned} \tag{2.153}$$

and

$$\omega^{(1)}\frac{\partial X_1^{(1)}}{\partial\tau}+F\sin(Y_2^{(1)})=0,$$
$$\omega^{(1)}X_1^{(1)}\frac{\partial Y_2^{(1)}}{\partial\tau}+sX_1^{(1)}-3\alpha(X_1^{(1)})^3+F\cos Y_2^{(1)}=0. \tag{2.154}$$

Choosing $x_0(\tau,\ t_0)$ as a solution of the equation

$$\frac{\partial x_0}{\partial t_0}+x_0=0 \tag{2.155}$$

we find

$$\frac{dA_0}{dt_0}+A_0=0,\ A_0\,(t_0)\ =\ A_0\,(0)\exp(-t_0) \tag{2.156}$$

and, therefore,

$$x_0(\tau,\ t)=A_0\,(0)\,\tau e^{-t_0}, \tag{2.157}$$

where $A_0(0)$ can be found from (2.147). Substitution of (2.156) into (2.146) yields:

$$\omega_0\,(t_0)=\frac{1}{2\sqrt{2}}\left|\frac{3}{2}\alpha A_0^2\,(0)\exp(-2t_0)-s\right| \tag{2.158}$$

Insertion $A_0(t_0)$ and $\omega_0(t_0)$ from (2.156, 2.158) into (2.150) and (2.151) determines $y_0(t_0)$ and $y_1(\tau,\ t_0)$. The approximation holds until ω_0 is of order unity.

As an example, we consider a system with the parameters

$$\gamma=0.1\,(or\,\gamma=0), s=0.2,\ \alpha=0.333;\ F=1, \theta=0 \tag{2.159}$$

Figure 2.16 demonstrates the shape of $\omega_0(t_0)$, $t_0=\gamma\tau$.

Figure 2.17a, b depict the LPT for the undamped system (2.115) and the solutions X_{10} for system (2.134) with parameters (2.159). As the frequency $\omega_0(t_0)$ decreases, the time to reach the first maximum of X_{10} in system (2.116) exceeds a similar interval in its undamped counterpart (3.1) but the maximum value of X_{10} in (2.115) is less than in (2.117). For the undamped system (2.117) we have the maximum $M={X^*}_{10}\approx 1.835$ at ${\tau_1}^*\approx 1.835$ in the leading order approximation, and $M\approx 1.686$ at ${\tau_1}^*\approx 2$ for the numerical solution; for system (2.115) with $\gamma=0.1$ we have the maximum

$$M=X_{10}^*\approx 1.46 \text{ at } {\tau_1}^*\approx 2.4. \tag{2.160}$$

Figure 2.17b proves that the correction term x_1 is negligibly small. In a similar way, one can evaluate the small term y_1. This implies that one can ignore the terms x_1, y_1 in the further analysis and let

$$X_{10}=x_0, Y_{20}=y_0 \tag{2.161}$$

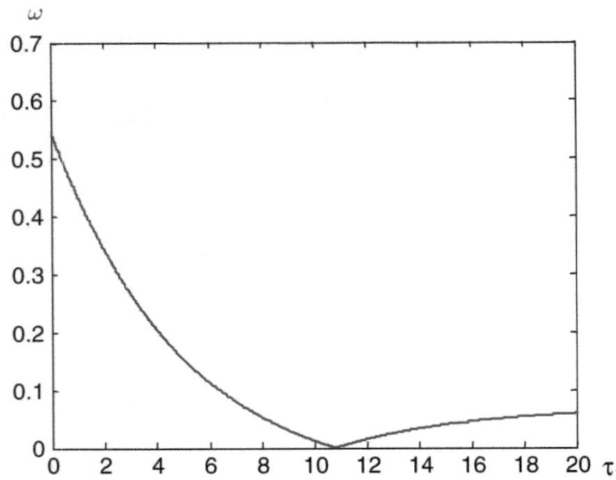

Fig. 2.16 Function $\omega_0(t_0)$

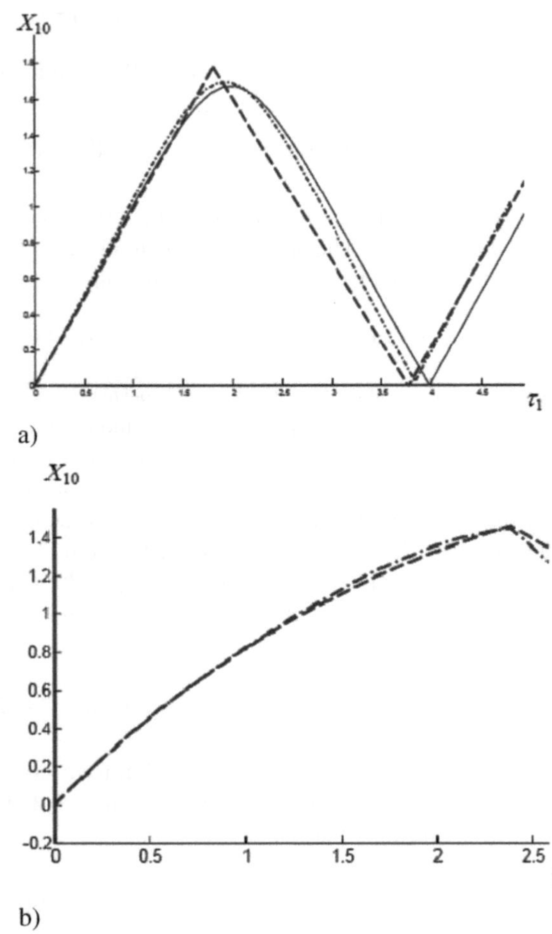

Fig. 2.17 (**a**) Numerical solution (*solid line*) and two analytical approximations (*dashed* – the leading order, *dash-and-dot* – the first order), corresponding to LPT; (**b**) Analytical solution for X_{10} (*dashed* – the leading order, *dash-and-dot* – the first order)

Explicit estimates of x_1 and y_1 can be found by the direct calculation by formulas (2.144) and (2.151). For brevity, we omit the calculation of the terms X_{11}, Y_{21} as well as the determination of higher order approximations.

Finally, we determine an instant $\tau_1{}^*$ at which the function X_{10} reaches its first maximum. Given $\omega_0(t_0)$, we obtain from Eq. (2.128)

$$\int_0^{\tau_1^*} \omega_0(\gamma s) ds = 1. \tag{2.162}$$

2.3.5 Quasi-Linear Oscillations

In this section we examine quasi-linear oscillations in the second interval of motion, $\tau > 1$. It is easy to see that an orbit of the dissipative system tends to its steady state as $\tau \to \infty$. The steady state O: (a_0, Δ_0) for system (2.115) is determined by the equality

$$a^2 \left[(s - 3\alpha a^2)2 + \gamma^2 \right] = F^2 \tag{2.163}$$

or, for sufficiently small γ,

$$\begin{aligned} &\gamma a_0 = -F \sin\Delta_0, \; s a_0 - 3\alpha a_0^3 = -F \cos \Delta_0 \\ &\Delta_0 \approx -\gamma a_0 / F + O(\gamma^3), \; a_0(s - -3\alpha a_0^2) = -\, F + O(\gamma^2). \end{aligned} \tag{2.164}$$

In fact, the smallness of γ is required only in the second set of Eq. (2.164).

We define deviations from the steady state as

$$\xi = a - a_0, \eta = \Delta_0. \tag{2.165}$$

In addition, we must impose the matching conditions

$$a_0 + \xi = x_0^*, \frac{d\xi}{d\tau_1} = 0 \text{ at } \tau_1 = \tau_1 \tag{2.166}$$

where $x_0{}^* = x_0(\tau_1{}^*)$. The instant $\tau_1{}^*$ is determined by formula (2.151).

As above, we construct an approximate solution by the multiple scale method. First, we suppose that the contribution of the nonlinear terms and dissipation force in the overall response is relatively small. Secondly, to account for the small effects, we introduce the parameter μ symbolizing small terms. As a result, we obtain

$$\begin{aligned} &\frac{d\xi}{d\tau_1} + F\eta = -\mu\gamma\xi \\ &\frac{d\eta}{d\tau_1} - \frac{k_1}{a_0}\xi = \mu\left(\frac{s}{a_0^2}\xi^2 + \frac{F}{2a_0}\eta^2\right) - \mu\gamma\eta. \end{aligned} \tag{2.167}$$

where

$$k_1 = 9\alpha a_0^2 - s, k_2 = 18\alpha a_0,$$

In the final solution we let $\mu = 1$. We introduce the fast and slow time scales as $\theta_0 = \tau_1, \theta_1 = \mu\theta_0$. As above, the solution is sought as the expansion

$$\xi = \xi_0 + \mu\xi_1 + \cdots, \eta = \eta_0 + \mu\eta_1 + \cdots, \tag{2.168}$$

and

$$\begin{aligned} \frac{d\xi}{d\tau_1} &= \frac{\partial\xi_0}{\partial\theta_0} + \mu\left(\frac{\partial\xi_0}{\partial\theta_1} + \frac{\partial\xi_1}{\partial\theta_0}\right) + \ldots \\ \frac{d\eta}{d\tau_1} &= \frac{\partial\eta_0}{\partial\theta_0} + \mu\left(\frac{\partial\eta_0}{\partial\theta_1} + \frac{\partial\eta_1}{\partial\theta_0}\right) + \ldots \end{aligned} \tag{2.169}$$

Using the procedure similar to that of Sect. 2.3.4, we obtain the equations of the leading order approximation

$$\frac{\partial\xi_0}{\partial\theta_0} + F\eta_0 = 0, \quad \frac{\partial\eta_0}{\partial\theta_0} - \frac{k_1}{a_0}\xi_0 = 0, \tag{2.170}$$

with the matching conditions

$$\xi_0 = x_0^* - a_0\frac{\partial\xi_0}{\partial} = 0 \,\text{at}\, \theta_0 = \tau_1^*. \tag{2.171}$$

If $k_1 > 0$, the solution of system (2.170, 2.171) is written as

$$\xi_0 = (\theta_0, \theta_1) = C_0(z_1)\cos(kz_0), \eta_0(\theta_0, \theta_1) = rC_0(z_1)\sin kz_0, \tag{2.172}$$

where we denote

$$z_0 = \theta_0 - \tau_1^* = \tau_1 - \tau_1^* > 0, z_1 = \varepsilon z_0, k^2 = Fk_1/a_0 > 0, r = k/F.$$

The slowly varying function $C_0(z_1)$ will be found at the next step of approximation.
The equations of the first order approximations take the form

$$\begin{aligned} &\frac{\partial\xi_1}{\partial\theta_0} + F\eta_1 = -\left(\frac{\partial\xi_0}{\partial\theta_1} + \gamma\xi_0\right) \\ &\frac{\partial\eta_1}{\partial\theta_0} - \frac{k_1}{a_0}\xi_1 = \left(\frac{s}{a_0^2}\xi_0^2 + \frac{F}{2a_0}\eta_0^2\right) - \left(\frac{\partial\eta_0}{\partial\theta_1} + \gamma\eta_0\right). \end{aligned} \tag{2.173}$$

Insertion of (2.172) into (2.173) and exclusion of the secular terms yield

$$C_0(z_1) = c_o e^{-\gamma Z1}, c_0 = x_0^* - a_0. \tag{2.174}$$

and

$$\begin{aligned}\xi_1(\theta_0,\theta_1) &= -\frac{1}{2\kappa^2}Fc_0^2e^{-2\gamma z_1}\left(K_1-\frac{K_2}{3}\cos 2kz_0\right)\\ \eta_1(\theta_0,\theta_1) &= -\frac{1}{F}\frac{\partial\xi_1}{\partial z_0}=-\frac{K_2}{3\kappa}c_0^2e^{-2\gamma z_1}\sin 2kz_0,\end{aligned} \tag{2.175}$$

where

$$K_1=\frac{s}{a_0^2}+\frac{F}{2a_0},\ K_2=\frac{s}{a_0^2}-\frac{F}{2a_0}$$

In particular, for system (2.173) with the parameters (2.159), we find

$$x_0^*=1.46, a_0=1.065, \Delta_0=0.1.k_1=3.2, k_2=6.39, K_1=0.645, K_2=-0.294$$

and, therefore,

$$c_0=0.395, k=\sqrt{3} \tag{2.176}$$

Since $\gamma << \kappa$; the assumption of small dissipation holds. Finally, we obtain

$$\begin{aligned}\xi_1(\theta_0,\theta_1) &= -c_0^2e^{-2\gamma z_1}(0.108+\ 0.098\cos 2kz_0)\\ \eta_1(\theta_0,\theta_1) &= 0.075c_0^2e^{-2\gamma z_1}\sin 2kz_0\end{aligned} \tag{2.177}$$

It is obvious that $|\xi_1(\theta_0,\ \theta_1)| << |\xi_0(\theta_0,\ \theta_1)|$, $|\eta_1(\theta_0,\ \theta_1)| << |\eta_0(\theta_0,\ \theta_1)|$, and, therefore, the linear approximation suffices to describe the second stage of motion (Fig. 2.18).

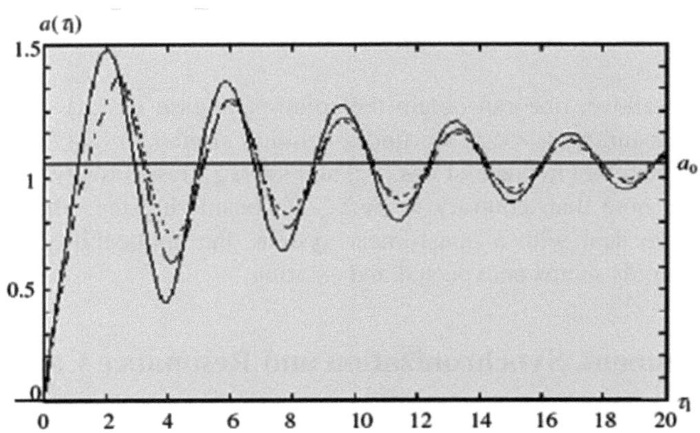

Fig. 2.18 Transient dynamics of system (2.115): *solid line* – numerical solution; *dot-and-dash line* – segment (2.157); *dashed line* – nonlinear solution (2.168); *dotted line* – linear solution (2.172)

Figure 2.18 demonstrates a good agreement between the numerical solution of Eq. (2.115) with the parameters (2.157) and the approximate solution found by matching the segment (2.157) (dot-and-dash line) with the approximate analytic solution of the linearized (dotted line) and nonlinear (dashed line) systems at the point $\tau_1{}^*$; the latter solution is calculated by (2.168, 2.172, 2.177) with $\mu = 1$. As seen in Fig. 2.18, small nonlinear terms (2.177) allow considering a weak asymmetry of the nonlinear solution. Despite a certain discrepancy in the initial interval of motion, the numerical and analytic solutions approach closely as τ_1 increases. In particular, this implies that a simplified model, expressed by (2.157) for $0 \leq \tau_1 \leq \tau_1{}^*$, and linear solution (2.172) for $\tau_1 \geq \tau_1{}^*$, suffices to describe a complicated near-resonance dynamics.

Results of numerical integration for the original Eq. (2.106) are given in Fig. 2.19

We now correlate numerical and analytic results. As seen in Fig. 2.19, the first maximum of the slowly varying envelope of the process $u(t,\varepsilon)$ equals $M_1 \approx 1.5$; it is reached at the instant $t^* \approx 20$, or $\tau_1{}^* \approx 2$; the second maximum $M_2 \approx 1.4$ is at $t^* \approx 60$, $\tau_1{}^* \approx 6$, the third maximum $M_3 \approx 1.3$ is at $t^* \approx 100$, $\tau_1{}^* \approx 10$, etc. When these results are compared with that of Fig. 2.18, it is apparent that the numerically constructed envelope is in a good agreement with the asymptotic approximations of the function $a(\tau_1)$.

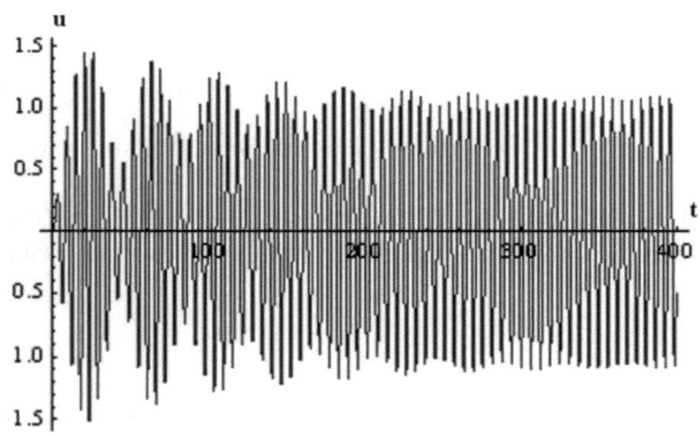

Fig. 2.19 Numerical integration of Eq. (2.106): $\varepsilon = 0.1$, $s = 0.2$, $\alpha = 0.333$, $F = 1$, and $\gamma = 0.1$

Arguing as above, one can obtain the solution in case $k_1 < 0$. Denoting $k^2 = F|k_1|/a_0$ and assuming $\gamma << k$, we find a solution similar to (2.172, 2.175), with cosh (kz) and sinh (kz) in place of cos (κz) and sin (κz), respectively.

Finally, we note that, contrary to Sect. 2.2 considering the vibro-impact processes, here we deal with a quasi-linear system; the applicability of so similar mathematical tools seems unexpected and exciting.

2.4 Entrainment, Synchronization and Resonance Capture

There exist many situations, in which the motion of oscillatory systems is affected by external forces. Perhaps, the simplest example is the motion of linear oscillator under harmonic forcing, considered above in Sect. 2.1. Still, if the linear system is asymptotically stable and subject to external harmonic forcing, it is possible to prove that the response regime will be unique.

Even for weakly nonlinear systems (to say nothing about the essentially nonlinear ones), this uniqueness is in general not the case. The forced and damped nonlinear system may exhibit different behavior, dependent on the initial conditions. In this case it is accepted to say that the system has multiple attractors. Without going into deeper mathematical details (some of them will be discussed below), it is possible to say that the attractor is the limit set of points in the state space of the dynamical system for $t \to \infty$. The knowledge of these attractors may help one to bring the nonlinear system to desired regime of motion or to avoid some undesired regimes.

Sometimes such control over the attractor of the oscillating system may be of vital importance – for instance, the role of heart pacemakers is to keep the heart of the patient close to the attractor associated with the normal beating and far from the attractor associated with potentially mortal fibrillations. The other example may be taken from the fields of mechanical and civil engineering – the designer would like to avoid attractors related with high-amplitude oscillations of the structure, which may bring about failure.

The current section describes a few simple basic models of nonlinear systems under external forcing and discusses possible dependence of the response regimes on parameters and initial conditions.

2.4.1 Pendulum with Constant External Torque

The first and perhaps the simplest model of essentially nonlinear system under effect of external forcing which demonstrates the idea of alternative attractors is the model of a pendulum with constant external torque. Without loss of generality, if the damping is absent, the governing equation for this model may be written as:

$$\ddot{u} + \sin u = c \tag{2.178}$$

where $c = \text{const}$, frequency of the pendulum is set to unity. We restrict ourselves only by the case of non-negative c – the opposite case is considered trivial by change of x sign.

Fixed points of Eq. (2.178) u_e are determined by the obvious equation:

$$\sin u_e = c \tag{2.179}$$

Equation (2.179) has solutions only if $c \leq 1$, therefore one should expect a qualitative change of the system behavior if c passes unity. In order to visualize and explain this change, it is instructive to plot the potential energy for Eq. (2.178) for both cases (Fig. 2.20a, b):

$$U(u) = -\cos u - cu \tag{2.180}$$

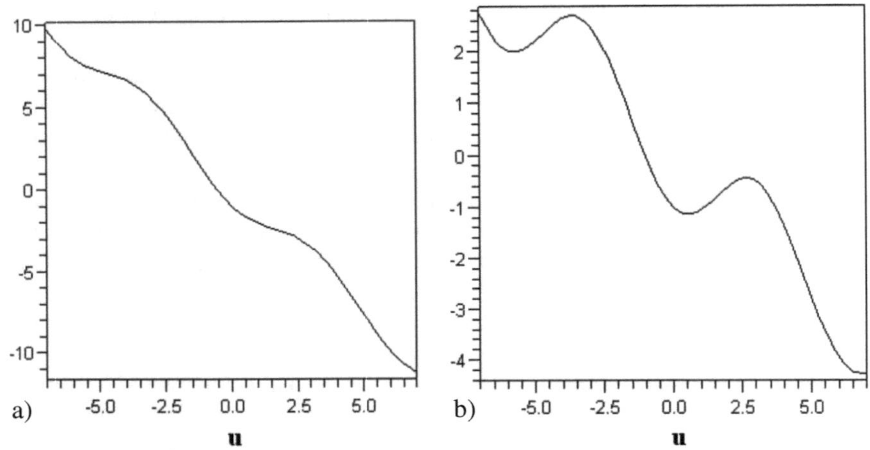

Fig. 2.20 Potential function (2.180) (**a**) c>1; (**b**) c<1

At Fig. 2.20a the case of c>1 is presented. It is easy to see that there are no equilibrium points and the system will inevitably rotate with growing velocity. On the contrary, the profile of the potential energy for $c < 1$ at Fig. 2.20b has infinitely many potential wells. Therefore, the system may either travel through the system with growing velocity, as in previous case, or oscillate in one of the wells. So, two qualitatively different types of motion are possible here. It should be mentioned that due to the presence of the first integral the system may be integrated in quadratures.

The regime of motion of the system is determined by its initial conditions. To clarify this point, we present the phase portraits of the system with $c > 1$ (Fig. 2.21a) and c < 1 (Fig. 2.21b).

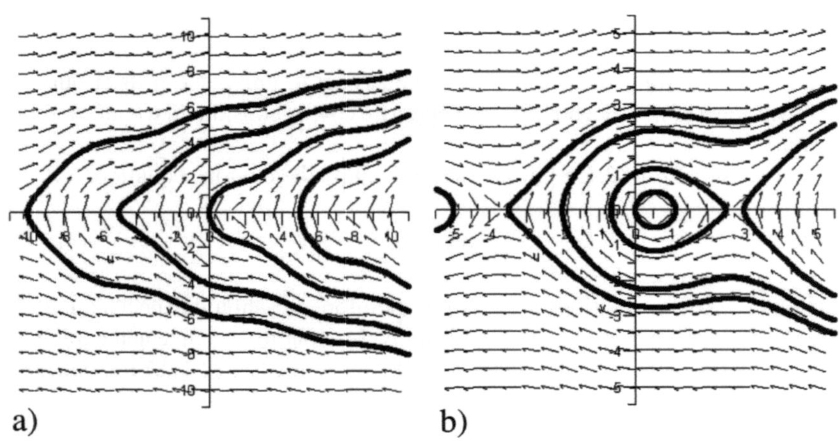

Fig. 2.21 Phase portrait of the oscillator with external constant torque, (**a**) c>1; (**b**) c<1

It is clear that all phase trajectories in Fig. 2.21a describe unlimited growth of the solution, whereas in Fig. 2.21b some of the solutions still grow without limit, whereas the amplitude of the others (situated inside the homoclinic loops) remains limited. In order to find the limited amplitude for all times one should start close enough to the stable equilibrium point – within the homoclinic loop. If the phase trajectory starts beyond this region, it will have an infinitely growing amplitude.

If viscous damping is introduced into the pendulum, Eq. (2.178) will be modified to the form

$$\ddot{u} + \lambda\dot{u} + \sin u = c \tag{2.181}$$

where λ is the damping coefficient. Unlikely Eq. (2.178), this equation does not possess the first integral and cannot be integrated in quadratures. Still, its fixed points are described by Eq. (2.179) and no solutions exist for c>1. In this case, the phase portrait will be very similar to one presented at Fig. 2.21a (the symmetry with respect to horizontal axis will be lost) and no solutions with limited amplitude will exist, despite the damping. For c<1, however, drastic change will occur. Phase portrait for the case $c = 0.5$ and $\lambda = 0.1$ is presented in Fig. 2.22.

One can see that the homoclinic loop is destroyed (stable and unstable manifolds are split) and therefore some phase trajectories coming "from infinity" will enter the vicinity of the fixed point and will eventually be attracted to it – such a situation is not possible in the conservative version of the system. It is acceptable to say that such phase trajectories of the system are *captured* by the fixed point. It is possible to demonstrate that a relative amount of these captured trajectories is proportional to λ.

One more remark will be in order in this section. In the conservative case, the trajectories within the homoclinic loops remain bounded but never approach the

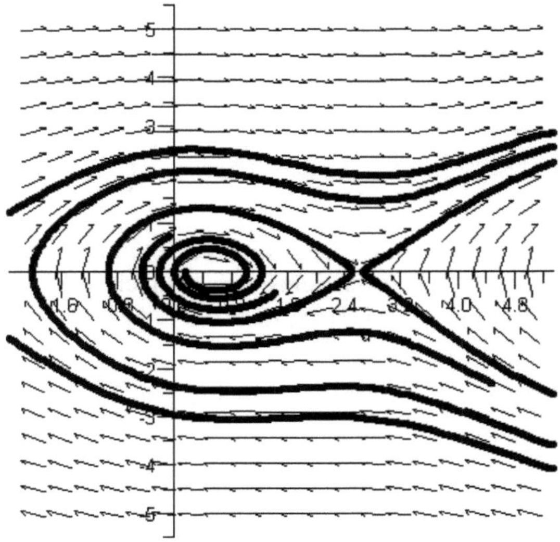

Fig. 2.22 Phase portrait of the damped system

fixed points. In the damped case, the captured trajectories are attracted to the fixed points, entering an arbitrarily small vicinity of the point at some finite time.

2.4.2 Entrainment of the Van der Pol Oscillator by External Harmonic Force

As an example of a system with self-excitation affected by a signal from outside, let us consider a standard Van-der-Pol oscillator with external harmonic forcing. Its equation of motion in dimensionless form, with frequency of external forcing and characteristic amplitude of the nonlinear term normalized to unity, may be written as:

$$\ddot{u} - \varepsilon(1 - u^2)\dot{u} + \Omega^2 u = \varepsilon A \cos t \tag{2.182}$$

It is supposed that the natural frequency of the oscillator is close to the forcing frequency: $\Omega = 1 + \Delta\varepsilon$. Values of A and Δ are considered to be of order unity, $\varepsilon \ll 1$.

Transfer to complex variables according to standard ansatz:

$$\phi(t)\exp(it) = \dot{u} + iu \tag{2.183}$$

with subsequent averaging yields the following condition for absence of the secular terms (up to order $O(\varepsilon)$):

$$\varphi' - i\varepsilon\Delta\varphi - \frac{\varepsilon}{2}\left(1 - \frac{|\varphi|^2}{4}\right)\varphi = \frac{\varepsilon A}{2} \tag{2.184}$$

The apostrophe denotes the derivative with respect to slow time after the averaging. Steady-state responses of the system correspond to fixed points φ_0 of Eq. (2.184) and may be computed from the following equation:

$$-i\Delta\varphi_0 - \frac{1}{2}\left(1 - \frac{|\varphi_0|^2}{4}\right)\varphi_0 = \frac{A}{2} \tag{2.185}$$

In order to investigate possible solution of Eq. (2.185), we split the complex function $\varphi(t)$ to modulus and argument $\varphi(t) = N(t)\exp(i\delta(t))$ and, accordingly, $\varphi_0 = N_0\exp(i\delta_0)$. Then, from Eq. (2.184) we can obtain two real equations:

$$\delta' - \varepsilon\Delta = -\frac{\varepsilon A}{2N}\sin\delta$$

$$N' - \frac{\varepsilon}{2}\left(1 - \frac{N^2}{4}\right)N = \frac{\varepsilon A}{2}\cos\delta$$

and from Eq. (2.185) for the fixed points:

$$\begin{aligned} \Delta &= \frac{A}{2N_0}\sin\delta_0 \\ -\frac{1}{2}\left(1-\frac{N_0^2}{4}\right)N_0 &= \frac{A}{2}\cos\delta_0 \end{aligned} \tag{2.186}$$

By putting $N_0^2 = Z$, dividing the second equation of (2.186) by 2, squaring up and summing, one obtains a single equation for Z:

$$\Delta^2 Z + \frac{1}{4}(1 - Z/4)^2 Z = A^2/4$$

Technical change of variables

$$\sigma = 2\Delta, \rho = Z/4, F = A/2 \tag{2.187}$$

reduces the equation above to the standard form:

$$\sigma^2\rho + \rho(1-\rho)^2 = F^2 \tag{2.188}$$

Our goal is to investigate how number and properties of solutions of Eq. (2.188) depend on values of parameters F and σ. First of all, let us investigate possible behavior of function $f(\rho) = \sigma^2\rho + \rho(1-\rho)^2$ (clearly, ρ cannot be negative). For $\rho = 0$ $f = 0$, for $\rho\to\infty$ $f(\rho)\to\infty$, therefore for every value of σ and F at least one solution of (11) exists. More than one solution can exist if f(ρ) is not monotonous for $\rho \geq 0$. To check that, we find extreme points of f(ρ):

$$f'(\rho_0) = \sigma^2 + 1 - 4\rho_0 + 3\rho_0^2 = 0 \tag{2.189}$$

Solutions of Eq. (2.189) are

$$\rho_0 = \frac{2}{3} \pm \frac{1}{3}\sqrt{1-3\sigma^2} \tag{2.190}$$

From (2.190) it is clear that for $|\sigma| < 1/\sqrt{3}$ the function f(ρ) will have a maximum and a minimum in the points with positive ρ, otherwise it will grow monotonously and only one solution of (2.188) will exist. These two possibilities are presented in Fig. 2.23a ($|\sigma| < 1/\sqrt{3}$) and 2.23b ($|\sigma| > 1/\sqrt{3}$).

From Fig. 2.23a it is clear that for given σ for $Q_1 > F^2 > Q_2$ there are three solutions; otherwise, there is only one. The boundaries correspond to merging of two solutions. Values of Q_1 and Q_2 are easily obtained from (1.188) and (1.190):

$$\begin{aligned} Q_1 &= \frac{2}{3}\sigma^2 + \frac{2}{27} + \left(\frac{2}{9}\sigma^2 - \frac{2}{27}\right)\sqrt{1-3\sigma^2} \\ Q_2 &= \frac{2}{3}\sigma^2 + \frac{2}{27} - \left(\frac{2}{9}\sigma^2 - \frac{2}{27}\right)\sqrt{1-3\sigma^2} \end{aligned} \tag{2.191}$$

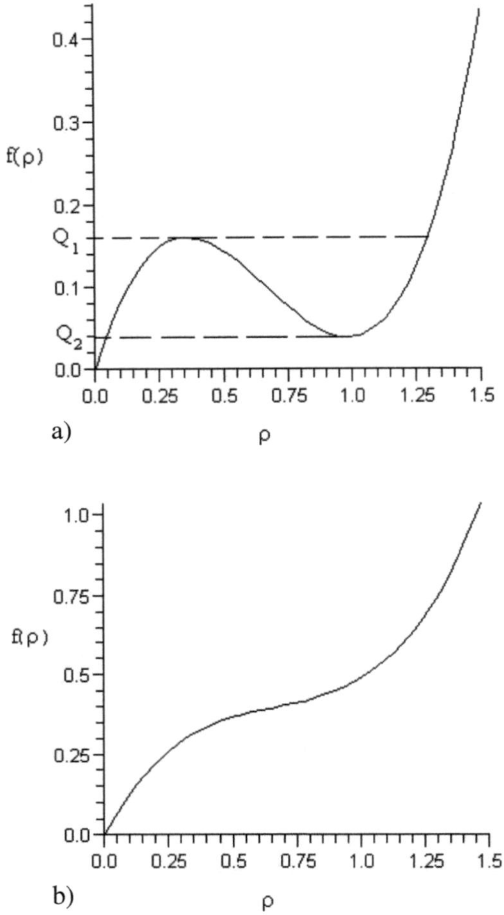

Fig. 2.23 Possible behaviors of function f(ρ)

The zone with three solutions at the plane of parameters (σ, F) is presented in Fig. 2.24 in accordance with Eq. (2.191).

It is easy to see that Eq. (2.188) can have three solutions only for $F < F_{crit} = \sqrt{8/27}$ and $\sigma < \sigma_{crit} = 1/\sqrt{3}$.

The result obtained here means that for certain combinations of parameters Eq. (2.188) can have more than one solution and therefore Eq. (2.182) in principle can have more than one response regime. However, in order to characterize these regimes one should describe the motion in the vicinity of the fixed points, i.e to determine their stability characteristics.

Easy way to do that is to consider small perturbation of the solution for the fixed point:

$$\varphi = \varphi_0 + \gamma(t), \gamma << 1 \tag{2.192}$$

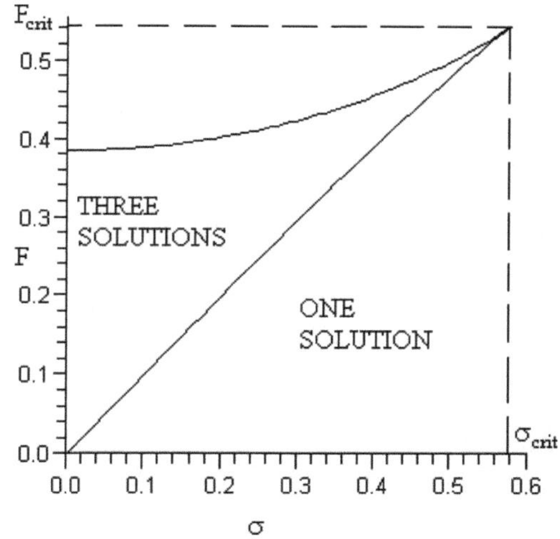

Fig. 2.24 Zones of multiplicity on the parametric plane

By substitution of (2.192) by (2.184) and keeping only linear terms, one obtains:

$$\gamma' - i\varepsilon\Delta\gamma + \frac{\varepsilon}{8}\left(\varphi_0^2\gamma^* + |\varphi_0|^2\gamma\right) - \frac{\varepsilon}{2}\left(1 - \frac{|\varphi_0|^2}{4}\right)\gamma = 0 \tag{2.193}$$

Solution of Eq. (2.193) is searched in a form

$$\gamma = \gamma_+ \exp(\mu t) + \gamma_- \exp(\mu^* t) \tag{2.194}$$

By substituting (2.194) to (2.193) and equating the coefficients of each of two exponents, one gets

$$\begin{aligned} &\left(\mu - i\varepsilon\Delta + \frac{\varepsilon|\varphi_0|^2}{4} - \frac{\varepsilon}{2}\right)\gamma_+ + \frac{\varepsilon|\varphi_0|^2}{8}\gamma_-^* = 0 \\ &\left(\mu^* - i\varepsilon\Delta + \frac{\varepsilon|\varphi_0|^2}{4} - \frac{\varepsilon}{2}\right)\gamma_- + \frac{\varepsilon|\varphi_0|^2}{8}\gamma_+^* = 0 \end{aligned} \tag{2.195}$$

Equations (2.195) are commensurate only if

$$\left(\mu + \frac{\varepsilon|\varphi_0|^2}{4} - \frac{\varepsilon}{2}\right)^2 + \varepsilon^2\Delta^2 = \frac{\varepsilon^2|\varphi_0|^4}{64} \tag{2.196}$$

or, in terms of new variables (2.187):

$$\mu = \frac{\varepsilon}{2}\left(1 - 2\rho \pm \sqrt{\rho^2 - \sigma^2}\right) \tag{2.197}$$

Stability of a given fixed point is determined by sign of real part of μ. Two cases should be distinguished:

(a) If $\sigma > \rho$, then the solution is stable for $\rho > 1/2$ and unstable for $\rho < 1/2$.
(b) If $\sigma < \rho$, then for $3\rho^2 - 4\rho + 1 + \sigma^2 > 0$ the fixed point is stable, otherwise – stable for $\rho > 1/2$ and unstable for $\rho < 1/2$.

Stability zones described above are presented in Fig. 2.25:

Partial information concerning stability of solutions on the plane of control parameters is summarized in Fig. 2.26. It should be mentioned again that this picture is only partial; especially, in the center of the picture there exists a very interesting region with complicated bifurcational structure. For detailed analysis, see (Rand, 2009).

It is remarkable that for relatively small values of detuning and relatively large forcing there exists only one stable steady-state solution. In other terms, the Van-der Pol oscillator oscillates with a frequency of external forcing rather than with its own frequency. Such behavior is referred to as *entrainment* – external force imposes its frequency on the oscillator.

For relatively large detuning and small forcing, the solution corresponding to entrainment is unstable. It is possible to demonstrate that the loss of stability occurs due to Hopf bifurcation at $\rho = 1/2$ and therefore stable limit cycle is formed for Eq. (2.184) – for initial forced Van-der-Pol equation, it corresponds to Neimark–Sacker bifurcation. It means that two frequencies are present in the solution – approximately frequency of the forcing and approximately natural frequency of the

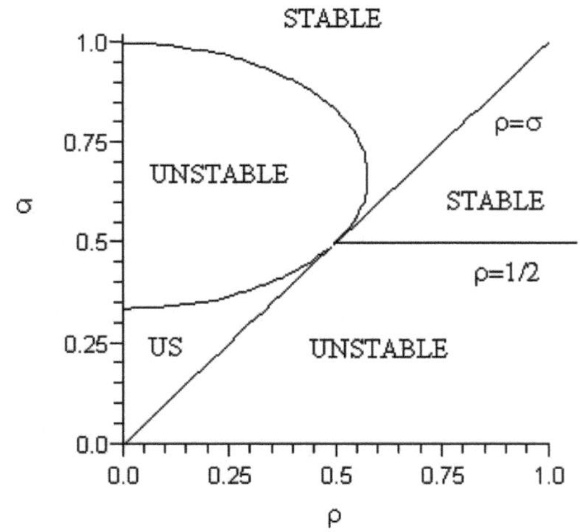

Fig. 2.25 Stability zones on the parametric plane ρ–σ

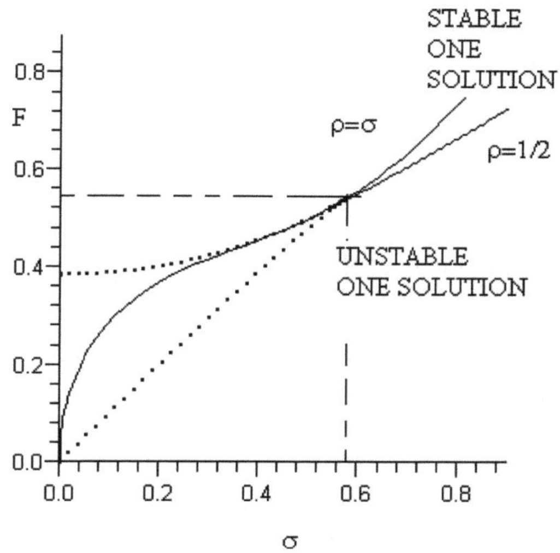

Fig. 2.26 Stability zones at the plane of parameters σ–F (partial view)

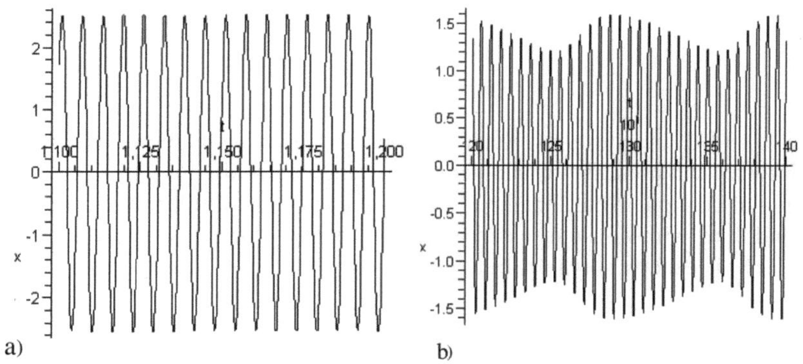

Fig. 2.27 Time series simulation of the forced Van-der-Pol oscillator. (**a**) Phase – locked response; (**b**) Quasiperiodic response

oscillator. Such behavior is referred to as *heterodyning*, beating, or *quasiperiodic oscillations*.

Numerical examples which illustrate these two types of behavior in initial Eq. (2.182) are presented in Fig. 2.27a ($A = 1.6$, $\Delta = 0.1$, $\varepsilon = 0.01$) and Fig. 2.27b ($A = 1.2$, $\Delta = 0.35$, $\varepsilon = 0.3$).

Figure 2.27a describes the regime of entrainment and Fig. 2.27b the regime of heterodyning or quasiperiodic response. The beatings in Fig. 2.27b are caused by the presence of two close frequencies – the natural frequency of the oscillator and the frequency of the external forcing.

The model of entrainment of the oscillator with self-excitation by external periodic force has many applications. Besides the already mentioned heart pacemakers which entrain the heart of the patient to beat with the prescribed frequency, the concept of entrainment appears useful in studies of circadean rhythms in living systems and even for the tuning of piano strings (Jackson, 1991).

It should be mentioned that the model of the Van-der-Pol oscillator presented above has certain advantages and shortcomings. The main advantage, of course, is that it is the classic (historically, the first) model of a self-excited system, which exhibits entrainment in the case of large forcing and small detuning, and does not exhibit it in the opposite case. However, there are some effects (multiplicity of solutions, complicated bifurcations as in the central part of Fig. 2.26) which occur due to the special shape of Eq. (2.182) rather than they represent essential common features of the entrainment process. In other terms, real systems with self-excitation and entrainment should exhibit the common features mentioned above but can have very different peculiarities of behavior from those predicted by the Van-der-Pol equation. Therefore, it might be instructive to develop a simpler model which will capture essential features of the phenomenon and will avoid irrelevant complications caused by special choice of the equation of self-excited oscillations.

The idea of such a model is given by the first equation of System (2.186). If one neglects variations of N, this equation may be presented in a form

$$\frac{d\delta}{dt} = \omega + G\sin\delta \tag{2.198}$$

where ω and G are constants. Such a model is referred to as *phase-only* model of entrainment, ω represents instantaneous frequency of oscillations, G – the intensity of external forcing. Note that Eq. (2.198) is somewhat similar to Eq. (2.198) of the pendulum with constant external torque, but is *of the first order*. Of course, Eq. (2.198) is exactly solvable. Fixed points exist in (2.198) if and only if

$$\left|\frac{\omega}{G}\right| \le 1 \tag{2.199}$$

Physically, this condition means that the forcing is strong enough. Fixed points correspond to the states with zero frequency, i.e. to the frequency of external force, and therefore describe the regime of entrainment. Moreover, exact solution of Eq. (2.198) for this case yields

$$\delta = 2\tan^{-1}\left(-\frac{G}{\omega} + \sqrt{\left(\frac{G}{\omega}\right)^2 - 1}\tanh\left(-\frac{t-t_0}{2}\sqrt{G^2-\omega^2}\right)\right) \tag{2.200}$$

It is easy to see that for any initial conditions and $t \to \infty$ solution (2.200) yields $\delta \to -\sin^{-1}(\omega/G)$, i.e. for any initial conditions the system will be finally attracted by the stable regime of entrainment. If condition (2.199) is not valid, then no fixed point exists and the entrainment does not occur – all solutions exhibit a *drift* with a never vanishing instantaneous frequency.

2.4.3 Synchronization of Oscillators and Related Models

The phenomenon of synchronization is a "close relative" of the phenomenon of entrainment discussed in the previous section. Historically, the concept of synchronization preceded that of entrainment by many years. Synchronization of a two pendulum clock with sufficiently close frequencies hung back to back on the same wall has been observed as early as in Seventeenth century by Christian Huygens. The physical reason for this synchronization is interaction between two self-excited oscillators – if it is strong enough and the frequencies are close enough, the oscillators will synchronize their motion.

In order to describe the synchronization in this case, we will take advantage of phase-only approximation, although historically this phenomenon was described by the treatment of coupled Van-der-Pol oscillators (see, e.g. Landa, 1996).

Two coupled phase-only oscillators are described by the following system of equations:

$$\begin{aligned} \frac{d\theta_1}{dt} &= \omega_1 + F_1(\theta_1, \theta_2) \\ \frac{d\theta_2}{dt} &= \omega_2 + F_2(\theta_1, \theta_2) \end{aligned} \tag{2.201}$$

Variables θ_i are phases, therefore functions F_i should be 2π-periodic with respect to both variables. At the beginning, for the sake of simplicity, we'll adopt that both functions F_i depend only on phase difference $\delta = \theta_1 - \theta_2$. Then (2.202) may be trivially reduced to the form

$$\frac{d\delta}{dt} = \omega_1 - \omega_2 + F(\delta),\ F = F_1 - F_2 \tag{2.202}$$

Function F is 2π-periodic. If we make the further assumption that only the first Fourier component of F is significant, then Eq. (2.202) is reduced to the form

$$\frac{d\delta}{dt} = \omega_1 - \omega_2 + \tilde{F}_1 \sin(\delta + \delta_0) \tag{2.203}$$

where $\tilde{F}_1$ is the amplitude of this significant Fourier component and δ_0 is the constant phase shift. Obviously, Eq. (2.203) is completely equivalent to Eq. (2.198) and therefore, phase variables θ_1 and θ_2 will be synchronized (i.e. δ will be constant) if and only if

$$\left| \frac{\tilde{F}_1}{\omega_1 - \omega_2} \right| > 1 \tag{2.204}$$

Physically, Condition (2.204) is rather understandable: in order to synchronize the oscillations, coupling between the oscillators should be strong enough and the difference between frequencies not too large.

Now is the time to refute the assumption that the interaction depends only on the phase difference. Generally speaking, 2π-periodicity of F_1 and F_2 with respect to both arguments implies that these functions may be presented as sums of a Fourier series including sines and cosines of all combinations $k\theta_1 + m\theta_2$, where k and m are whole numbers. Let us admit that one such Fourier component is significant and all others can be neglected. Then, it is again possible to reduce the system to a single first-order differential equation:

$$\begin{aligned} \frac{d\vartheta}{dt} &= k\omega_1 + m\omega_2 + \tilde{F}_{km}\sin(\vartheta + \vartheta_0) \\ \vartheta &= k\theta_1 + m\theta_2 \end{aligned} \tag{2.205}$$

This equation, again, is formally equivalent to (2.198) and the system is attracted to a fixed point with $\vartheta = const$ if and only if

$$\left| \frac{\tilde{F}_{km}}{k\omega_1 + m\omega_2} \right| > 1 \tag{2.206}$$

If Condition (2.206) is valid and k≠–m, then the oscillators are not synchronized; instead, the ratio of their frequencies is a rational number, or, in other terms, the frequencies are *commensurate*. This situation is referred to as *phase locking* in the system of coupled oscillators. The other common notion is that the oscillators are engaged in the *resonance k: m*. In the state space, the motion of the system is periodic, with period equal to the least common multiple of k and m.

Physically, Condition (2.206) means that either appropriate Fourier component is large, or the value $k\omega_1 + m\omega_2$ is small. The former condition may be the result of some peculiar form of the coupling. The latter condition implies that the ratio of the frequencies is close to some rational number. In fact, *any* frequency ratio is arbitrarily close to some rational number, but for "well-behaved" functions F, the coefficients of Fourier components decay very rapidly (in fact, exponentially) for large k and m. That is why usually the phase locking is observed for not very large absolute values of k and m; however, in principle it is possible for every rational number.

The analysis presented above is by no means rigorous. Mathematical treatment of the phase locking phenomenon involves consideration of so-called *circular map* and is presented in many books (see, e.g. Arnold et al., 2006). The final conclusion is similar: the locking can occur at any frequency ratio k:m, provided that the combination $k\omega_1 + m\omega_2$ is small enough or the relevant Fourier-component of the coupling is large enough. In the space of parameters, the zones of phase locking exist in the vicinity of any rational frequency ratio – they are known as *Arnold tongues*. Their relative width decreases drastically with growth of k and m.

Concepts of synchronization and phase locking have been broadly used in the last decades, for consideration of various systems of coupled oscillators, for the study of synchronization in chaotic systems and for the design of chaos control. The interested reader may consult recent books (Blekhman, 2000 and references therein, Pikovsky et al., 2003) to find out necessary details.

2.4.4 Resonance Capture

Resonance motions (i.e. appearance of commensurate frequencies) are ubiquitous in dynamical systems around us. The most common example, of course, is related to the motion of the Moon: we always see one side only and this means that the frequency of the rotation around the axis is equal to the frequency of rotation around the Earth. It is possible to say that we deal here with a 1:1 resonance. A less known example is related to the motion of Mercury around the sun: the two frequencies are related as 3:2. There are a lot of other resonances in the sun system (Arnold et al., 2006).

It seems that when the sun system was formed, resonant relationships between different rotation frequencies did not exist then – the question is really why the complicated and sometimes stochastic (hypothetically) process of planet formation should give rise to commensurate rotation frequencies characterized by ratios of relatively small numbers? Maybe, it is more reasonable to suggest that the frequency relations under discussion are the result of further evolution of the dynamical system; in other terms, the systems were *captured* into the resonance. This suggestion was presented (and the term *capture* was coined) in a paper by P. Goldreich and S. Peale (1966).

Before a brief presentation of their model, we would like to discuss briefly the physics of the involved phenomena. Quite obviously, if the planet rotating around the sun would be ideally spherical and absolutely rigid, then the spin motion and the orbital motion of the planet would be completely decoupled. The reason is that the gravitational interaction of two spheres according to Newton's law is exactly equivalent to the interaction of two points and rotation of the sphere is therefore not important, up to relativistic corrections which are irrelevant in our case. However, a real planet is neither ideally spherical nor absolutely rigid. Due to the eccentricity, the gravitational force between the Sun and the planet depends on the orientation of the planet with respect to the plane of its orbit. Absence of absolute rigidity is even more important – it gives rise to tidal forces which are capable of damping the energy of planetary spin out.

The model of Goldreich and Peale is based on the equation of planetary motion taking into account the eccentricity of the planet:

$$C\ddot{\theta} + \frac{3}{2}(B - A)\frac{GM}{r^3}\sin 2\psi = 0 \tag{2.207}$$

where A, B, C are main moments of inertia of the planets, C it the moment around the spin axis, M is the mass of the sun, G – the gravitational constant, r – instantaneous radius, θ – angular position of the planet's longer axis with respect to the longer axis of its orbit, ψ – the angle between the axis of the planet and the center of the orbit. It is easy to see that if there is no eccentricity of the planet (A=B), then $\dot{\theta} = const$, in accordance with the second Kepler law. This law is based on conservation of the angular momentum. Therefore, we see that the eccentricity of the planet gives rise to coupling between the spin and the orbital motion.

If the system is close to the resonance, one can take $\gamma = \theta - qL$ small, where L is the mean anomaly (the phase variable determining the position of the celestial body at its Kepler orbit) and q is the resonance ratio – rational number, which defines the ratio of periods in the state of exact resonance. The central idea here is that both L and θ change significantly at each period of orbit, but their difference changes only slightly due to the resonance relationship. Therefore one can average over period and get the following equation for γ:

$$C\ddot{\gamma} + Q\sin 2\gamma = 0 \tag{2.208}$$

where Q is certain constant. Needless to say, Eq. (2.208) describes a regular pendulum!

It is rather difficult to take into account the tidal force in an exact manner, but it can have a very significant effect on the motion. One of the suggestions made by Goldreich and Peale was to proceed empirically and to add the following terms to the right-hand side of Eq. (2.208):

$$C\ddot{\gamma} + Q\sin 2\gamma = a - b\dot{\gamma} \tag{2.209}$$

The term a on the right-hand side describes the constant torque which appears as a result of change of the planet eccentricity due to the tidal effects. The second term describes the dissipation of energy (for instance, due to viscous forces acting in the ocean). Again, the exact shape of these terms should not be taken too seriously – they are only empirical terms required to take into account the main features of the phenomenon rather than to describe it exactly.

It is easy to recognize Eq. (2.209) – it is the equation of the pendulum with constant external torque and viscous damping. One can immediately conclude that if the coefficient Q/C is large enough, then the system has a chance to be captured into the domain of attraction of the fixed point. This fixed point satisfies the equation $\gamma = const$, which implies the exact resonance between the orbital and the spin motions. Now it should be recalled that $Q \sim |A - B|$; therefore, the chances of the planet to be captured into such a resonance grow together with its eccentricity. The characteristic values of the tidal forces are also of great importance.

A very similar model in a more abstract framework (related to problem of averaging and formulated in the action-angle variables) has been proposed by A.I. Neishtadt (Neishtadt, 1975). In the book (Arnold et al., 2006) it was also demonstrated that capture into the resonance may occur if a dynamical system moves into the vicinity of any resonant hypersurface (and generically, such hypersurfaces fill the whole state space) and the local problem is qualitatively similar to the model of the pendulum with the external torque and damping.

The other well-known example of capture into the resonance is also related to the motion of eccentric systems in the rotating machinery. As it is presented in the paper of R. Rand (1998):

> Imagine a slightly unbalanced wheel attached to an elastic support. If a constant torque is applied to the wheel, then we would expect that the wheel would begin to spin faster and faster. A surprising thing happens, however, when the wheel's angular speed gets close to the natural frequency of the elastic support. The wheel may cease to spin up and the energy of the applied torque will instead produce large motions of the support. The wheel is said to be captured into the resonance.

The mechanism of the process mentioned above is described in details in papers (Quinn et al., 1995; Rand and Quinn, 1995). It is in fact very similar to the models inferred from celestial mechanics, but instead of the separatrix splitting due to the damping terms it involves the slow evolution of the separatrix, allowing the phase trajectories to cross it and to enter the capture region.

All models of the resonance capture mentioned here have one interesting common feature. Only a certain share of the phase trajectories is captured into the resonance. It is rather easy to understand from Fig. 2.22 that only the trajectories which enter the space between the split branches of the separatrix will be eventually captured, so the fate of the trajectory depends on the initial conditions. Moreover, it is possible to prove (Arnold et al., 2006) that in certain classes of systems the points in the state space which correspond to captured and non-captured trajectories are mixed in a such way that in every "ball" of size ε in the space of initial conditions one can find the origins for both captured and non-captured trajectories. If the accuracy of the initial condition is ε or less, then one cannot distinguish between them and therefore the resonance capture should be treated as *probabilistic phenomenon*, despite the completely deterministic nature of the system.

Such probabilistic models for resonance capture cannot explain the ubiquity of dynamical systems where the capture really occurs – they usually provide very low estimations for the capture probabilities. However, there exist more refined models which describe the deterministic resonance capture for large regions in the space of initial conditions. Interestingly enough (and unlike the model analyzed in this section above), all such models known to authors involve more than two time scales. It seems that the first such model has been introduced by Burns and Jones (2003). It provides a somewhat formal generalization of the model of A.I. Neishtadt mentioned earlier (Neishtadt, 1975), with introduction of an additional super-slow time scale. We do not present it here in detail, since a somewhat similar model which yields deterministic capture into the resonance for relatively large regions in the space of initial conditions will be discussed in Sect. 2.6: a targeted energy transfer in a linear oscillator with essentially nonlinear attachment.

2.4.4.1 Exercises

1. Consider the Neishtadt system in action-angle variables:

$$\dot{I} = \varepsilon\left(1 + a\sin\gamma - \frac{1}{4}I\right)$$
$$\dot{\gamma} = I$$
$$a > 0$$

Demonstrate that this model describes capture into the resonance and find the conditions for the capture. What characteristic share of phase trajectories will be captured?

2. Consider phase-only system of coupled oscillators:

$$\text{(A)}\quad \begin{aligned}\dot{\theta}_1 &= 1 + a\sin\theta_1\cos\theta_2\\ \dot{\theta}_2 &= -1.1 + a\cos\theta_1\sin\theta_2\end{aligned}$$

$$\text{(B)}\quad \begin{aligned}\dot{\theta}_1 &= 1.2 + a\sin\theta_1\cos\theta_2 + 0.2\sin\theta_2\\ \dot{\theta}_2 &= 0.85 + a\cos\theta_1\sin\theta_2 + 0.1\sin\theta_1\end{aligned}$$

$$\text{(C)}\quad \begin{aligned}\dot{\theta}_1 &= 1.01 + a\sin^2\theta_1\cos\theta_2\\ \dot{\theta}_2 &= 0.47 + a\cos^2\theta_1\sin\theta_2\end{aligned}$$

For what values of parameters will the motion of the oscillators be synchronized or phase-locked? At what frequency ratio? Verify your analytic findings by numeric simulations.

Solution

(A)

$$\dot{\theta}_1 = 1 + a\sin\theta_1\cos\theta_2$$
$$\dot{\theta}_2 = -1.1 + a\cos\theta_1\sin\theta_2$$

Right-hand sides of the equations may be easily presented in a form of Fourier series:

$$\dot{\theta}_1 = 1 + \frac{a}{2}(\sin(\theta_1 + \theta_2) + \sin(\theta_1 - \theta_2))$$

$$\dot{\theta}_2 = -1.1 + \frac{a}{2}(\sin(\theta_1 + \theta_2) - \sin(\theta_1 - \theta_2))$$

Obviously, only phase locking with $k = m = 1$ and $k = 1$, $m = -1$ is possible. By introducing new variables:

$$\Phi = \theta_1 + \theta_2$$

$$\Psi = \theta_1 - \theta_2$$

the problem is reduced to a trivial system of equations

$$\dot{\Phi} = -0.1 + a\sin\Phi$$
$$\dot{\Psi} = 2.1 - a\sin\Psi$$

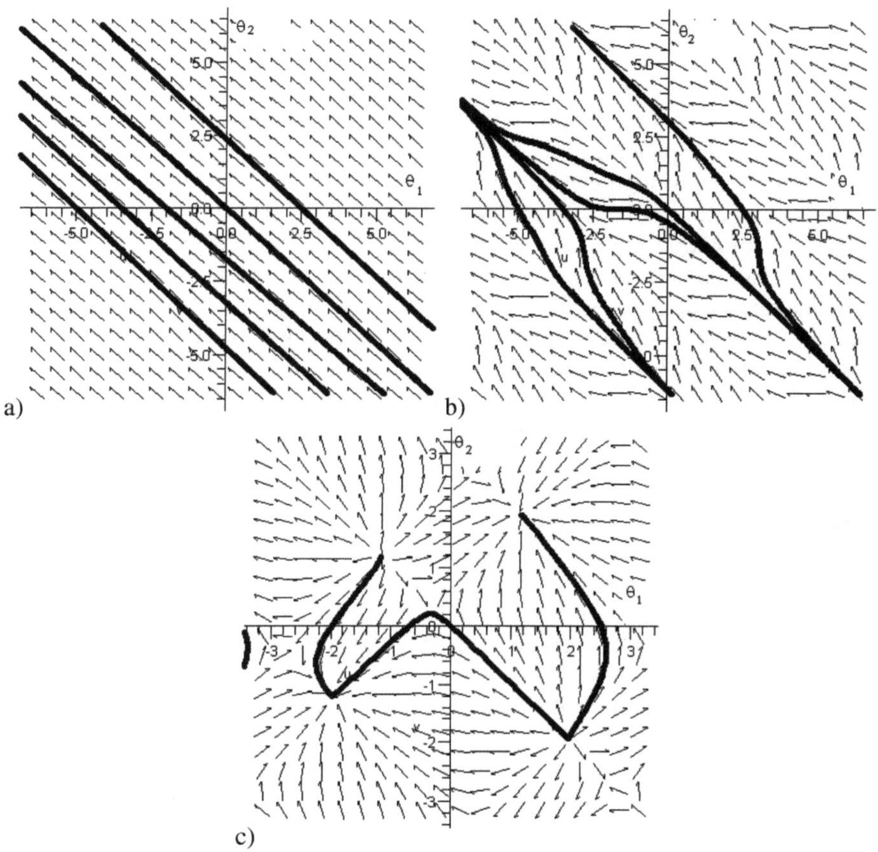

Fig. 2.28 Illustration to Problem 2A – phase portraits

Therefore, one should conclude that for $|a| < 0.1$ the system will not be phase-locked, for $0.1 < |a| < 2.1$ the system will be phase-locked with $\theta_1 + \theta_2 = const$. For $2.1 < |a|$ the condition $\theta_1 - \theta_2 = const$ also must be valid. This can be only if both variables are constants, i.e. the system is attracted to the fixed point for any initial conditions. Evolution of the phase portrait of the system is demonstrated in Fig. 2.28a (a = 0.05; no phase locking), Fig. 2.28b (a = 1, phase locking at $\theta_1 + \theta_2 = const$) and Fig. 2.28c ($a = 3$, appearance of attractors – fixed points).

2.4.5 Forced Oscillator with Multiple States of Equilibrium

Let us consider the motion of a simple symmetric oscillator with two states of equilibrium subject to external harmonic forcing and linear viscous damping (Guckenheimer and Holmes, 2002):

$$\ddot{u} - u + 2u^3 + \varepsilon(\lambda\dot{u} - \gamma\cos\omega t) = 0 \tag{2.210}$$

For $\varepsilon = 0$ the phase portrait of this system is presented in Fig. 2.10. It is characterized by two centres and one saddle point. Characteristic trajectories include two families: one of them corresponds to low energies and surrounds only one centre and the other corresponds to high energies and encircles both centres. These two families are divided by the separatrix trajectory.

If the perturbation in (2.210) is small enough, one should expect that the state of the system can still be qualitatively characterized on the base of Fig. 2.10. In particular, the saddle point will be preserved, as well as the two-well structure. If the frequency of the external forcing allows, one can expect nonlinear resonances in one of two potential wells or above two wells, giving rise to periodic or quasiperiodic trajectories. Such regimes can be considered with the help of the methods presented in the previous sections; the presence of multiple states of equilibrium is not very significant for this case.

The situation drastically changes if the perturbed motion occurs in the vicinity of the separatrix. In this case the trajectory can "switch" between encircling either one of the two wells or both, thus giving rise to chaotic behaviour. This problem has been thoroughly studied since 1960s, starting from pioneering works of V.K. Melnikov (1963). If the homoclinic orbit is perturbed, then the stable and the unstable manifolds of the saddle point do not coincide anymore and this highly degenerate structure is destroyed (Fig. 2.29).

In the case of general small perturbation, one can use the so-called Melnikov function to establish whether the stable and the unstable manifolds of the homoclinic

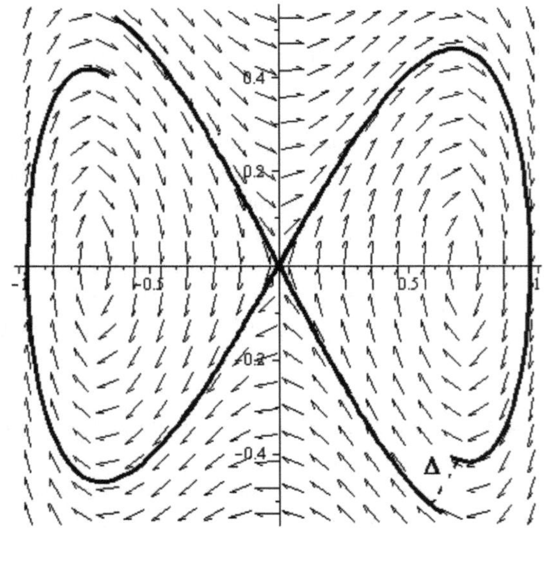

Fig. 2.29 Perturbation of the homoclinic orbit and splitting of the stable and unstable manifolds of the *saddle point*

point intersect. This function is proportional to the split between the manifolds. In the particular case of Eq. (2.210), this function is defined as:

$$M(t_0) = \int_{-\infty}^{\infty} \dot{u}_s(t)(\gamma \cos \omega(t + t_0) - \lambda \dot{u}_s(t)) = -\frac{2\lambda}{3} + \gamma \pi \omega \frac{\sin(\omega t_0)}{\cosh(\frac{\pi \omega}{2})} \tag{2.211}$$

where u_s is determined by Eq. (2.105). The intersection occurs if $M = 0$. Thus, the following criterion for the intersection of the manifolds can be formulated:

$$\gamma \geq \gamma_{crit} = \lambda \frac{2 \cosh\left(\frac{\pi \omega}{2}\right)}{3\pi \omega} \tag{2.212}$$

It is proven in (Guckenheimer and Holmes, 2002) that if single intersection occurs, then an infinite number of such intersections will occur, giving rise to the special structure of a Smale horseshoe. Still, it should be mentioned that criterion (2.212) can strongly underestimate the critical amplitude of the forcing required for transition to chaos (see Fig. 2.30)

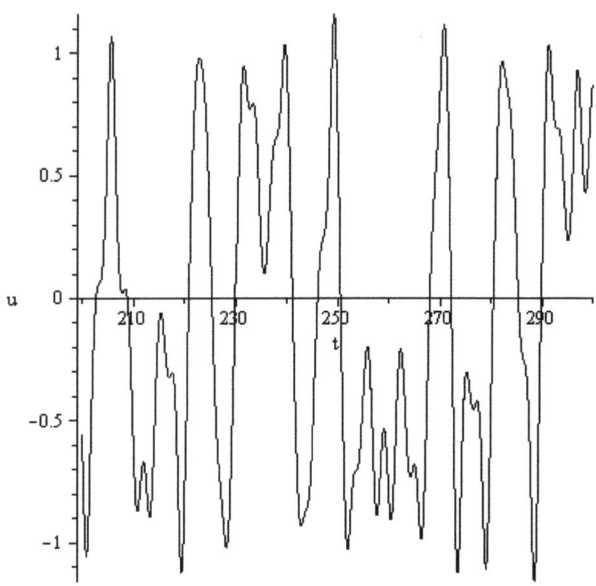

Fig. 2.30 Chaotic response of equation (), ε=0.1, ω=2, λ=1, γ=6.15. For this case, $\gamma \sim 5\gamma_{crit}$. For $\gamma \sim 3\gamma_{crit}$ or lower, no chaos is observed in simulations at these values of the parameters

2.5 Symmetric Systems of Coupled Nonlinear Oscillators Beating Phenomena

Beating in a linear system with two degrees of freedom (2 DOF) having close eigenfrequencies is one of the most well-known dynamic phenomena. Because of the validity of the superposition principle, their analytical description presents no difficulties and was considered above in Sect. 2.1.4. Several decades ago, close phenomena in nonlinear systems became the subject of growing interest because of their significant role in nonlinear optics (Akhmeriev and Ankiewicz, 1992; Uzunov et al., 1995). A similar problem arises in nonlinear mechanics (Kosevitch and Kovalyov, 1989; Manevich and Manevitch, 2005). Besides nonlinear optics and mechanics, intensive energy transfer may be important in all physical systems described by coupled Klein–Gordon equations (Khasnutdinova and Pelinovsky, 2003) as well as in the nonlinear dynamics of coupled polymer chains and deoxyribonucleic acid (DNA) double helices.

Let us first consider the simplest nonlinear problem of energy transfer in two weakly coupled equal nonlinear oscillators with cubic restoring forces. This problem can be described by the following system of two nonlinear equations (in dimensionless form):

$$
\begin{aligned}
&\frac{d^2U_1}{d\tau_0^2} + U_1 + 2\beta\varepsilon(U_1 - U_2) + 8\alpha\varepsilon U_1^3 = 0,\\
&\frac{d^2U_2}{d\tau_0^2} + U_2 + 2\beta\varepsilon(U_2 - U_1) + 8\alpha\varepsilon U_2^3 = 0,
\end{aligned}
\tag{2.213}
$$

where

$$
U_j = \frac{u_{j0}}{L_0}, \tau_0 = \sqrt{\frac{c_1}{m}}\, t, 8\alpha\varepsilon = \frac{c_3 L_0^2}{c_1}; \quad 2\beta\,\varepsilon = \frac{c_{12}}{c_1} \tag{2.214}
$$

L_0 is the length of unloaded nonlinear spring c_1 and c_3 are the linear and nonlinear stiffnesses of the oscillators respectively, and c_{12} is the stiffness of the coupling spring. Introducing the complex variables:

$$
\begin{aligned}
\varphi_1 &= e^{-i\tau_0}\left(\frac{dU_1}{d\tau_0} + i\,U_1\right) \qquad & \varphi_0^* &= e^{i\tau_0}\left(\frac{dU_1}{d\tau_0} - i\,U_1\right)\\
\varphi_2 &= e^{-i\tau_0}\left(\frac{dU_2}{d\tau_0} + i\,U_2\right) \qquad & \varphi_2^* &= e^{i\tau_0}\left(\frac{dU_2}{d\tau_0} - i\,U_2\right)
\end{aligned}
\tag{2.215}
$$

and slow time $\tau_1 = \varepsilon\tau_0$ (along with the fast time τ_0), one can use the following two-scale expansions

$$
\varphi_j(\tau_0, \tau_1) = \sum_n \varphi_{j,n}(\tau_0, \tau_1)\varepsilon^n, j = 1, 2. \tag{2.216}
$$

After corresponding calculations that take into account (2.213–2.216) we arrive at the equations of the principal asymptotic approximation:

$$\begin{aligned} &\frac{df_1}{d\tau_1} + i\beta f_2 - 3i\alpha\,|f_1|^2 f_1 = 0,\\ &\frac{df_2}{d\tau_1} + i\beta f_1 - 3i\alpha\,|f_2|^2 f_2 = 0, \end{aligned} \tag{2.217}$$

$$\varphi_j = e^{i\beta\tau_1} f_j, j = 1,\ 2, \tag{2.218}$$

which describe a number of interesting model systems, including optic couplers (Kosevitch and Kovalyov, 1989). This system is completely integrable and has two independent integrals of motion:

$$H = \beta(f_2 f_1^* + f_1 f_2^*) - \frac{3}{2}\alpha(|f_1|^4 + |f_2|^4) \tag{2.219}$$

$$N = |f_1|^2 + |f_2|^2 \tag{2.220}$$

The best way to handle these integrals is to use (2.219) to use coordinates θ and Δ defined below and to reduce system (2.217) to two real equations:

$$\begin{aligned} &f_1 = \sqrt{N}\cos\theta e^{i\delta_1}, f_2 = \sqrt{N}\sin\theta e^{i\delta_2}, \Delta = \delta_1 - \delta_2\\ &\frac{d\theta}{d\tau_1} = \beta\sin\Delta,\ \ \sin 2\theta\frac{d\Delta}{d\tau_1} = 2\beta\cos 2\theta\cos\Delta + \frac{3}{2}\alpha N\sin 4\theta \end{aligned} \tag{2.221}$$

Integral (2.219) is reduced to the form

$$H = (\cos\Delta + k\sin 2\theta)\sin 2\theta,\ k = \frac{3\alpha N}{4\beta}, \alpha > 0 \tag{2.222}$$

The latter condition is equivalent to the requirement of a hardening nonlinearity. One should mention that system (2.221) is strongly nonlinear, even in the case of an initially linear problem.

Before further analysis, let us present plots of the phase trajectories for different values of k in Fig. 2.31. Because of the phase plane periodicity one only has to consider the two lower quadrants.

We can see that two dynamic transitions are clearly distinguished when the nonlinearity parameter k increases. The first transition consists of the appearance of two additional stationary points, corresponding to nonlinear normal modes (NNMs); their number changes from 2 (if $k < 1/2$) to 4 (for $k > 1/2$). This transition is distinctly seen in Fig. 2.31c. The second transition, which occurs when $k = 1$, is connected to the behavior of the limiting phase trajectory (LPT) corresponding to a complete energy transfer between the oscillators. Namely, this trajectory transforms into separatrix at this value of k (Fig. 2.31e). This means that the characteristic time for complete energy transfer diverges. For $k > 1$ such a transfer becomes impossible (see Fig. 2.31f). Simultaneously, energy localization on the excited particle becomes

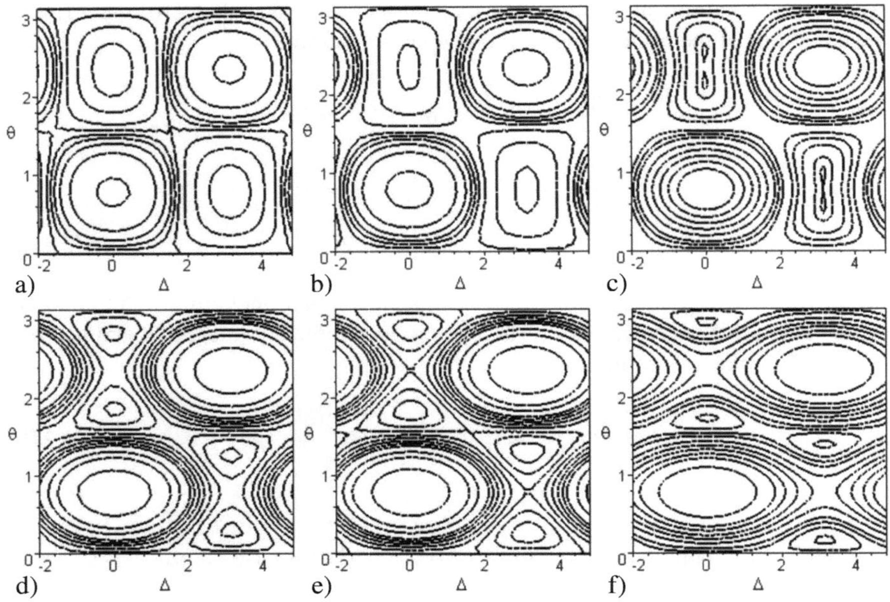

Fig. 2.31 Phase trajectories in the $\theta - \Delta$ plane for: (**a**) $k = 0.2$, (**b**) $k = 0.4$, (**c**) $k = 0.55$, (**d**) $k = 0.9$, (**e**) $k = 1$, (**f**) $k = 1.5$

possible. It is easy to check that the values $H = 1 + k^2$, $H = -1 + k^2$ correspond to the stationary points for in-phase and out-of-phase cooperative modes, respectively. Regimes of this kind are synchronized motions that can be presented as straight or curved lines in the configuration space of the initial variables (The existence of such normal modes in strongly nonlinear systems was first shown in Rosenberg, 1966). Efficient techniques for their construction even in the case when they are not straight may be developed by applying the principle of least action in Jacobi's form and the corresponding equations for the trajectories in the configuration space. Such techniques allow one to find the NNMs using power expansions in the independent variable (i.e., one of the unknowns in this case) in the framework of a nonlinear boundary problem (Manevitch et al., 1989; Vakakis et al., 1996).

For the discussed problem there is no need for such expansions because the NNMs are represented here by stationary points; this advantage is widely used in the papers devoted to NNMs and their bifurcations, as well as when searching the close-to-NNM regimes in damped and forced weakly coupled systems (Manevitch et al., 1989; Vakakis et al., 1996). The regimes close to stationary points in the $\theta - \Delta$ plane in the conservative system under consideration are beats with weak energy transfer between two oscillators. The equations of motions can be linearized in the vicinity of the stable stationary points and the solutions present small-amplitude oscillations of both θ and Δ around their values, corresponding to the NNMs. If one linearizes the Eq. (2.221) after the transformation $\theta_1 = \theta - \pi/4$, $\Delta_1 = \Delta - \pi$ (the latter for the case $H = -1 + k^2$), one arrives at the equations of linear oscillators, which are valid for initial conditions close to those for the normal modes themselves

$$\frac{d^2\theta_1}{d\tau_1^2} + \alpha_1^2\,\theta_1 = 0, \quad \frac{d^2\theta_1}{d\tau_1^2} + \alpha_2^2\,\theta_1 = 0, \tag{2.223}$$

where $\alpha_1^2 = 4\beta^2(1+2k)$, $\alpha_2^2 = 4\beta^2(1-2k)$, $k = 3\alpha N/4\beta$.

These equations contain a contribution that depends on the nonlinear terms of the initial system. Moreover, they lead to the conclusion that instability is possible if $k > 1/2$, which corresponds to instability of the out-of-phase nonlinear normal mode (if $\alpha > 0$).

This transition, which leads to the appearance of two new out-of-phase modes, does not noticeably influence the behavior of the LPT, which is far from the stationary points and describes a complete energy transfer between the oscillators. Certainly, one can extend the range of validity of the Eq. (2.223) by taking into account the nonlinear terms, which depend on θ_1 and Δ_1. However, one cannot attain LPT in this way. Let us show that we can consider the LPT as another type of fundamental solution (similarly to the NNMs) whose behavior determines the second dynamic transition in the behavior of the oscillatory system. The LPT, which is far from the stationary points, can then be used as a generating solution to construct close trajectories with strong energy transfer.

The value of H in integral (2.222), corresponding to the LPT, is zero. Therefore the variables θ and Δ in this case are connected by the equation

$$\cos\Delta = -k\sin 2\theta \tag{2.224}$$

so that $\sin\Delta = \pm\sqrt{1 - k^2\sin^2 2\theta}$. Then the first of the Eq. (2.221) can be written as:

$$\frac{d\theta}{d\tau_1} = \pm\beta\sqrt{1 - k^2\sin^2 2\theta} \tag{2.225}$$

Solution of Eq. (2.225) for the plus sign is the Jacobi elliptic function $\theta = (1/2)\,am\,(2\beta\tau_1, k)$. Because $0 \le \theta \le \pi/2$ by definition, one can use the negative sign for $(2n-1)\,\pi < 2\beta\tau_2 < 2n\pi$, $n = 1, 2, 3\ldots$ Then, one obtains the solution

$$\theta = \frac{1}{2}\left|am\,(2\beta\tau_1, k)\right|, \Delta = \pm\arccos\left[k\mathrm{sn}\,(2\beta\tau_1, k)\right] \tag{2.226}$$

with period $K(k)$, i.e., the complete elliptic integral of the first kind (for the in-phase oscillations). The solution for the out-of-plane oscillations is

$$\theta = \frac{1}{2}\left|am\,(2\beta\tau_1, k)\right|, \Delta = \pi \pm \arcsin\left[n\,(2\beta\tau_1, k)\right] \tag{2.227}$$

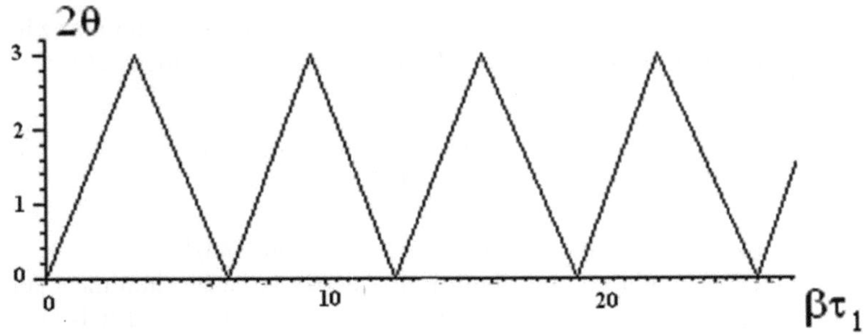

Fig. 2.32 The function $2\theta = |am\,(2\tau_1, k)|$ for k=0.5

Fig. 2.33 The function $2\theta = |am\,(2\tau_1, k)|$ for k=0.9

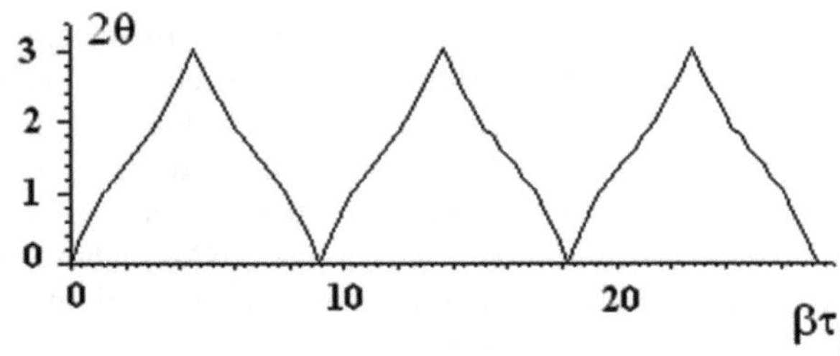

Periodic functions (2.226, 2.227) are not smooth; $\Delta\,(2\beta\tau_1)$ has discontinuities at points $2\beta\tau_1 = (2n-1)\,\pi$, $n = 0,1,\ldots$ and $\theta\,(2\beta\tau_1)$ has a discontinuous derivative at these points (in terms of distributions $d\theta/d\tau_1 = (2\beta/\pi)\,\Delta$). Plots of $\theta\,(2\beta\tau_1)$ for two values of parameter k are presented in Figs. 2.32 and 2.33. The value $k = 0.5$ corresponds exactly to the first dynamic transition. However, the solution for the LPT (Fig. 2.32) is still close to that of the linear case, except for a small change of period. Only for values of k that are close to the unit the deflections form an exact saw-tooth profile and the change of period becomes noticeable.

The second dynamic transition occurs when $k = 1$. In this case one can find a simple analytical solution corresponding to the LPT:

$$2\beta\tau_1 = \int\limits_0^{2\theta} \frac{d\,(2\theta)}{\cos 2\theta} \Rightarrow \theta = \frac{1}{2}\arcsin\frac{1-e^{-2\beta\tau_1}}{1+e^{-2\beta\tau_1}} \tag{2.228}$$

One can see from (2.228) that the LPT actually coincides with a separatrix if $k \to 1{:}\theta \to \pi/4$ when $\tau_1 \to \infty$. Here it is also convenient to use two non-smooth functions $\tau(\tau_1)$, $e(\tau_1)$ (Fig. 2.7).

It is natural to apply these non-smooth basic functions when describing the beatings (in terms of the variables θ and Δ) and closed trajectories with strong energy transfer. Actually, in the case $k = 0$ (the linearized system) the definitions in System (2.227) can be rewritten in the form $\theta = (\pi/2)\,\tau, \Delta = (\pi/2)\,e,\ \tau = \tau\left(\tau_1/a\right)$,

$e = e\left(\tau_1/a\right)$ where $a = \pi/2\beta$ (exactly as in a vibro-impact process with velocity $\Delta = \pi/2$). After introducing the basic functions $\tau\,(\tau_1/a)$, $e(\tau_1/a)$, we can present the solution as

$$\theta = X_1(\tau) + Y_1(\tau)\,e\left(\frac{\tau_1}{a}\right),\ \Delta = X_2(\tau) + Y_2(\tau)\,e\left(\frac{\tau_1}{a}\right) \qquad (2.229)$$

where the smooth functions $X_i(\tau)$, $Y_i(\tau)$ satisfy Eq. (2.221):

$$\begin{aligned}
&\frac{\partial}{\partial \tau}\begin{Bmatrix} X_1 \\ Y_1 \end{Bmatrix} = \frac{1}{2}a\beta[\sin(X_2 + Y_2)\ \sin(X_2 - Y_2)] \\
&\frac{\partial}{\partial \tau}\begin{Bmatrix} X_2 \\ Y_2 \end{Bmatrix} = a\beta[ctg2(X_1 + Y_1)\cos(X_2 + Y_2)\mp \\
&\mp ctg2(X_1 - Y_1)\cos(X_2 - Y_2)] \\
&+\frac{3a}{2}\alpha N[\cos 2(X_1 + Y_1) \mp \cos 2(X_1 - Y_1)]
\end{aligned} \qquad (2.230)$$

Then, we can search for the solution of Eq. (2.221) in the form of power expansions in the independent variable τ:

$$X_i = \sum_{l=0}^{\infty} X_{i,\,l}\tau^l,\ Y_i = \sum_{l=0}^{\infty} Y_{i,\,l}\tau^l,\ i = 1, 2 \qquad (2.231)$$

where the generating solution is the linear beating:

$$X_{1,0} = 0, \quad X_{1,1} = \frac{\pi}{2}, Y_{1,0} = 0, X_{2,0} = 0, Y_{2,0} = \frac{\pi}{2}, \qquad (2.232)$$

satisfying exactly the $\theta - \Delta$ equations for the case of the strongest beating. It can be proven that the presentation (2.229), taking into account (2.231), actually recovers the exact solution of the nonlinear problem for the most intensive energy transfer between the oscillators. As this takes place, the expansions (2.231) restore the exact local representation of the corresponding elliptic function (near $\tau = 0$), but the expressions (2.229) allow the prediction of the exact global behavior of the system. It is important to note that, even for large enough values of k, the solution appears close to that of linear beatings; the only differences are a minor curvature of the lines that are straight for linear beatings, and a change of the period.

One can find corresponding corrections by considering the next order of approximations, namely, $X_{1,0} = 0$, $X_{1,1} = \alpha\,\beta$, $X_{1,3} = -(2/3)\,(\alpha\beta)^3\,k^2$, $Y_{2,0} = \pi/2$, $Y_{2,1} = 2\alpha\,\beta\,k$ which coincide with those in the expansions of the exact solution.

Contrary to previous applications of non-smooth transformations, in the considered case there is no need to formulate boundary problems to compensate for singularities arising due to the substitution of non-smooth functions into the equations of motion (2.221) in order to derive these equations in terms of the smooth functions X_i and Y_i. Singularities arising due to the substitution of non-smooth functions into the second of the equations of motion (2.221) are exactly compensated for LPT since $\sin 2\theta = 0$ at the singular points.

The most important feature of the proposed technique is the unification of the local and global approaches. The local approach is invoked using power expansions, with unusually good results even in the lowest-order approximation. For global characteristics such as the period of oscillations $T = 2a$, its expansion in the parameter k can be found, separately after construction of the analytical form of the solution (with the period still unknown). The key point for the solution of this problem is a preliminary knowledge of the amplitude values of the θ and Δ functions (in particular, $\theta(a) = \pi/2$). This is another important distinction from previous applications of non-smooth transformations, in which the problem was solved step by step; in the lowest-order approximation $a = \pi/2\beta$.

Here is the place to discuss the behavior of the arising power series. The zeroth approximation turns out to be efficient even for large values of the nonlinearity parameter (i.e., far from the first bifurcation point, which corresponds to a qualitative change of the phase plane). However, the convergence of these expansions is slow and practically does not depend on the magnitude of the nonlinearity parameter, that is, on the modulus of the elliptic integral of the first kind. This situation resembles the behavior of asymptotic series where the first few terms normally give a reliable representation. Then one can use these terms to construct the Pade approximation, which yields the approximation with an extended applicability range. For example, the /5, 2/ Pade approximation in the case $k = 0.62$, $\beta = 0.58$ yields the value $T = 6.18$ for the period, which is close to the numerical value $T = 6.14$. When keeping nine terms in the power expansion, one obtains $T = 4.28$.

This procedure may be applied if we study processes in systems with internal resonance, which are far from their stationary states and consequently close to a beat with complete energy transfer. We underline that LPT in systems that are linearized in terms of displacements but strongly nonlinear in terms of $\theta - \Delta$ is a good approximation for the LPT in nonlinear system. It is important that consideration of LPTs enables one to recognize the second dynamic transition that occurs in a system when the nonlinearity parameter N increases, caused by the transformation of the LPT into a separatrix. This means that complete energy transfer from the first mass to the second one becomes impossible, as mentioned above. When $k > 1$ the structure of the phase plane changes drastically and unbounded trajectories appear. Simultaneously, the role of the two stable asymmetric normal modes, which have appeared due to the bifurcation of the initial out-of-phase mode, becomes more important. They represent vibrations concentrated predominantly on one of the masses. The result of direct numerical integration of the initial system confirms that complete energy transfer does not exist when $k > 1$. When only the first particle is initially excited, one can see that for $k = 2$ the system oscillates in the attractive region of the localized nonlinear normal mode with energy concentrated predominantly on the first particle (Fig. 2.34).

In the case of a linear system ($k = 0$) one can find an exact analytical solution using linear normal modes. However, the proposed description of the beating

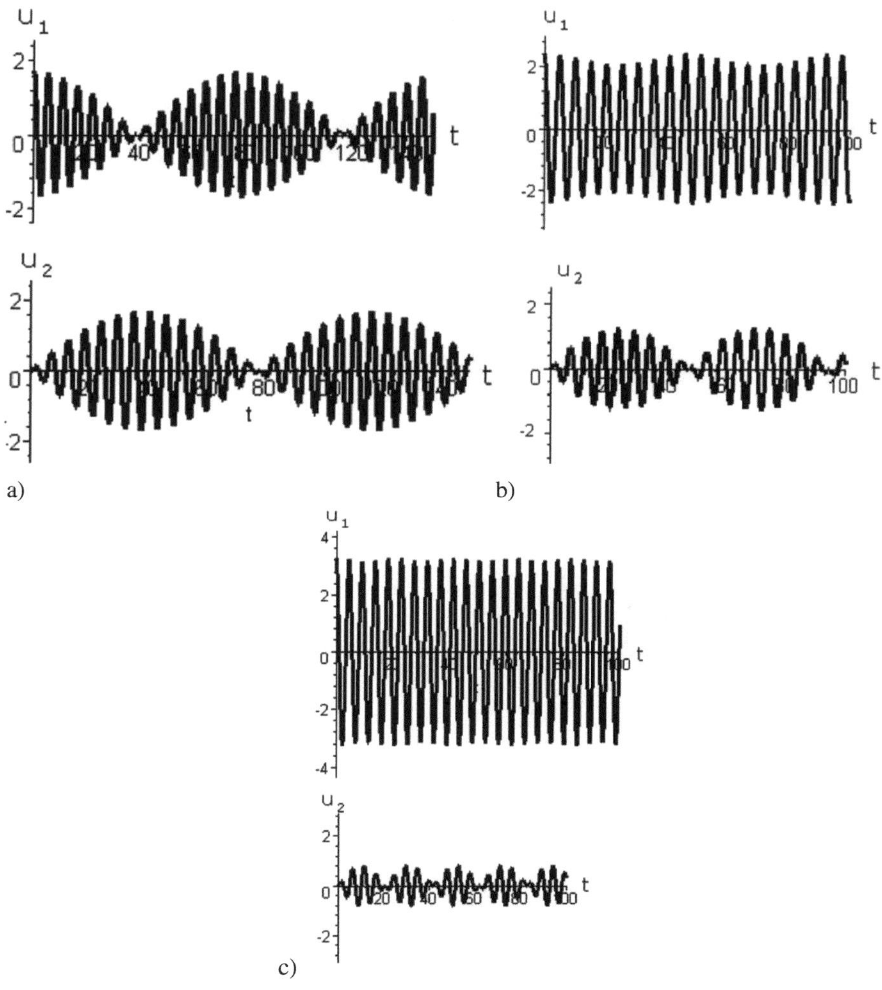

Fig. 2.34 Free oscillations with ($\alpha = 0.125$, $\beta = 0.5$, $\varepsilon = 0.1$) for: (**a**) $k = 0.55$, $N = 2.933$, $u_1(0) = 1.7127$, $u_{1,t}(0) = u_2(0) = u_{2,t}(0) = 0$; (**b**) $k = 1.1$, $N = 5.867$, $u_1(0) = 2.422$, $u_{1,t}(0) = u_2(0) = u_{2,t}(0) = 0$; (**c**) $k = 2.0$, $N = 10.67$, $u_1(0) = 3.266$, $u_{1,t}(0) = u_2(0) = u_{2,t}(0) = 0$

phenomena via non-smooth basic functions has the advantage of being physically adequate in both linear and nonlinear cases. It is clear that both linear and nonlinear beating close to the LPT can be more adequately described in terms of the basic functions τ and e than in terms of trigonometric functions. We choose this rather simple system to illustrate the main ideas. Now let us discuss briefly some examples that demonstrate the applicability of power-expansions techniques in problems relating to trajectories close to LPTs in more-complicated conservative, dissipative and forced 2 DOF systems as well as in weakly coupled oscillatory chains.

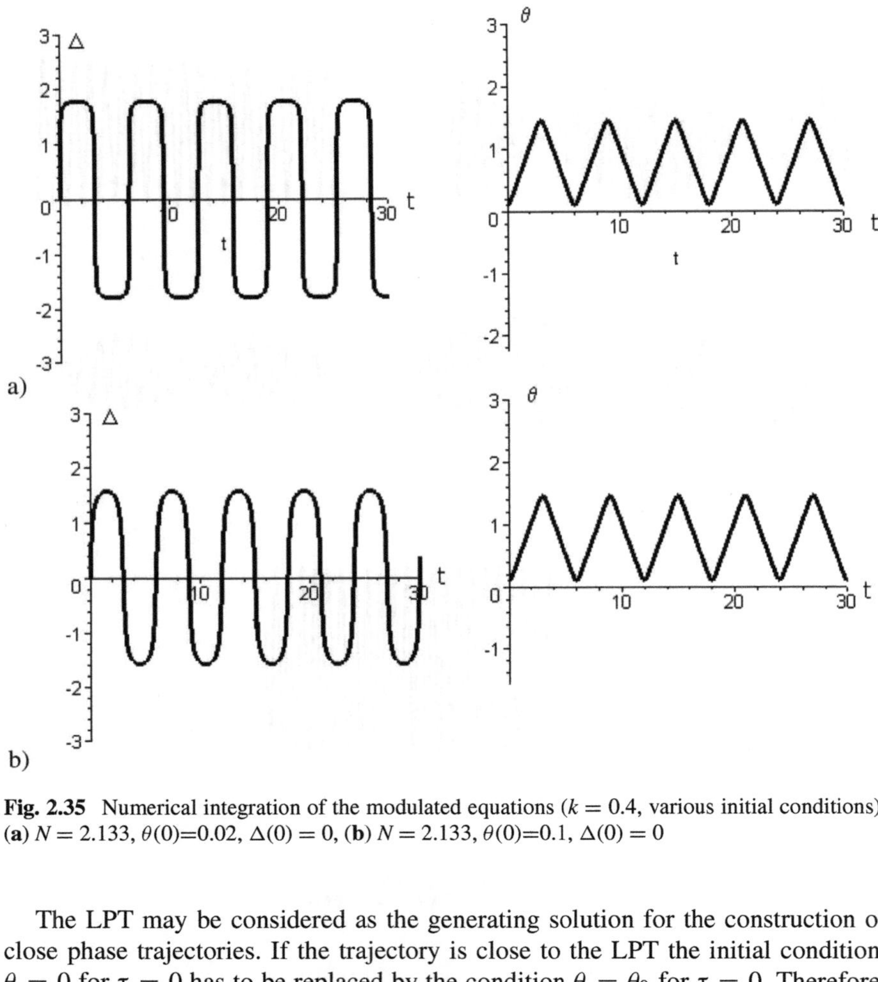

Fig. 2.35 Numerical integration of the modulated equations ($k = 0.4$, various initial conditions), (**a**) $N = 2.133$, $\theta(0)$=0.02, $\Delta(0) = 0$, (**b**) $N = 2.133$, $\theta(0)$=0.1, $\Delta(0) = 0$

The LPT may be considered as the generating solution for the construction of close phase trajectories. If the trajectory is close to the LPT the initial condition, $\theta = 0$ for $\tau = 0$ has to be replaced by the condition $\theta = \theta_0$ for $\tau = 0$. Therefore, we can use the expansions (2.228) again, but $X_{1,0} = \theta_0$ is not equal to zero. The corresponding solution is shown in Fig. 2.35 for $k = 0.4$ and two different values of $\theta(0)$. In both cases $\Delta(0) = 0$.

One can add the dissipative terms $2n\varepsilon \left(dU_j / d\tau_0\right)$ to the left part of the governing equations of motion. The final complex equations of motion then have the form (Manevitch, 2001):

$$\frac{\partial \Phi_{j,0}}{\partial \tau_1} + i\beta\, \Phi_{3-j,0} - i\alpha\, e^{-2\gamma\tau_2} \left|\Phi_{j,0}\right|^2 \Phi_{j,0}, \;\; j = 1, 2 \tag{2.233}$$

where $\Phi_{j,0} = e^{\gamma\tau_1} f_{1,0}, \varepsilon^2\gamma = n / \sqrt{c_1 m}$ and the second integral (2.220) remain valid if we replace $f_{1,0}$ with $\Phi_{j,0}$; $2n$ is the coefficient of damping in the initial equations of motion. The final equations following from (2.233), which describe the phase trajectories in the $\theta - \Delta$ plane can be written as follows

$$\begin{aligned} \frac{\partial \theta}{\partial \tau_1} &= \beta \sin \Delta \\ \sin 2\theta \frac{\partial \Delta}{\partial \tau_1} &= 2\beta \cos 2\theta \cos \Delta + \frac{3}{2}\alpha N e^{-2\gamma \tau_2} \sin 4\theta \end{aligned} \tag{2.234}$$

One can again present the solution of (2.234) as (2.229) where the smooth functions $X_i(\tau)$, $Y_i(\tau)$ satisfy the equations

$$\begin{aligned} &\frac{\partial}{\partial \tau}\begin{Bmatrix} X_1 \\ Y_1 \end{Bmatrix} = \frac{1}{2}a\beta[\sin(X_2+Y_2) \mp \sin(X_2-Y_2)] \\ &\frac{\partial}{\partial \tau}\begin{Bmatrix} X_2 \\ Y_2 \end{Bmatrix} = a\beta[ctg\, 2(X_1+Y_1)\cos(X_2+Y_2) \mp \\ &\mp ctg\, 2(X_1-Y_1)\cos(X_2-Y_2)] \\ &+\frac{3a}{2}\alpha N e^{-2\gamma\tau}[\cos 2(X_1+Y_1) \pm \cos 2(X_1-Y_1)] \end{aligned} \tag{2.235}$$

Using the procedure described above while taking into account (2.235) one can find the following representation for θ and Δ:

$$\theta = \alpha\beta\tau - \frac{2}{3}(\alpha\beta)^3\, k^2\tau^3 + \cdots \tag{2.236}$$

$$\Delta = e\left[\frac{\pi}{2} + 2\alpha\beta\, k\tau - \frac{4}{3}\alpha\beta\gamma k\tau^2 - \frac{4}{3}(\alpha\beta)^3\, k\tau^3 + \cdots\right] \tag{2.237}$$

So, in the given approximation we find not only the corrections to θ and Δ caused by the nonlinearity, but also a correction to Δ due to damping; the corresponding decrease in the amplitude is taken into account by multiplying the complex functions $fj,0$ by $e^{-2\gamma\tau_1}$. This means that the trajectories in the phase plane (Fig. 2.31) will preserve their form except for a small deformation caused by the damping.

The conditions for the LPT excitation are fulfilled in the important particular case when one of two different masses is exposed to the action of an initial impulse. If we take viscous friction into account, we can formulate the problem of almost irreversible energy transfer from an initially excited mass to the second, smaller one, known as the energy pumping problem discussed in the next section.

2.5.1 Exercise

Calculate the second-order correction to functions X_i, Yi describing the beatings

2.6 2DOF Systems of Nonlinear Oscillators with Essential Asymmetry Targeted Energy Transfer (TET)

2.6.1 Targeted Energy Transfer in an Unforced 2DOF System

We shall consider a system of two oscillators with essential asymmetry in the context of a so-called "targeted energy transfer (TET)", or "energy pumping", dealing with passive irreversible energy transfer from a linear system to a strongly nonlinear defending element (or nonlinear energy sink (NES). Detailed studies of this complicated problem are presented in a recent book (Vakakis et al., 2008).

As a preliminary example of TET, we consider a two degree-of-freedom (DOF) dissipative unforced system described by the following equations:

$$\begin{aligned} \ddot{y}_1 + \lambda_1 \dot{y}_1 + y_1 + \lambda_2(\dot{y}_1 - \dot{y}_2) + k(y_1 - y_2)^3 &= 0 \\ \varepsilon \ddot{y}_2 + \lambda_2(\dot{y}_2 - \dot{y}_1) + k(y_2 - y_1)^3 &= 0 \end{aligned} \tag{2.238}$$

Physically, these equations describe a damped linear oscillator (LO) with mass and natural frequency normalized to unity, and viscous damping coefficient λ_1, and an essentially nonlinear attachment with normalized mass ε, normalized nonlinear stiffness coefficient k, and viscous damping coefficient λ_2. Note that system (2.238) can not be regarded as a small perturbation of a linear system due to the strongly nonlinear coupling terms.

We simulate system (2.238) numerically for the parameter values $\varepsilon = 0.1$, $k = 0.1$, $\lambda_1 = 0.01$ and $\lambda_2 = 0.01$. The selected initial conditions correspond to the impulse $F = A\,\delta(t)$ applied to the linear oscillator (where $\delta(t)$ is Dirac's delta function – this impulsive forcing is equivalent to imposing the initial velocity $\dot{y}_1(0+) = A$) with the system being initially at rest, i.e., $y_1(0) = y_2(0) = \dot{y}_2(0) = 0$ and $\dot{y}_1(0+) = A$. Hence, the initial energy is stored only in the LO. The instantaneous transfer of energy from the LO to the nonlinear attachment can be monitored by computing the nondimensional energy ratio κ, which denotes the portion of instantaneous total energy stored in the nonlinear attachment,

$$\kappa = \frac{E_2}{E_1 + E_2}, \quad E_1 = \frac{1}{2}\left(y_1^2 + \dot{y}_1^2\right), \quad E_2 = \frac{\varepsilon}{2}\dot{y}_2^2 + \frac{k}{4}(y_1 - y_2)^4 \tag{2.239}$$

where E_1 and E_2 are instantaneous energies of the LO and the attachment, respectively. Of course, all quantities in relations (2.239) are time-dependent.

In Figs. 2.36 and 2.37 we depict the evolution of the energy ratio κ for impulse strengths $A = 0.5$ and $A = 0.7$, respectively. From Fig. 2.36 it is clear that only a small amount of energy (of order of 7%) is transferred from the LO to the nonlinear attachment. However, for a slightly higher impulse the energy transferred climbs to almost 95% (cf. Fig. 2.37), and within a rather short time (t~15, much less than characteristic time of viscous energy dissipation in the LO) almost the entire impulsive energy is passively transferred from the LO to the nonlinear attachment, which acts

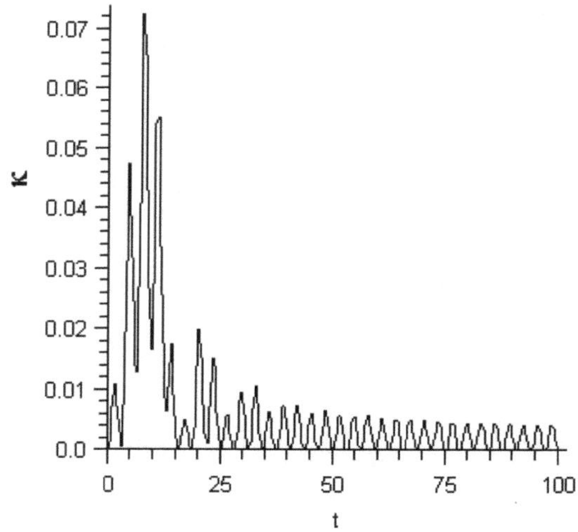

Fig. 2.36 Evolution of the energy ratio κ for impulse strength $A = 0.5$

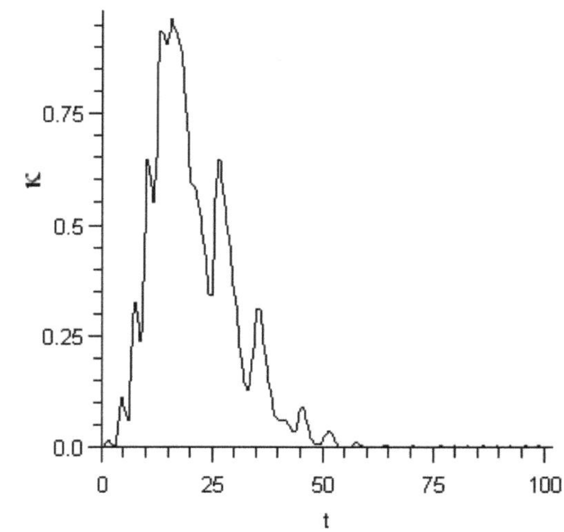

Fig. 2.37 Evolution of the energy ratio κ for impulse strength $A = 0.7$

as nonlinear energy sink. It should be mentioned that the mass of the attachment in this particular example is just 10% of the mass of the LO.

The phenomenon of TET in damped strongly nonlinear autonomous system (Gendelman, 2001; Vakakis and Gendelman, 2001; Gendelman et al., 2001, 2005; Gendelman, 2004) puts forward a series of problems of principal importance. One of them is the possibility of optimization of energetic sink parameters. Efficient analytical description of the TET process in a strongly non-homogeneous 2DOF system proposed in (Manevitch et al., 2007) turned out to be appropriate for the statement and solution of the optimization problem in application to a "cubic" type sink studied in previous papers (Gendelman, 2001; Gourdon and Lamarque, 2005). Meanwhile, together with this type of sink, other strongly nonlinear sinks were also discussed in the framework of the general TET problem. However, isolated numerical estimations and the absence of a general criterion of relative sink efficiency make their comparison difficult. Such a general criterion follows from the viewpoint proposed in (Manevitch et al., 2007) when considering energy pumping process as a damped beating.

On this basis, we consider and compare the sinks which are systems with several equilibrium states and the sinks close to vibro-impact systems. Some general regularity determining the efficiency of an energy sink are discussed in detail.

The following system of coupled oscillators is considered:

$$\begin{aligned} &M\frac{d^2x_1}{dt^2}+\tilde{\mu}_1\frac{dx_1}{dt}+\tilde{\eta}\left(\frac{dx_1}{dt}-\frac{dx_2}{dt}\right)+k_1\,x_1+\\ &+k_3\,(x_1-x_2)^{2n-1}\pm D\,(x_1-x_2)=0\\ &m\frac{d^2x_2}{dt^2}-\tilde{\eta}\left(\frac{dx_1}{dt}-\frac{dx_2}{dt}\right)-k_3\,(x_1-x_2)^{2n-1}\pm D\,(x_2-x_1)=0\\ &n\geq 2 \end{aligned} \tag{2.240}$$

Nonlinear coupling with multiple states of equilibrium corresponds to the symbol "–". The linear primary structure is excited by an impulse, so we consider free oscillations of structures with the following initial conditions:

$$t=0: x_1=x_2=0,\ \frac{dx_2}{dt}=0,\ \frac{dx_1}{dt}=C_I.$$

System (2.240) can be analyzed by using the perturbation theory. The following change of variables $\tilde{U}_1$=x_1, $\tilde{U}_2$= x_2–x_1 is considered. Then system (2.240) is reduced to the form

$$\begin{aligned} &(M+m)\,\frac{d^2\tilde{U}_1}{dt^2}+m\frac{d^2\tilde{U}_2}{dt^2}+\tilde{\mu}_1\frac{d\tilde{U}_1}{dt}+k_1\,\tilde{U}_1=0\\ &m\frac{d^2\tilde{U}_2}{dt^2}+m\frac{d^2\tilde{U}_1}{dt^2}+\tilde{\eta}\frac{d\tilde{U}_2}{dt}+k_3\tilde{U}_2^{2n-1}+D\tilde{U}_2=0 \end{aligned} \quad n\geq 2 \tag{2.241}$$

To clarify the equations, dimensionless coefficients and displacements are used, and Eq. (2.241) and the initial conditions are rewritten as:

$$\begin{aligned} &(1+\varepsilon)\,\frac{d^2U_1}{d\tau^2}+\varepsilon\frac{d^2U_2}{d\tau^2}+\varepsilon\mu_1\frac{dU_1}{d\tau}+U_1=0\\ &\varepsilon\frac{d^2U_2}{d\tau^2}+\varepsilon\frac{d^2U_1}{d\tau^2}+\varepsilon\eta\frac{dU_2}{d\tau}+cU_2^{2n-1}+\varepsilon\alpha U_2=0\\ &\tau=0: U_1=U_2=0;\ \frac{dU_1}{d\tau}=-\frac{dU_2}{d\tau}=\frac{C_I}{\omega}, \end{aligned} \quad n\geq 2 \tag{2.242}$$

where

$$\omega=\sqrt{\frac{k_1}{M}}, U_1=\frac{\omega}{C_I}\tilde{U}_1, U_2=\frac{\omega}{C_I}\tilde{U}_2, \varepsilon=\frac{m}{M}, \tau=\omega t, \varepsilon\mu_1=\sqrt{\frac{1}{k_1 M}}\,\tilde{\mu}_1, \varepsilon\eta=\sqrt{\frac{1}{k_1 M}}\,\tilde{\eta}, \varepsilon\alpha=\sqrt{\frac{1}{k_1 M}}\,D,$$

$$c=\frac{C_I^{2n-2}k_3}{\omega^{2n-2}k_1}.$$

Here and below ε is a small parameter, representing a mass ratio which has to be very small.

For small U_1, U_2, the contribution of nonlinear term in the second of Eq. (2.242) is much lesser than the contribution of linear terms in the same equation. The following change of variables: $u_1=\varepsilon^{-1/(2n-2)}U_1$, $u_2=\varepsilon^{-1/(2n-2)}U_2$ is introduced. Then Eq. (2.242) and the initial conditions are transformed into the following system

$$\begin{aligned}
&(1+\varepsilon)\frac{d^2u_1}{d\tau^2}+(1+\varepsilon)\,u_1+\varepsilon\left(\frac{d^2u_2}{d\tau^2}+\mu_1\frac{du_1}{d\tau}-u_1\right)=0\\
&\frac{d^2u_2}{d\tau^2}+u_2+\left[-u_2+\frac{d^2u_1}{d\tau^2}+\eta\frac{du_2}{d\tau}+c\,u_2^{2n-1}\pm\alpha u_2\right]=0\\
&\tau=0: u_1=u_2=0;\ \frac{du_1}{d\tau}=-\frac{du_2}{d\tau}=\varepsilon^{-1/(2n-2)}\frac{C_I}{\omega}.
\end{aligned}\tag{2.243}$$

We assume further that the oscillations occur near the resonance at frequency close to unity. Then we should suppose that the sum of terms in the square brackets in the second of Eq. (2.243) is a small quantity of $\boldsymbol{O(\varepsilon)}$. To accomplish this, let us enter the "bookkeeping" factor $\delta=1/\varepsilon$ which can be nominally taken as being equal to 1 during the further asymptotic analysis (actually the sum of terms in square brackets is assumed to be small). We shall take into account the true value of factor δ in numerical calculations (such a procedure, as we can see, is fully justified by detailed numerical analysis). Then Eq. (2.243) look as follows

$$\begin{aligned}
&(1+\varepsilon)\frac{d^2u_1}{d\tau^2}+(1+\varepsilon)\,u_1+\varepsilon\left(\frac{d^2u_2}{d\tau^2}+\mu_1\frac{du_1}{d\tau}-u_1\right)=0\\
&\frac{d^2u_2}{d\tau^2}+u_2+\varepsilon\delta\left(-u_2+\frac{d^2u_1}{d\tau^2}+\eta\frac{du_2}{d\tau}+cu_2^{2n-1}\pm\alpha u_2\right)=0
\end{aligned}\tag{2.244}$$

Introducing the change of variables

$$\begin{aligned}
\varphi_1&=e^{-i\tau}\left(\frac{du_1}{d\tau}+i\,u_1\right) & \varphi_1^*&=e^{i\tau}\left(\frac{du_1}{d\tau}-i\,u_1\right)\\
\varphi_2&=e^{-i\tau}\left(\frac{du_2}{d\tau}+i\,u_2\right) & \varphi_2^*&=e^{i\tau}\left(\frac{du_2}{d\tau}-i\,u_2\right)
\end{aligned}\tag{2.245}$$

and performing multiple scale analysis:

$$\tau_0=\tau, \tau_1=\varepsilon\tau, \tau_2=\varepsilon^2\tau, \ldots\tag{2.246}$$

$$\varphi_1 = \varphi_{10} + \varepsilon\varphi_{11} + \varepsilon^2\varphi_{12} + \dots \tag{2.247}$$

$$\varphi_2 = \varphi_{20} + \varepsilon\varphi_{21} + \varepsilon^2\varphi_{22} + \dots \tag{2.248}$$

one obtains the following equations

$$\frac{\partial\varphi_1}{\partial\tau_1} + \frac{i}{2}(\varphi_1 + \varphi_2) + \frac{\mu_1}{2}\varphi_1 = 0 \tag{2.249}$$

and:

$$\frac{\partial\varphi_2}{\partial\tau_1} + \delta\left[\frac{i}{2}(\varphi_1 + \varphi_2) + \frac{i\eta}{2}\varphi_2\left|\varphi_2\right|^{2n-1} \pm \frac{i\alpha}{2}\varphi_2\right] = 0 \tag{2.250}$$

Multiplying Eqs. (2.249 and 2.250) by $\varphi_1{}^*$ and $\varphi_2{}^*$ respectively, and combining these equations and the complex conjugate ones we get:

$$\frac{\partial\left|\varphi_2\right|^2}{\partial\tau_1} + \delta\frac{\partial\left|\varphi_1\right|^2}{\partial\tau_1} + \eta\delta\left|\varphi_2\right|^2 + \delta\mu_1\left|\varphi_1\right|^2 = 0 \tag{2.251}$$

If there is no damping in the system (2.240), i.e. $\eta = \mu_1 = 0$, then Eq. (2.251) is the conservation law of quantity $H =\mid \varphi^2 + \delta\left|\varphi_1\right|^2$ relative to time τ_1. One can consider relation (2.251) as an ordinary differential equation with respect to function $\left|\varphi_2\right|^2$ the term $\delta\frac{\partial\left|\varphi_1\right|^2}{\partial\tau_1} + \delta\mu_1\left|\varphi_1\right|^2$ being the right-hand member. Directly applying the Laplace transformation to Eq. (2.251), we obtain its solution in the form

$$\Psi(s) = \frac{G(s) + \left|\varphi_2\right|^2(0)}{s + \delta\eta}$$

where $\Psi(s)$ is a Laplace representation of function $|\phi|^2(\tau_1)$, $G(s)$ is a Laplace representation of function $-\delta\frac{\partial\left|\varphi_1\right|^2}{\partial\tau_1} - \delta\mu_1\left|\varphi_1\right|^2$. After the application of an inverse Laplace transformation to this equation we can find the following representation for function $H(\tau_1)$

$$H(\tau_1) = \exp(-\delta\eta\tau_1)[H(0) + \delta(\delta\eta - \mu_1)\int\limits_0^{\tau_1}\exp(\delta\eta z)\left|\varphi_1\right|^2(z)dz] \tag{2.252}$$

To find a solution we expand the integral in the right-hand member of Eq. (2.252) in a Taylor series in a vicinity of point $\tau_1 = 0$. It allows us to calculate function $H(\tau_1)$ avoiding the solution of Eqs. (2.250 and 2.251) or the initial Eq. (2.2400). Then Eq. (2.252) looks like

$$\begin{aligned} &H(\tau_1) = \exp(-\delta\eta\tau_1)[H(0)+ \\ &+\delta(\delta\eta - \mu_1)\{\tau_1\left|\varphi_1\right|^2(0) + \tfrac{\tau_1^2}{2}(\delta\eta\left|\varphi_1\right|^2(0) + \tfrac{\partial}{\partial\tau_1}(\left|\varphi_1\right|^2)(0)) + \dots\}] \end{aligned} \tag{2.253}$$

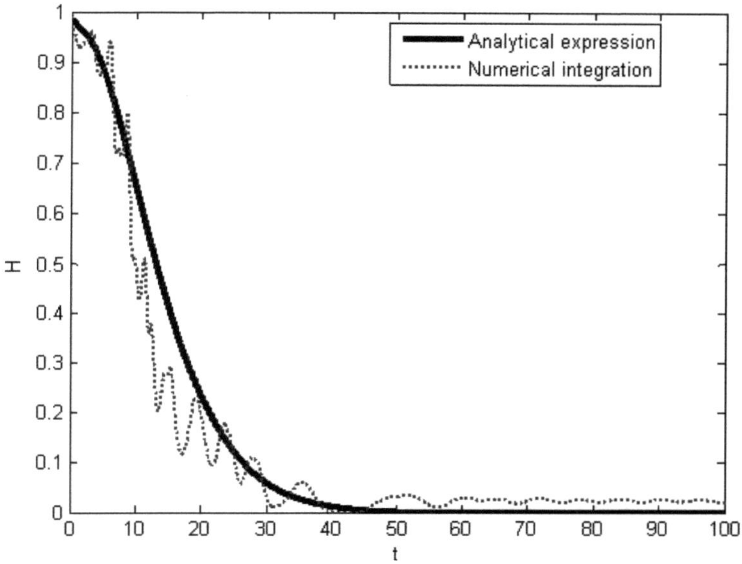

Fig. 2.38 Time dependence of function H(t). The *solid line*, taking into account the Taylor series up to the terms of the fifth order on τ_1, inclusive, depicts solution (2.253). The *dashed line* depicts the numerical solution of System (2.240)

Quantities $H\,(0)$ and $|\varphi_1|^2(0)$ are known from the initial conditions. The derivative $\dfrac{\partial\,|\varphi_1|\,(0)}{\partial\tau_1}$ and higher-order derivatives of function $|\varphi_1|$ at the same point $\tau_1 = 0$ can be found from the initial conditions and equations of motion (2.240).

When energy pumping occurs, the analytical approximation (2.253) is good as shown in Fig. 2.38 where the analytical solution H of (2.253), taking into account the Taylor series up to the terms of the fifth order on τ_1 and the numerical integration of System (2.240) have been compared ($n = 2$, $\omega = 1$, $\mu_1 = 0$, $\varepsilon = 0.1$, $\eta = 0.2$, $c = 0.8$, $\alpha = 0.2$, $\frac{dx_1}{dt}(t = 0) = 0.3$ and sign "–" is considered in Eq. (2.240).

In this case, energy pumping occurs as shown in Fig. 2.39 where the numerical solutions of the System (2.240) have been plotted with and without coupling.

Not only is the analytical approximation (2.253) good, but the different φ_1, φ_2 introduced are also a good approximation as shown in Fig. 2.40, where those analytical approximations are compared with results of integrating initial system (2.240). Initial conditions are:

$$\frac{dx_1}{dt}(t = 0) = 0.5,\ \frac{dx_2}{dt}(t = 0) = x_1(t = 0) = x_2(t = 0) = 0$$

parameters of the system: $n = 2$, $\omega = 1$, $\mu_1 = 0$, $\varepsilon = 0.1$, $\eta = 0.2$, $c = 0.8$, $\alpha = 0.2$ and sign "–" remain in Eq. (2.240).

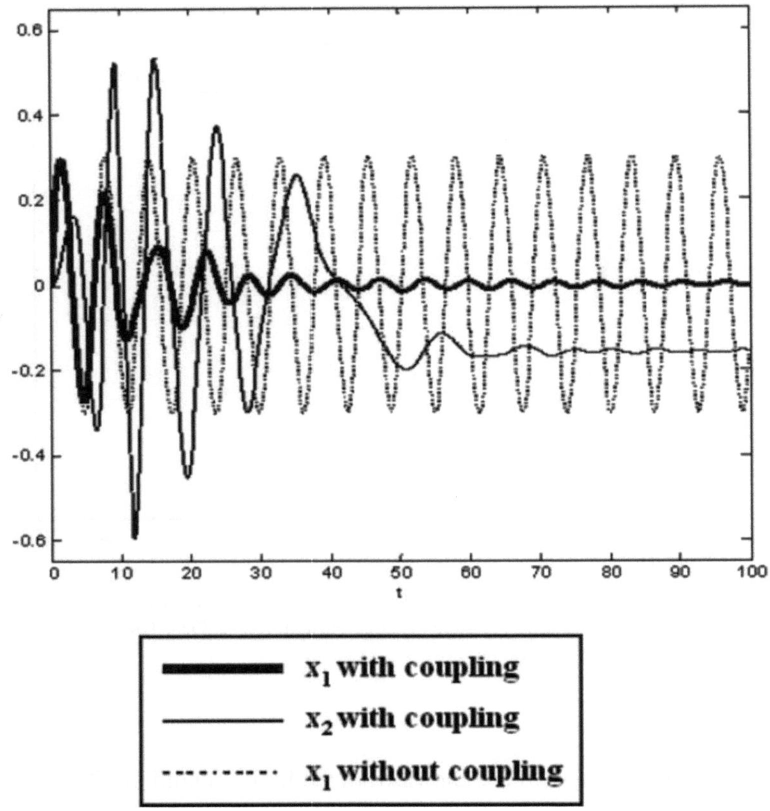

Fig. 2.39 Responses with numerical integration of (2.240) with and without coupling

If sign "+" is considered in Eq. (2.240), then the analytical expression (2.253) also provides satisfactory accuracy, as shown in Fig. 2.41, where

$$\frac{dx_1}{dt}(t=0) = 0.3, \ \frac{dx_2}{dt}(t=0) = x_1(t=0) = x_2(t=0) = 0,$$

parameters of the system: $n = 2$, $\omega = 1$, $\mu_1 = 0$, $\varepsilon = 0.1$, $\eta = 0.5$, $c = 0.8$, $\alpha = 0.2$.

So it is now possible to try to design the optimal energy sink owing to the calculation of H. Indeed, we can see that if sign "+" is considered in Eq. (2.240), then energy pumping appears to be more efficient since the decrease of energy H is more abrupt. The energy decreases faster with sign "+" in Eq. (2.240) than with sign "–" (if all other parameters are fixed) as shown in Fig. 2.42 where

$$\frac{dx_1}{dt}(t=0) = 0.3, \ \frac{dx_2}{dt}(t=0) = x_1(t=0) = x_2(t=0) = 0$$

and $n = 2$, $\omega = 1$, $\mu_1 = 0$, $\varepsilon = 0.1$, $\eta = 0.5$, $c = 0.8$, $\alpha = 0.2$.

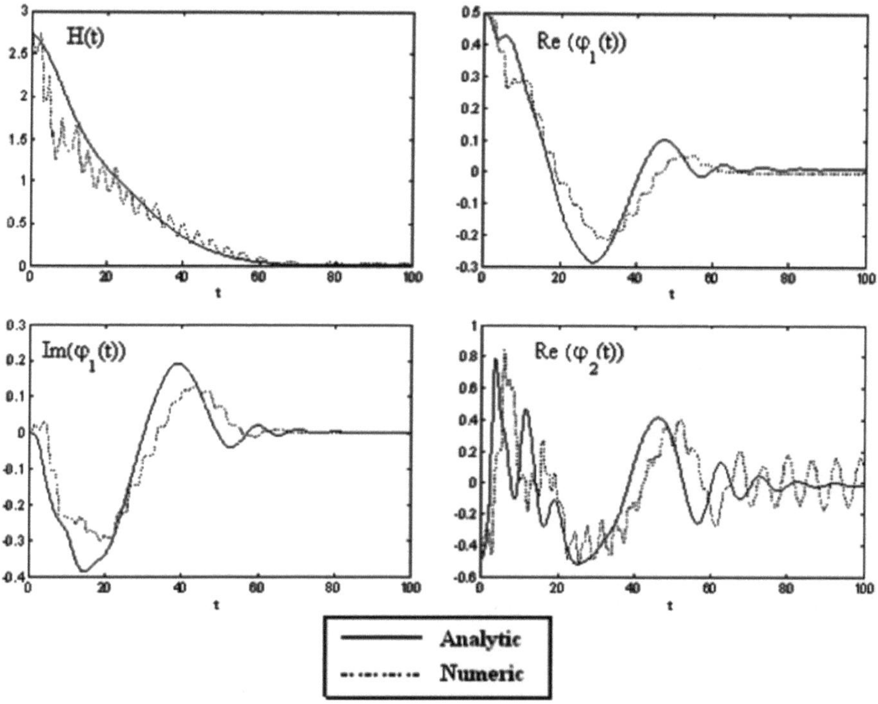

Fig. 2.40 Function *H(t)*, Imag $\varphi_1(t)$, Re $\varphi_1(t)$, Re $\varphi_2(t)$, compared with results of integrating initial System (2.240)

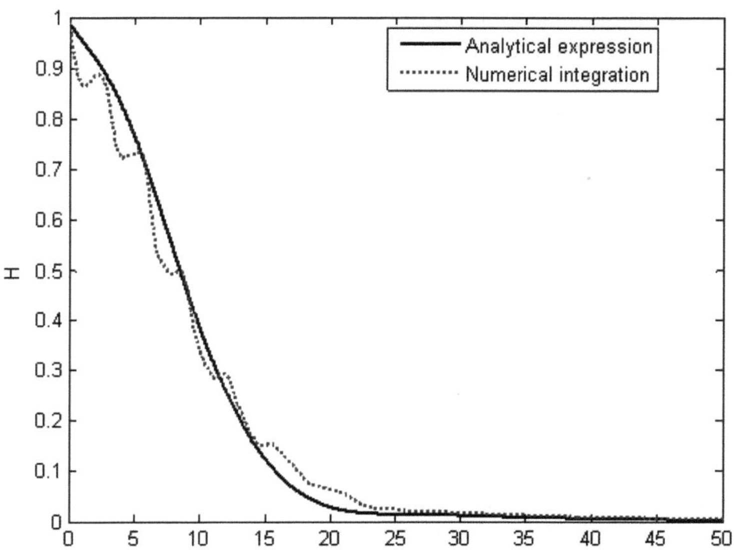

Fig. 2.41 Function *H(t)*. The *solid line*, taking into account the Taylor series up to the terms of the fifth order on τ_1, inclusive, depicts solution (2.253). The *dashed line* depicts the numerical solution of System (2.240)

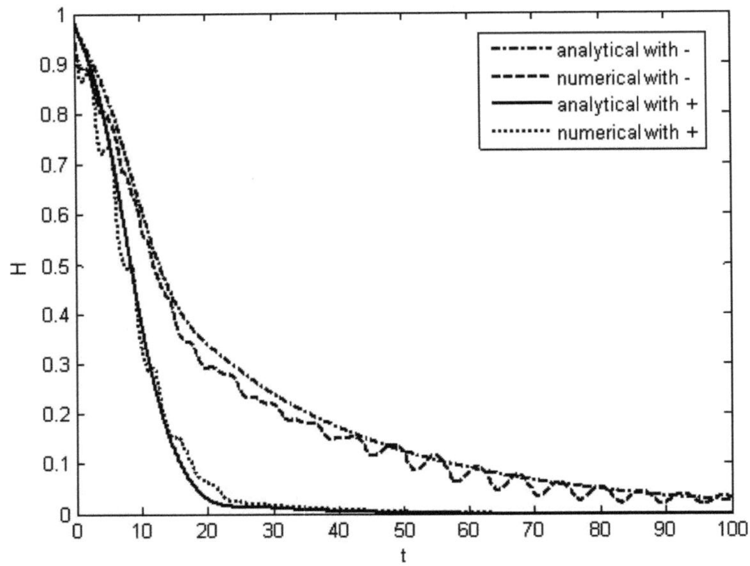

Fig. 2.42 Comparison of function *H*(*t*) with consideration of sign + or – in Eq. (2.240)

Thus, energy pumping is more efficient when sign "+" is considered in Eq. (2.240) as shown in Fig. 2.43 with the numerical integration of System (2.240) with the same values of parameters as previously. In this figure, it clearly appears that the vibrations are almost completely suppressed at t = 20 s when sign "+" is considered in Eq. (2.240).

Moreover, we can also consider the influence of the degree n of the nonlinearity on the efficiency of the sink. Indeed, for a given set of parameters, there exists an optimal value of n for which the efficiency of energy pumping is optimal. For this study we now consider the case of sign "+" in Eq. (2.240) since the efficiency in this case seems better. For example, if

$$\frac{dx_1}{dt}(t=0) = 0.4, \frac{dx_2}{dt}(t=0) = 0, x_1(t=0) = x_2(t=0) = 0$$

and n = 2,3,4, ω =1, μ_1=0, $\varepsilon = 0.1$, $\eta = 0.2$, $c = 0.8$, $\alpha = 0.2$ and sign "+" in Eq. (2.240), then the optimal value of n is 3 (the degree of the nonlinearity is 5) as shown in Fig. 2.44. In this figure, we can also see that for $n = 4$ the analytical approximation is less good after $t = 30$ s since after that the energy pumping does not occur and there is no resonance anymore.

As we will see, important information about sink efficiency can be extracted from the analysis of a corresponding conservative system. If η=μ_1=0, the Eqs. (2.249 and 2.250) look like this:

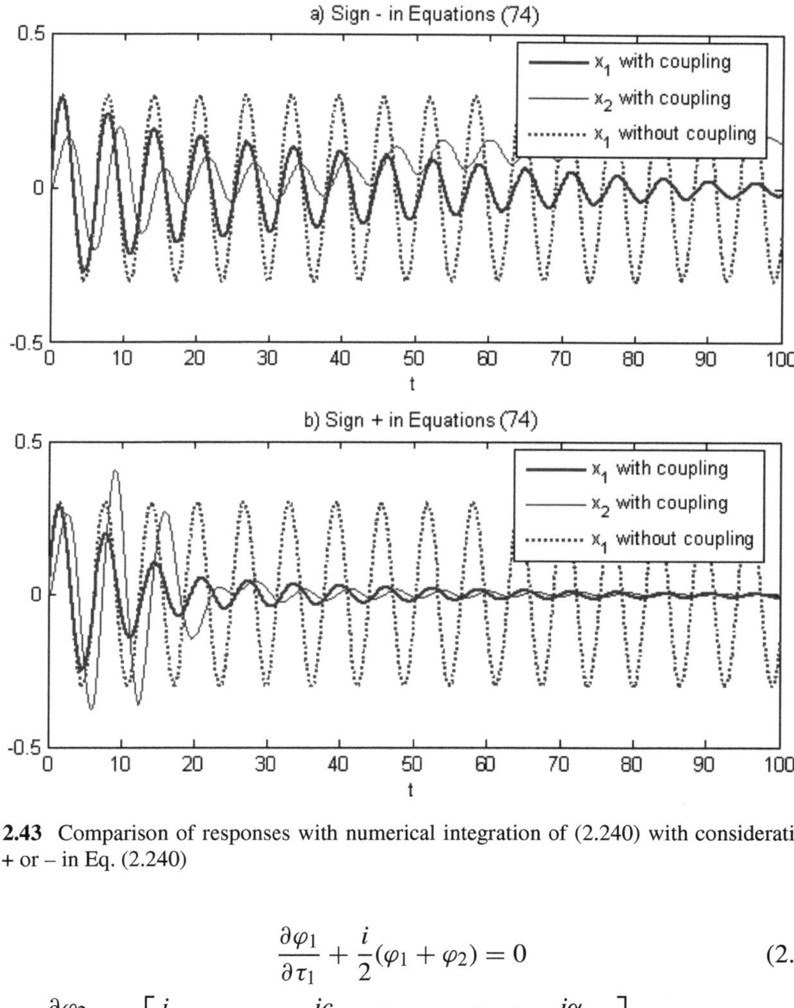

Fig. 2.43 Comparison of responses with numerical integration of (2.240) with consideration of sign + or – in Eq. (2.240)

$$\frac{\partial \varphi_1}{\partial \tau_1} + \frac{i}{2}(\varphi_1 + \varphi_2) = 0 \tag{2.254}$$

$$\frac{\partial \varphi_2}{\partial \tau_1} + \delta \left[\frac{i}{2}(\varphi_1 + \varphi_2) - \frac{ic}{2} C_{2n-1}^{n-1} \varphi_2 \left|\varphi_2\right|^{2n-1} \pm \frac{i\alpha}{2\varphi}\varphi_2 \right] = 0 \tag{2.255}$$

Introducing the change of variables:

$$\varphi_1 = f_1, \ \varphi_2 = \sqrt{\delta} f_2 \tag{2.256}$$

the equations of motion can be rewritten as follows

$$\frac{\partial f_1}{\partial \tau_1} + \frac{i}{2}(f_1 + \sqrt{\delta} f_2) = 0 \tag{2.257}$$

$$\frac{\partial f_2}{\partial \tau_1} + \sqrt{\delta} \left[\frac{i}{2}(f_1 + \sqrt{\delta} f_2) - \frac{ic}{2^{2n-1}} \delta^{\frac{2n-1}{2}} C_{2n-1}^{n-1} f_2 \left|f_2\right|^{2n-1} \pm \frac{i\alpha}{2\varphi} \sqrt{\delta} f_2 \right] = 0 \tag{2.258}$$

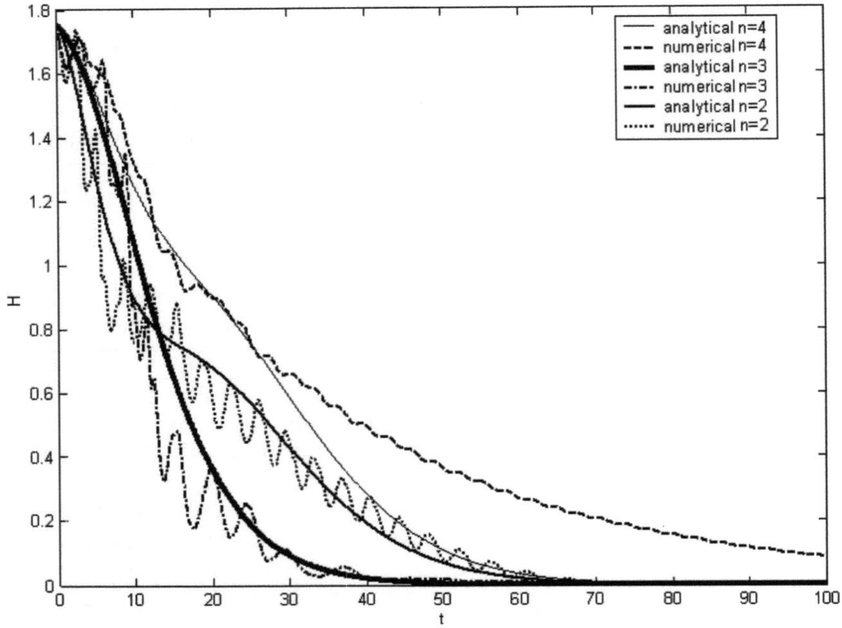

Fig. 2.44 Comparison of $H(t)$ for different values of n

The system is now completely integrable with the two first integrals of motion:

$$H = \frac{i}{2}(|f_1|^2 + \sqrt{\delta}\,|f_2|^2) - \frac{i}{2}(f_1 f_2^* + f_2 f_1^*) + ic\delta \frac{C_{2n-1}^{n-1}}{2^{2n-1}} |f_2|^{2n} \pm \frac{i\alpha}{2}\delta\,|f_2|^2 \quad (2.259)$$

$$N = |f_1|^2 + |f_2|^2 \quad (2.260)$$

Now we can introduce the following changes of variables:

$$f_1 = \sqrt{N}\cos\theta e^{i\delta_1}, f_2 = \sqrt{N}\sin\theta e^{i\delta_2} \quad (2.261)$$

$$\Delta = \delta_1 - \delta_2 \quad (2.262)$$

Finally we obtain:

$$\frac{\partial\theta}{\partial\tau_1} - \frac{\sqrt{\delta}}{2}\sin\Delta = 0 \quad (2.263)$$

$$\frac{\partial\Delta}{\partial\tau_1} - \frac{\delta-1}{2} \pm \frac{\alpha\delta}{2}\sqrt{\delta}\cos\Delta\cot 2\theta + c\delta N^{n-1}\frac{C_{2n-1}^{n-1}}{2^{2n-1}}\sin^{2n-2}\theta = 0 \quad (2.264)$$

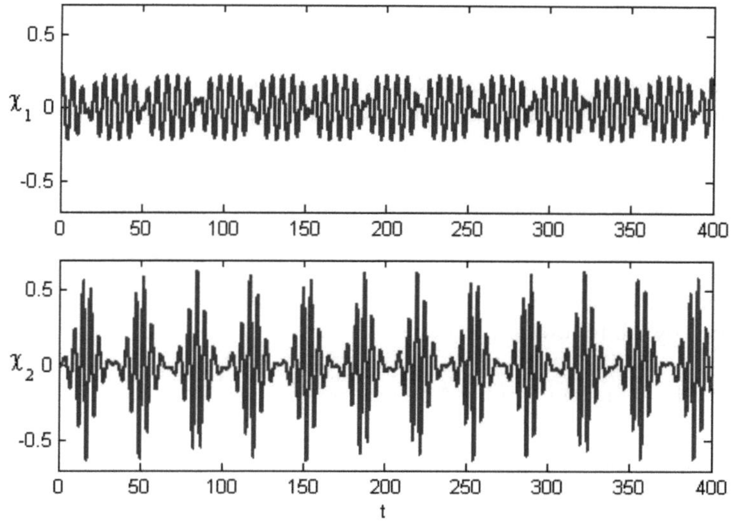

Fig. 2.45 Nonlinear beating

Then, the following two cases can be distinguished:

- If $-\frac{\delta-1}{2} \pm \frac{\alpha\delta}{2} > 0$ then for all values of N there exist only 2 NNMs;
- If $-\frac{\delta-1}{2} \pm \frac{\alpha\delta}{2} < 0$ then under a certain value of N there exist 2 NNMs and above a certain value of N, there exist 4 NNMs.

When all 4 NNMs appear, the energy pumping phenomenon occurs with the beating phenomenon. This nonlinear beating can also be seen in the two displacements x_1 and x_2 as shown in Fig. 2.45 with the same values as previously.

2.6.2 Targeted Energy Transfer in Forced 2DOF System

In the previous section it was demonstrated that the addition of a relatively lightweight strongly nonlinear attachment to a primary (discrete or continuous) linear structure under shock excitation can drastically modify its transient dynamic response and bring about the TET phenomenon. Hence, it is not unreasonable to expect that similar salient dynamical behavior will also be revealed for the case of external periodic excitation. The transition from shock (broadband) to periodic (narrowband) excitation, however, is not trivial, and the application of nonlinear energy sinks (NESs) to structures under narrowband excitation deserves special consideration. For, example, it is not obvious that the capacity for TET of an NES under conditions of shock excitation of a primary structure can be extended to the case of periodic excitation.

We aim to demonstrate that the steady state response of a primary system under harmonic excitation with an attached NES exhibits not only common steady-state and weakly modulated responses, but also very special types of responses characterized by large modulations of the resulting oscillations; this response type is referred to as *Strongly Modulated Response* (*SMR*), and may be regarded as the extension of the TET phenomenon to structures under periodic (narrowband) excitation. Moreover, we demonstrate that SMRs are related to relaxation oscillations of the corresponding averaged dynamical flows (the slow-flows of the dynamics), and in fact, one can regard SMRs as a form of repetitive TETs under the action of persisting periodic forcing.

The system considered here is comprised of a linear oscillator and strongly non linear attachment (pure cubic nonlinearity) and is forced harmonically. The system is described by the following equations:

$$\begin{aligned}&\ddot{y}_1 + \varepsilon\lambda(\dot{y}_1 - \dot{y}_2) + (1+\varepsilon\sigma)y_1 + \frac{4}{3}\varepsilon(y_1 - y_2)^3 = \varepsilon A\cos t\\&\varepsilon\ddot{y}_2 + \varepsilon\lambda(\dot{y}_2 - \dot{y}_1) + \frac{4}{3}\varepsilon(y_2 - y_1)^3 = 0\end{aligned} \tag{2.265}$$

where y_1 and y_2 are the displacements of the linear oscillator and the attachment respectively, $\varepsilon\lambda$ is the damping coefficient, εA is the amplitude of external force and σ is the frequency detuning parameter. $\varepsilon << 1$ is a small parameter which establishes the order of magnitude for coupling, damping, amplitude of the external force, detuning and mass of the attachment.

Coefficients: A, λ, σ are adopted to be of order unity. Rigidity of the nonlinear spring is adopted to be equal to $\frac{4}{3}\varepsilon$ and linear frequency of the primary oscillator – close to unity. The latter adoption does not affect the treatment below, since it may be changed independently by proper rescaling of the dependent variables.

We are interested in the motion of the system in the vicinity of a 1:1:1 resonance manifold, where all variables oscillate with a frequency close to that of the external force. Applicability and technicalities of the averaging procedure for this kind of essentially nonlinear systems are discussed elsewhere (Gendelman, 2004; Gendelman et al., 2008). Successive changes of variables:

$$\begin{aligned}v &= y_1 + \varepsilon y_2\\w &= y_1 - y_2\end{aligned} \tag{2.266}$$

(transition to coordinates "center of mass-internal displacement") and

$$\begin{aligned}\varphi_1\exp(it) &= \dot{v} + iv\\\varphi_2\exp(it) &= \dot{w} + iw\end{aligned} \tag{2.267}$$

where the complex variables φ_j, j = 1,2 describe a slow modulation of variables v and w respectively, yield the following equations for the modulation amplitudes:

$$\begin{aligned}&\dot{\varphi}_1 + \frac{i\varepsilon}{2(1+\varepsilon)}(\varphi_1 - \varphi_2) - \frac{i\varepsilon\sigma\,(\varphi_1 + \varepsilon\varphi_2)}{2(1+\varepsilon)} = \frac{\varepsilon A}{2}\\&\dot{\varphi}_2 + \lambda(1+\varepsilon)\frac{\varphi_2}{2} + \frac{i}{2(1+\varepsilon)}(\varphi_2 - \varphi_1) -\\&-\frac{i\varepsilon\sigma\,(\varphi_1 + \varepsilon\varphi_2)}{2(1+\varepsilon)} - \frac{i(1+\varepsilon)}{2}|\varphi_2|^2\varphi_2 = \frac{\varepsilon A}{2}\end{aligned} \tag{2.268}$$

The fixed points of Eq. (2.268) correspond to periodic responses of the system described by Eq. (2.265). For the case $\sigma = 0$, these were investigated in details in a paper (Starosvetsky and Gendelman, 2008). The investigation of these fixed points and their stability is beyond the scope of this paper – it can be performed by standard methods and will be published elsewhere. System (2.268) has a somewhat special form – the time derivative in the first equation is proportional to the small parameter and thus the time evolution of variable φ_1 can be considered as slow compared to φ_2. This peculiarity means that the dynamics of System (2.268) in a 4-dimensional real state space may be presented in terms of two "slow" and 2 "super-slow" real variables, thus giving a chance of a tractable global description. The term "fast" is kept for the oscillations with close to unit frequency, which are averaged out. So, the problem requires the analysis of three time-scales.

By simple manipulations, System (2.268) may be reduced to a single second-order ODE:

$$\begin{aligned}&\frac{d^2\varphi_2}{dt^2}+\frac{d}{dt}\left[\alpha\varphi_2-\frac{i(1+\varepsilon)}{2}\left|\varphi_2\right|^2\varphi_2+\frac{i\varepsilon}{2(1+\varepsilon)}(1-\sigma)\varphi_2\right]+\\&+\frac{i\varepsilon}{2(1+\varepsilon)}(1-\sigma)\left[\alpha\varphi_2-\frac{i(1+\varepsilon)}{2}\left|\varphi_2\right|^2\varphi_2-\frac{\varepsilon A}{2}\right]-\\&-\frac{i\varepsilon\beta}{2(1+\varepsilon)}\left[1+\varepsilon\sigma\right]\varphi_2=\frac{\varepsilon A\beta}{2}\end{aligned} \tag{2.269}$$

where

$$\alpha=\frac{\lambda(1+\varepsilon)^2+i-i\varepsilon^2\sigma}{2(1+\varepsilon)}\;;\;\beta=\frac{i}{2(1+\varepsilon)}(1+\varepsilon\sigma)$$

Multiple scale expansion of the differential operators is introduced as:

$$\begin{aligned}&\varphi_2=\varphi_2(\tau_0,\tau_1,\ldots)\;;\;\tfrac{d}{dt}=\tfrac{\partial}{\partial\tau_0}+\varepsilon\tfrac{\partial}{\partial\tau_1}+\ldots\\&\tau_k=\varepsilon^k t,k=0,1,\ldots\end{aligned} \tag{2.270}$$

Substituting (2.270) into (2.269) and equating the like powers of ε, one obtains equations for zero and the first order approximations:

$$\begin{aligned}&\varepsilon^0:\;\frac{\partial^2\varphi_2}{\partial\tau_0^2}+\frac{\partial}{\partial\tau_0}\left[\frac{\lambda\varphi_2}{2}+\frac{i\varphi_2}{2}-\frac{i}{2}\left|\varphi_2\right|^2\varphi_2\right]=0\\&\varepsilon^1:\;2\frac{\partial^2\varphi_2}{\partial\tau_0\partial\tau_1}+\frac{\partial}{\partial\tau_1}\left[\frac{\lambda\varphi_2}{2}+\frac{i\varphi_2}{2}-\frac{i}{2}\left|\varphi_2\right|^2\varphi_2\right]+\\&+\frac{\partial}{\partial\tau_0}\left[\frac{\lambda\varphi_2}{2}+\frac{i(1-\sigma)\varphi_2}{2}-\frac{i}{2}\left|\varphi_2\right|^2\varphi_2\right]+\\&+\frac{1-\sigma}{4}\left|\varphi_2\right|^2\varphi_2+\left[\tfrac{\sigma}{4}+\frac{i\lambda(1-\sigma)}{4}\right]\varphi_2-\frac{iA}{4}=0\end{aligned} \tag{2.271}$$

The first equation of (2.271) describes a "slow" evolution of the averaged system. It can be trivially integrated:

$$\frac{\partial}{\partial \tau_0}\varphi_2 + \left(\frac{i}{2}\varphi_2 + \frac{\lambda}{2}\varphi_2 - \frac{i}{2}\left|\varphi_2\right|^2 \varphi_2\right) = C(\tau_1, \ldots) \tag{2.272}$$

where C is an arbitrary function of higher-order time scales. Approximations of higher orders are not used in the current analysis. Then, for the sake of brevity, only dependence on time scales τ_0 and τ_1 will be explicitly denoted below. The fixed points $\Phi(\tau_1)$ of Eq. (2.272) depend only on time scale τ_1 and obey the algebraic equation:

$$\frac{i}{2}\Phi + \frac{\lambda}{2}\Phi - \frac{i}{2}\left|\Phi\right|^2 \Phi = C(\tau_1) \tag{2.273}$$

Equation (2.273) is easily solved by taking $\Phi(\tau_1) = N(\tau_1)\exp(i\theta(\tau_1))$ and performing trivial calculations:

$$\lambda^2 N^4 + (N^2 - N^4)^2 = 4\left|C(\tau_1)\right|^2 N^2$$

or, equivalently

$$\begin{aligned} &\lambda^2 Z(\tau_1) + Z(\tau_1)(1 - Z(\tau_1))^2 = 4\left|C(\tau_1)\right|^2 \\ &Z(\tau_1) = (N(\tau_1))^2 \end{aligned} \tag{2.274}$$

The expression for the argument of the fixed point may be written as

$$\theta(\tau_1) = \arg C(\tau_1) - \tan^{-1}\frac{1 - Z(\tau_1)}{\lambda} \tag{2.275}$$

where $Z(\tau_1)$ is solution of Eq. (2.274).

The number of solutions of Eq. (2.274) depends on $|C(\tau_1)|$ and λ. The function on the left-hand side can be monotonous or have a maximum and a minimum. In the first case, the change of $|C(\tau_1)|$ has no effect on the number of solutions – Eq. (2.274) will have one positive solution. In the latter case, the change of $|C(\tau_1)|$ will bring about a pair of saddle-node bifurcations. In order to distinguish between different cases we should check whether the derivative of the left-hand side of (2.274) has roots:

$$1 + \lambda^2 - 4Z + 3Z^2 = 0 \,\text{or}\, Z_{1,2} = \frac{2 \mp \sqrt{1 - 3\lambda^2}}{3} \tag{2.276}$$

Therefore, two roots and pair of saddle-node bifurcations exist for $\lambda < {}^{1}\!/\!_{\sqrt{3}}$ and do not exist otherwise. At the critical value $\lambda = {}^{1}\!/\!_{\sqrt{3}}$ two saddle-node bifurcation points coalesce, thus forming the typical structure of a cusp catastrophe.

It is easy to see from Eq. (2.272) whether only one solution of (2.274) exists, and whether it is stable with respect to time scale τ_0. If there are three solutions, two of them are stable (nodes) and one unstable (saddle). Therefore, at time scale

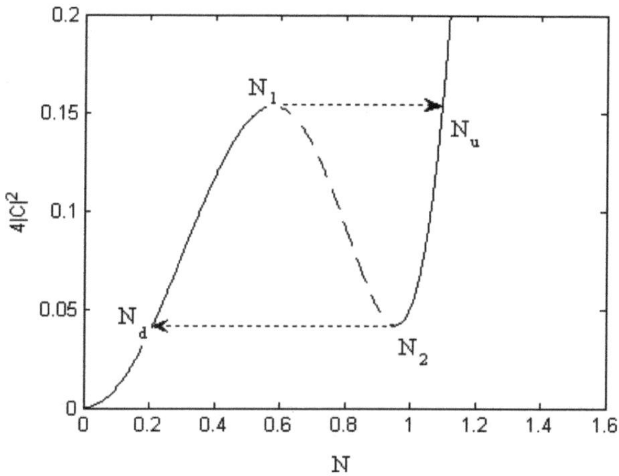

Fig. 2.46 Projection of the slow invariant manifold of the system in accordance with Eq. (10), $\lambda = 0.2$. The unstable branch is denoted by the *dashed line*. *Arrows* denote hypothetic "jumps" in the regime of the relaxation oscillations. N_1 and N_2 denote the fold lines, N_u and N_d – final points of the "jumps"

τ_0 the phase point will be attracted to one of the nodes. In fact, Eq. (2.273) defines a slow invariant manifold (SIM) of the problem. In the case $\lambda < 1/\sqrt{3}$ the fold lines $N_{1,2} : N(\tau_1) = Z_{1,2}^{1/2}$, $\theta(\tau_1) \in (0, 2\pi)$ divide stable and unstable branches of the SIM. Figure 2.46 demonstrates the projection of the two-dimensional SIM on the plane (N, C). The fold lines correspond to the points of maximum and minimum.

It is well-known (Arnold et al., 1994; Guckenheimer et al., 2005, 2006) that the structure of the SIM may give rise to relaxation-type oscillations of the system – the hypothetic "jumps" between the stable branches are denoted by arrows in Fig. 2.46. N_u and N_d denote the final points of jumps on the upper and lower stable branches of the SIM respectively. Still, such motion is possible only if the system can reach the fold lines $N_{1,2}$ while moving on the SIM with respect to the slow time scale. In order to assess this possibility, one should investigate the behavior of $\Phi(\tau_1)$. For this sake, we consider the ε^1 term of multiple-scale expansion, namely the second equation of (2.271). We are interested in the behavior of the solution on the stable branches of the SIM $\Phi(\tau_1) = \lim_{\tau_0 \to +\infty} \varphi_2(\tau_0, \tau_1)$. Taking the limit $\tau_0 \to \infty$ into the second equation of System (2.271) and taking into account the asymptotic stability of the points of the stable branches with respect to time scale τ_0, one obtains:

$$\frac{\partial}{\partial \tau_1}\left[\frac{\lambda\Phi}{2} + \frac{i\Phi}{2} - \frac{i}{2}|\Phi|^2\Phi\right] + \frac{1-\sigma}{4}|\Phi|^2\Phi + \left[\frac{\sigma}{4} + \frac{i\lambda(1-\sigma)}{4}\right]\Phi - \frac{iA}{4} = 0$$

This equation can be written in the more convenient form:

$$\begin{aligned}
&\left[\frac{\lambda}{2}-\frac{i}{2}+i\,|\Phi|^2\right]\frac{\partial\Phi}{\partial\tau_1}-\frac{i}{2}\Phi^2\frac{\partial\Phi}{\partial\tau_1}=G\\
&G=-\frac{1-\sigma}{4}\,|\Phi|^2\,\Phi-\left[\frac{\sigma}{4}+\frac{i\lambda(1-\sigma)}{4}\right]\Phi+\frac{iA}{4}
\end{aligned}\tag{2.277}$$

By taking a complex conjugate of (2.277), it is possible to extract the derivative $\dfrac{\partial\Phi}{\partial\tau_1}$:

$$\frac{\partial\Phi}{\partial\tau_1}=\frac{2\left[\left(\lambda-i+2i\,|\Phi|^2\right)G+i\Phi^2G^*\right]}{\lambda^2+1-4\,|\Phi|^2+3\,|\Phi|^4}\tag{2.278}$$

Splitting the variable Φ to modulus and argument $\Phi(\tau_1)=\mathrm{N}(\tau_1)\exp\left(\mathrm{i}\theta(\tau_1)\right)$, one obtains the equations of the reduced flow in polar coordinates:

$$\begin{aligned}
\frac{\partial N}{\partial\tau_1}&=\frac{-\lambda N-AN^2\cos\theta+\lambda A\sin\theta+A\cos\theta}{2(\lambda^2+1-4N^2+3N^4)}\\
\frac{\partial\theta}{\partial\tau_1}&=\frac{\begin{bmatrix}(1-4\sigma)N^2+(\sigma-\lambda^2(1-\sigma))-\\-3(1-\sigma)N^4+3AN\sin\theta+A(\lambda\cos\theta-\sin\theta)/N\end{bmatrix}}{2(\lambda^2+1-4N^2+3N^4)}
\end{aligned}\tag{2.279}$$

Denoting the numerators and denominator of the right hand side of system (2.279) by $f_1(N,\theta)$ for the first equation, $f_2(N,\theta)$ for the second equation and $g(N)$ for the denominator, system (15) is presented in the following form

$$\begin{aligned}
\frac{\partial N}{\partial\tau_1}&=\frac{f_1(N,\theta)}{g(N)}\\
\frac{\partial\theta}{\partial\tau_1}&=\frac{f_2(N,\theta)}{g(N)}
\end{aligned}\tag{2.280}$$

Rescaling the time by the function $g(N)$ yields the equations for a "desingularized" flow:

$$\begin{aligned}
N'&=f_1(N,\theta)\\
\theta'&=f_2(N,\theta)
\end{aligned}\tag{2.281}$$

Regular points of the SIM are defined as those that satisfy the inequality $g(N)\neq 0$. The *fold lines* $N_i, i=1,2$ of the SIM are, by the definition above, the sets of points (N,θ) where $g(N)=0$.

Let us start from the particular case $\mathrm{A}=0$. In this case, Eq. (2.279) are reduced to

$$\begin{aligned}
\frac{\partial N}{\partial\tau_1}&=\frac{-\lambda N}{2(\lambda^2+1-4N^2+3N^4)}\\
\frac{\partial\theta}{\partial\tau_1}&=\frac{(1-4\sigma)N^2+(\sigma-\lambda^2(1-\sigma))-3(1-\sigma)N^4}{2(\lambda^2+1-4N^2+3N^4)}
\end{aligned}\tag{2.282}$$

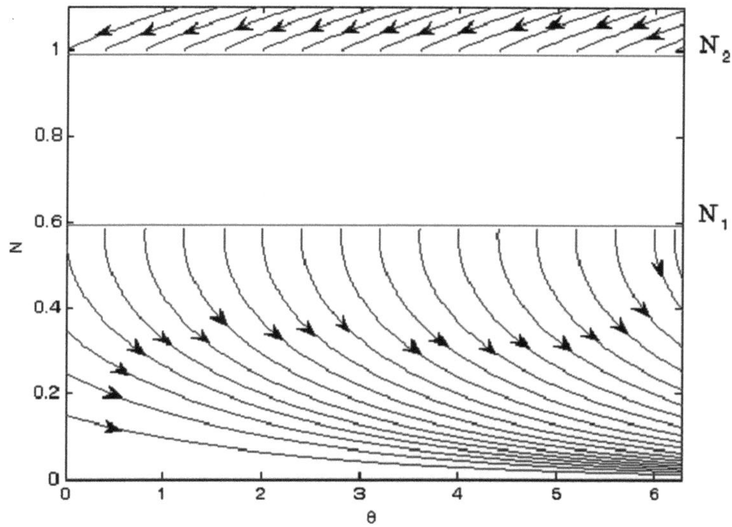

Fig. 2.47 Phase portrait of the slow invariant manifold for the case $A = 0, \lambda = 0.2, \sigma = 0$

Phase portrait of system (2.282) is presented in Fig. 2.47. The system parameters are: $\lambda = 0.2$, $\sigma = 0$. Fold lines N_1, N_2 are marked on the phase portrait as dashed lines. It is clear from the first equation of system (2.282) that the phase trajectories on the upper stable branch are directed towards the fold line, whereas the trajectories at the lower stable branch cannot bring the slow flow to the fold line. It means that the trajectory can "jump" from the upper stable branch to the lower one, but cannot jump back. It is trivial, since in the absence of forcing the system should be damped out.

In order to allow the jumps from the lower stable branch (and, therefore, to provide the necessary condition for the relaxation oscillations) the slow flow in the vicinity of the lower fold line should undergo some bifurcations. Namely, the N' value for some points on the lower fold should change its sign from a negative to a positive one. Consequently, we can state that for some point or points on the lower fold the *normal switching condition* (Guckenheimer et al., 2005) should be violated in the course of the bifurcation. In order to investigate this mechanism, we first compute the fixed points of the slow-flow Eq. (2.281) for arbitrary A.

Before this, we proceed with the calculation of the equilibrium points of (2.281) which let us define the two different types of these points. The first type is referred to as *ordinary fixed point*. These are equilibrium points of slow flow (2.281) which satisfy $N' = \theta' = 0$ and $g(N) \neq 0$. The second type is referred to as *folded singularities*. Folded singularities satisfy both $N' = \theta' = 0$ and $g(N) \neq 0$. They can be classified as equilibrium points of the two dimensional flow (2.281) belonging to the fold lines.

Equilibrium points of the slow-flow system are found from (2.281) by setting both time derivatives equal to zero, thus providing

$$\begin{aligned} f_1(N,\theta) &= 0 \\ f_2(N,\theta) &= 0 \end{aligned} \tag{2.283}$$

System (2.283) can be presented in the following matrix form

$$\begin{pmatrix} \alpha_{11} & \alpha_{12} \\ \alpha_{21} & \alpha_{22} \end{pmatrix} \begin{pmatrix} \cos\theta \\ \sin\theta \end{pmatrix} = \begin{pmatrix} \beta_1 \\ \beta_2 \end{pmatrix} \tag{2.284}$$

where

$$\alpha_{11} = \frac{1}{4}\lambda A;\ \alpha_{12} = -\frac{1}{4}A + \frac{3}{4}N^2 A;\ \alpha_{21} = \frac{1}{4}A - \frac{1}{4}N^2 A;\ \alpha_{22} = \frac{1}{4}\lambda A$$
$$\beta_1 = \frac{1}{4}N\sigma + \frac{1}{4}N^3 + \frac{1}{4}N\lambda^2\sigma - \frac{3}{4}N^5 - N^3\sigma - \frac{1}{4}N\lambda^2 + \frac{3}{4}N^5\sigma;\ \beta_2 = -\frac{1}{4}N\lambda$$

System (2.284) has two different types of solutions. The first type is obtained by solving (2.284) and assuming that the α matrix determinant does not vanish ($\alpha_{11}\alpha_{22} - \alpha_{21}\alpha_{12} \neq 0$). Thus the first type of solution is calculated from:

$$\begin{aligned} &\left[\lambda^2 + \frac{\sigma^2}{(1-\sigma)^2}\right] N_0^2 + \frac{2\sigma}{1-\sigma} N_0^4 + N_0^6 = \frac{A^2}{(1-\sigma)^2}; \\ &\theta_0 = \tan^{-1}\left(\frac{\sigma}{\lambda(1-\sigma)} + \frac{N_0^2}{\lambda}\right) \end{aligned} \tag{2.285}$$

It is easy to derive that

$$\alpha_{11}\alpha_{22} - \alpha_{21}\alpha_{12} = \frac{A^2}{16}(1 + \lambda^2 - 4\,N^2 + 3\,N^4) = \frac{A^2}{32} g(N);$$

then, nullification of g(N) brings us to the simultaneous nullification of $\alpha_{11}\alpha_{22} - \alpha_{21}\alpha_{12}$. Therefore, solution (2.285) describes the ordinary fixed points. This solution also coincides with the solution for fixed points of initial equation (2.269). This is rather obvious since fixed points of the global flow quite naturally belong to the slow invariant manifold.

The second type of solution obeys the following condition:

$$g(N) = 3N^4 - 4N^2 + 1 + \lambda^2 = 0 \Rightarrow \alpha_{11}\alpha_{22} - \alpha_{12}\alpha_{21} = 0 \tag{2.286}$$

Combination of (2.286) and (2.284) yields the following equality:

$$\frac{\alpha_{11}}{\alpha_{21}} = \frac{\alpha_{12}}{\alpha_{22}} = \frac{\beta_1}{\beta_2} \tag{2.287}$$

The second type of solution is generated by (2.286) and by one of the equations of (2.284) since another equation is satisfied automatically due to (2.287). Thus, picking the first equation of (2.284), one obtains the following solution for the folded singularities:

$$
\begin{aligned}
\Theta_{1,2} &= \gamma_{01} \pm \cos^{-1} \frac{\lambda N_1}{A\sqrt{(1-N_1^2)^2+\lambda^2}}, \\
\Theta_{3,4} &= \gamma_{02} \pm \cos^{-1} \frac{\lambda N_2}{A\sqrt{(1-N_2^2)^2+\lambda^2}} \\
\gamma_{0k} &= \sin^{-1} \frac{\lambda}{\sqrt{(1-N_k^2)^2+\lambda^2}}, k = 1, 2
\end{aligned}
\tag{2.288}
$$

The first pair of the folded singularities exists on the lower fold and is given by(N_1, Θ_1), (N_1, Θ_2). The second pair exists on the upper fold and is given by (N_2, Θ_3), (N_2, Θ_4). The first pair of the folded singularities exists if the following conditions hold:

$$
\left| \frac{\lambda}{\sqrt{(1-N_k^2)^2+\lambda^2}} \right| \le 1 \tag{2.289a}
$$

$$
\left| \frac{\lambda N_1}{A\sqrt{(1-N_1^2)^2+\lambda^2}} \right| \le 1 \tag{2.289b}
$$

Condition (2.289a) holds for arbitrary values of λ. However, condition (2.289b) holds only if

$$
A \ge A_{1\mathrm{crit}} = \frac{\lambda N_1}{\sqrt{(1-N_1^2)^2+\lambda^2}} \tag{2.290}
$$

Similarly, for the second pair of folded singularities the solvability condition reads

$$
A \ge A_{2\mathrm{crit}} = \frac{\lambda N_2}{\sqrt{(1-N_2^2)^2+\lambda^2}} \tag{2.291}
$$

Therefore, it is easy to see that if the external forcing is relatively small,

$$
A < A_{1\mathrm{crit}} = \frac{\lambda N_1}{\sqrt{(1-N_1^2)^2+\lambda^2}} \tag{2.292}
$$

there are no folded singularities at the SIM. Consequently, the slow flow in the vicinity of both fold lines remains qualitatively similar to that in Fig. 2.47, providing no possibility for the relaxation oscillations.

The bifurcation diagram of the periodic response regime is presented in Fig. 2.48 as the amplitude of the response vs. the amplitude of excitation. System parameters are $\lambda = 0.2$, $\sigma = 0$. For this set of parameters only a single periodic response exists.

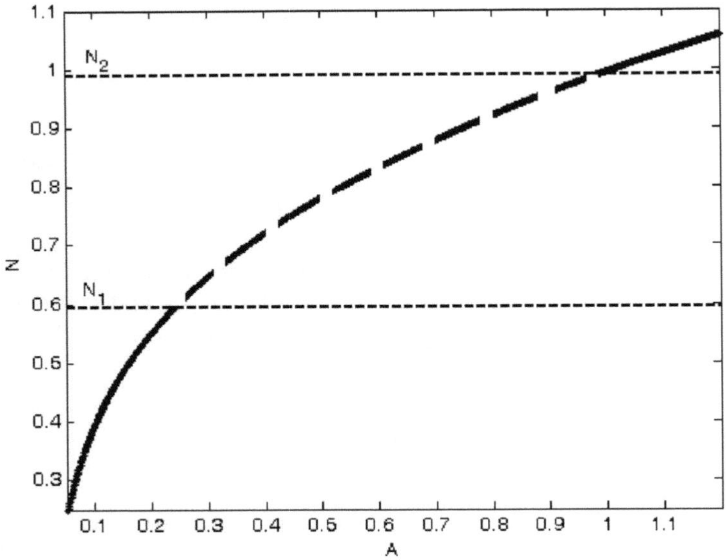

Fig. 2.48 Amplitude of excitation – system response, bifurcation diagram. *Dashed lines* refer to the unstable region of the periodic solution when the *solid lines* refer to the stable one. *Lower* and *upper folds* of the SIM are marked on the diagram with the *thin dashed lines*. System parameters: $\lambda = 0.2, \sigma = 0$

If the amplitude of external excitation (A) is varied, one can see qualitative changes (possible bifurcations) in the stability of the folded singularities. To track the behavior of the folded singularities and regular fixed points with variation of A parameter, we plot the positions of the ordinary fixed point and the folded singularities vs. the forcing amplitude in a 3D diagram (Fig. 2.49a). The parameters of the system are: $\lambda = 0.2$, $\sigma = 0$.

One observes transcritical bifurcations of the ordinary fixed point and folded singularities. Both transcritical bifurcations (on the lower fold and on the upper fold within large circles in Fig. 2.49a) are zoomed and presented in Fig. (2.49b, c) respectively. These bifurcations predict a change of stability of the folded singularities when the regular fixed point crosses the fold lines.

In order to illustrate the changes in the reduced flow dynamics for the various forcing amplitudes, we construct several phase portraits for System (2.279). These phase portraits are plotted only for the case of a single ordinary fixed point. It is convenient to pick zero frequency detuning $\sigma = 0$. Only the flow at two stable branches of the SIM is presented. We start with the case $0 < A < A_{1crit}$ ($A = 0.1, \lambda = 0.2$).

An ordinary fixed point is marked on the phase portrait (lower stable branch of SIM) with a rectangle. As it comes from the phase portrait of Fig. 2.50 there are no folded singularities for this case ($0 < A < A_{1crit}$), therefore we can see that all trajectories are finally attracted to the ordinary fixed point and there are no possibilities for relaxation oscillation. However, as the forcing approaches the value

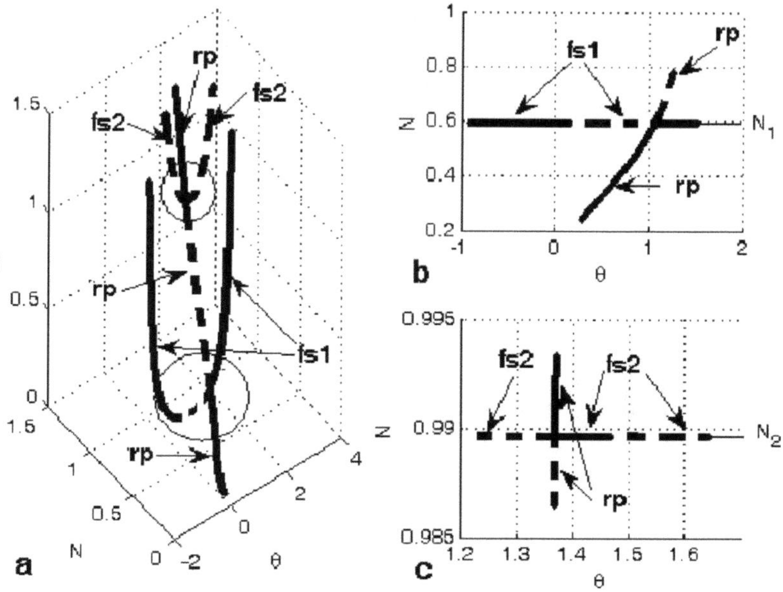

Fig. 2.49 (**a**) Three dimensional bifurcation diagram “fs1” refers to the first pair of folded singularities (*lower fold*); “fs2” – to the second pair of folded singularities (*upper fold*); “rp” – to the regular fixed point. The bifurcation regions bounded by the circles are zoomed (**b**) Zoom of transcritical bifurcation on the *lower fold* projected on the N,θ plane. (**c**) Zoom of transcritical bifurcation on the *upper fold* projected on the N,θ plane $\lambda = 0.2, \sigma = 0$

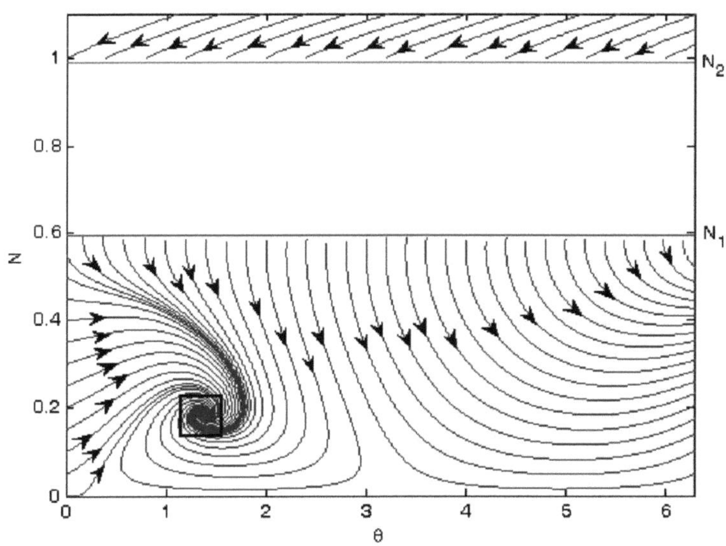

Fig. 2.50 Phase portrait of the slow invariant manifold for the case $0<A<A_{1crit}$ (only stable branches of the SIM are shown). System parameters: $A = 0.1, \lambda = 0.2, \sigma = 0$. Ordinary fixed point is marked (on the lower stable branch of SIM) with *rectangle*

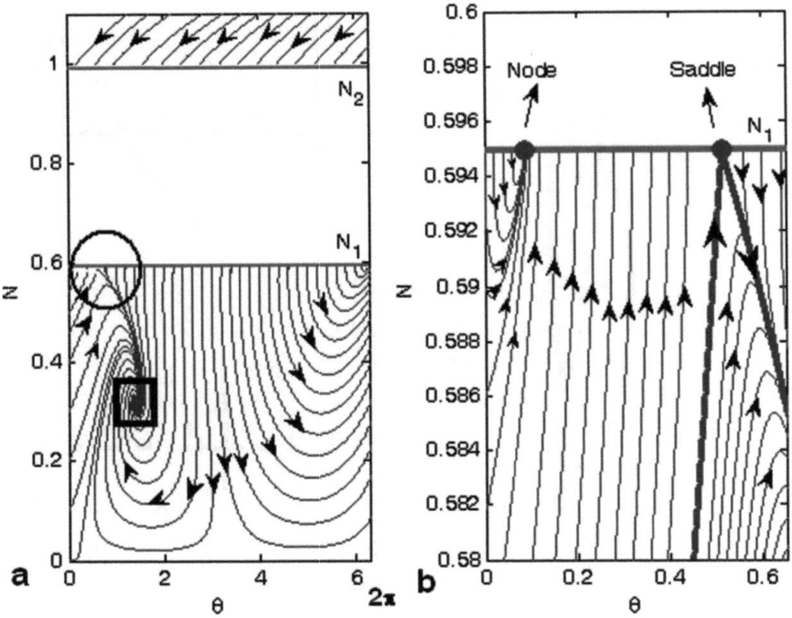

Fig. 2.51 (**a**) Phase portrait of the slow invariant manifold for the case A_{1crit} <A< A_{2crit} (only stable branches of the SIM are shown). System parameters: $A = 0.18, \lambda = 0.2, \sigma = 0$. Ordinary fixed point is marked (on the lower stable branch of SIM) with *rectangle*. (**b**) Zoomed part of the phase portrait (marked with a *circle* on Fig. 2.51a) which contains folded singularities (*saddle* and *node*)

$A = A_{1\mathrm{crit}}$, the "saddle-node" bifurcation occurs at the lower fold line at $\theta = \gamma_{01}$. The phase portrait of the SIM for the case $A_{1crit} < A < A_{2crit} = \frac{\lambda N_2}{\sqrt{(1-N_2^2)^2+\lambda^2}}, A = 0.18$ is presented in Fig. 2.51.

The phase portrait presented in Fig. 2.51 contains both an ordinary fixed point on the lower branch of the SIM (marked with *rectangle* on the figure) and folded singularities (of saddle and node types). The region on the phase portrait (Fig. 2.51a) bounded with a circle is zoomed and illustrated in Fig. 2.51b. Folded singularities marked with bold dots on the fold line are clearly observed in Fig. 2.51b. The trajectory which comes close to the separatrix of the saddle point is marked with the bold solid line.

The following phase portrait (Fig. 2.52) is drawn for the same region ($A_{1crit} < A < A_{2crit}, A = 0.5$), but for the increased value of forcing.

As we can see in Fig. 2.52, the system dynamics undergoes some qualitative changes. Before we list all the changes it is essential to note that by increasing the value of forcing the folded singularities gradually become distant from another. It is important to emphasize that the folded saddle propagates in the right direction along the folded line and the folded node propagates in the left direction. Recalling

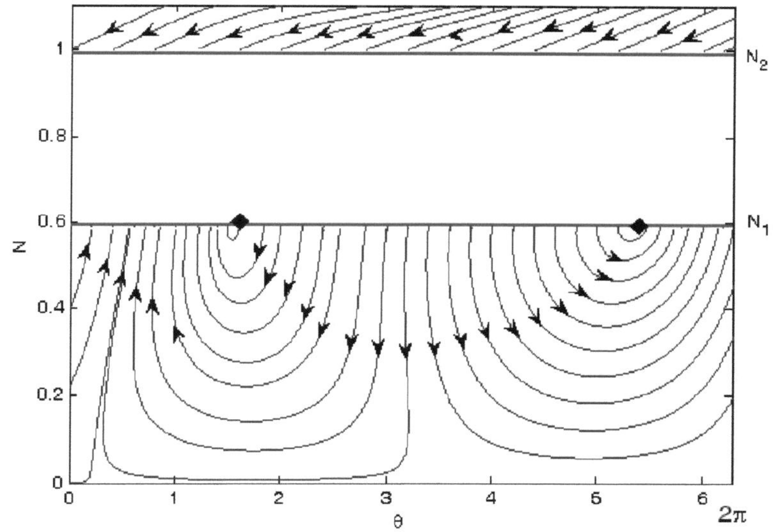

Fig. 2.52 Phase portrait of the slow invariant manifold for the case $0<A<A_{1crit}$ (only stable branches of the SIM are shown). System parameters: $A = 0.5, \lambda = 0.2, \sigma = 0$. Ordinary fixed point is absent on the stable branches of SIM. Folded singularities are marked with *diamonds*

that the θ coordinate is 2π periodic, then, if one of the folded singularities crosses the interval $[0,2\pi]$ it will appear from the other side. Therefore, the folded node crosses the left boundary of θ and appears from the right of the folded saddle when folded saddle continues to move in the same direction. Thus we need to identify first each one of the folded singularities in Fig. 2.52, namely to relate each one of them to the previously illustrated folded saddle node pair. It is clear from the previous discussion that the folded focus at the left side in Fig. 2.52 is related to the folded saddle in Fig. 2.51 and the folded focus from the right is related to the folded node in Fig. 2.51. Hence, two qualitative changes of the folded singularities are observed.

Bifurcation of the folded node to focus can be explained in terms of the eigenvalues of the linearized slow flow. Thus, for some critical value of forcing two real negative eigenvalues (related to the folded node) hit each other on the left half plane. Increasing the forcing value above the critical threshold leads to the symmetrical divergence of the eigenvalues across imaginary axes still staying in the left half plane; the fixed point becomes the stable focus.

The second folded focus in 2.2.51 requires one more step for its formation. At the beginning, the stable folded node appears due to the transcritical bifurcation described above. Then, this node turns to the folded focus as two negative real eigenvalues turn to the conjugated pair with a negative real part.

Slightly increasing the value of the forcing amplitude above the value of A_{2crit} ($A_{2crit} < A = 0.987$), we obtain the next phase portrait (Fig. 2.53). Observing the zoomed area denoted by a circle in Fig. 2.53a (Fig. 2.53b) one can notice the creation

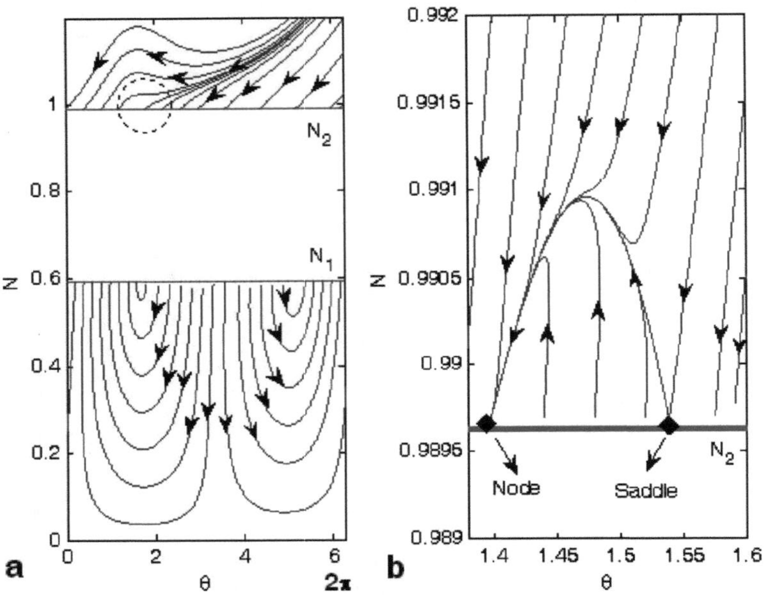

Fig. 2.53 (**a**) Phase portrait of the slow invariant manifold for the case A_{2crit} <A (only stable branches of the SIM are shown). System parameters: $A = 0.987, \lambda = 0.2, \sigma = 0$. (**b**) Zoomed part of the phase portrait (marked with a *circle* on (**a**)), which contains folded singularities (*saddle* and *node*) marked with *diamonds* on the *fold line*

of an additional saddle-node pair on the upper fold $N = N_2$. The regular fixed point has not reached the upper stable branch of the SIM yet.

Further increase of the forcing amplitude ($A_{2crit} < A = 1$) results in the passage of the regular fixed point to the upper stable branch of SIM. The phase portrait for that case is presented in Fig. 2.54.

Observing Fig. 2.54b related to the zoomed area bounded by the circle in Fig. 2.54a, one can easily recognize the appearance of the stable focus on the upper stable branch of SIM. This stable focus is situated within a rectangle in Fig. 2.54b; besides, pair of folded saddles is present. One of these saddles is formed at the fold when the regular stable point passes the fold line and additional transcritical bifurcation occurs. With further (very small) increase of the bifurcation parameter, the regular fixed point switches from the stable node to the stable focus. This process is partially demonstrated in Fig. 2.49c.

Summarizing the results of the present subsection, we would like to emphasize the changes in the slow flow due to the increase in forcing amplitude. From Figs. 2.51 and 2.52 it is easy to see that once the bifurcations occur, some phase trajectories on the SIM will bring the flow to the lower fold line $N = N_1$, thus providing a possibility for the jump to the upper stable branch. Then the flow can arrive to the upper fold line and jump down, thus closing the loop of the relaxation oscillation. It is interesting to note that the values of A_{crit} do not depend on the detuning

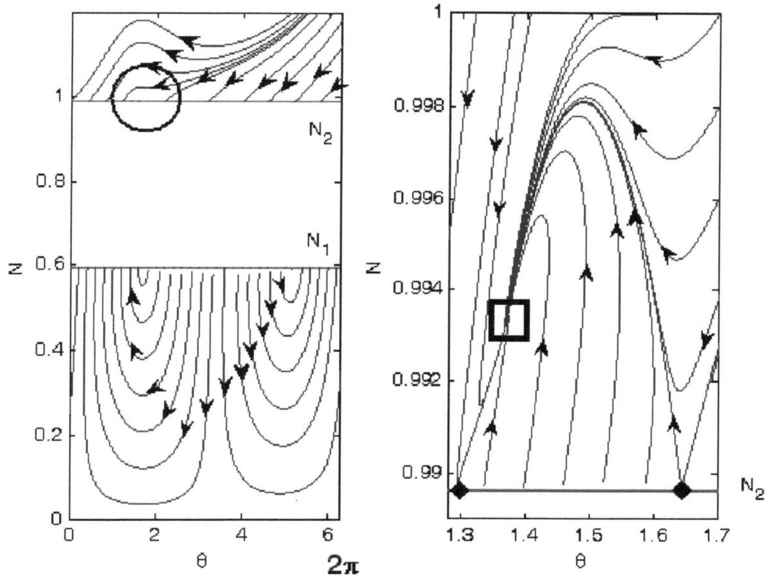

Fig. 2.54 (**a**) Phase portrait of the slow invariant manifold for the case A_{2crit} <A (only stable branches of the SIM are shown). System parameters: $A = 1, \lambda = 0.2, \sigma = 0$. (**b**) Zoomed part of the phase portrait (marked with a *circle* on (**a**)), which contains folded singularities (*two saddles*) marked with *diamonds* on the *fold line* and stable focus bounded with *rectangle*

parameter σ. One should keep in mind that all the considered bifurcation diagrams presented in the section are derived from the singular limit as it comes from the asymptotic analysis brought above; therefore one should expect that the results will be correct only for ε small enough.

From numerical simulations (Starosvetsky and Gendelman, 2008) it is known that the SMR related to the relaxation oscillations exists only in comparatively narrow vicinity of an exact 1:1 resonance between the external force and the natural frequency of the linear oscillator. It is therefore clear that the condition (2.290) is necessary, but by no means sufficient to provide the relaxation oscillations. In order to obtain the missing sufficient conditions, one should investigate more delicate details of the system dynamics.

Observing the phase portrait presented in Fig. 2.52 we can see that there is an interval of θ (on the lower fold line N_1) to which all the phase trajectories can arrive and jump from N_1. This interval is bounded by the folded singularities (for the case illustrated in Fig. 2.52 these folded singularities are stable foci). We denote this interval by $R = [\Theta_1, \Theta_2]$, where Θ_1, Θ_2 are the folded singularities which constitute the boundaries of the interval of the possible jump. In the regime of the relaxation oscillations, the phase trajectory jumps from a point of this interval to the upper branch of the SIM, then it moves along the line of the slow flow to the upper fold line, then jumps back to the lower branch and moves to the lower fold line, commencing within the interval R in order to enable the next jump. Therefore it is natural to consider this regime as mapping of the interval R into itself – the regime of the relaxation oscillations will correspond to the attractor of this one-dimensional map. Existence of this attractor is therefore necessary and a precondition for the existence of the SMR for System (2.268), or, equivalently, Eq. (2.269), when the mass ratio ε is small enough.

In order to construct the relevant map, we should separately consider the "slow" and the "fast" parts of the mapping cycle. As for the "slow" parts on the lower and the upper branches of the SIM, we can use Eq. (2.283) and directly connect the "entrance" and "exit" points. Due to the complexity of the equations, this part of the mapping should be accomplished numerically. As for the "fast" parts, the function ϕ_2 should be continuous at the points of contact between the "fast" and the "slow" parts. Therefore, for "fast" parts of the motion one obtains the complex invariant $C(\tau_1)$, defined by Eq. (2.273). If one knows its value at the point of "start", it is possible co compute N and θ for the point of "finish" unambiguously and thus to complete the mapping. Denoting the point of finish by (N_u, θ_u) and the start point as (N_1, θ_{01}) we are interested to provide the closed form formulae for the fast jump. The value of N_u may be easily calculated from the polynomial (2.274) by exploiting the invariance of $C(\tau_1)$

$$\begin{aligned} &\lambda^2 Z_1 + Z_1(1 - Z_1)^2 = \lambda^2 Z_u + Z_u(1 - Z_u)^2 = \\ &= \frac{2}{27}(1 + \sqrt{1 - 3\lambda^2}) + \frac{2\lambda^2}{9}(3 - \sqrt{1 - 3\lambda^2}) \\ &Z_u = N_u^2 = \frac{2}{3}(1 + \sqrt{1 - 3\lambda^2}) \end{aligned} \tag{2.293}$$

Then, from (2.273) and (2.293), one obtains the explicit expression for the phase variable at the "landing" point:

$$\begin{aligned} \theta_u &= \tan^{-1}\left(\frac{(N_u^2 - N_1^2)\lambda}{\lambda^2 - (1 - N_1^2)(N_u^2 - 1)}\right) + \theta_{01} = \\ &= \theta_{01} + \tan^{-1}\frac{9\lambda\sqrt{1 - 3\lambda^2}}{-1 + 15\lambda^2 - \sqrt{1 - 3\lambda^2}} \end{aligned} \tag{2.294}$$

Then, the part of mapping which corresponds to the "jump" from the lower fold line N_1 to the upper stable branch of the SIM is very simple – the amplitude switches to N_u and the phase rotates by a constant angle. Similarly, the jump from the point of the upper fold line N_2 with phase θ_{02} to the point N_d, θ_d is described by the following map:

$$\begin{aligned} &N_2 \to N_d = \sqrt{\tfrac{2}{3}(1 - \sqrt{1 - 3\lambda^2})} \\ &\theta_{02} \to \theta_d = \tan^{-1}\left(\frac{(N_d^2 - N_2^2)\lambda}{\lambda^2 - (1 - N_d^2)(N_2^2 - 1)}\right) + \theta_{02} = \\ &= \theta_{02} - \tan^{-1}\frac{9\lambda\sqrt{1 - 3\lambda^2}}{-1 + 15\lambda^2 + \sqrt{1 - 3\lambda^2}} \end{aligned} \tag{2.295}$$

The slow motion on the stable branches of the SIM should be simulated numerically. The procedure of numerical integration should be performed twice – for each of the two branches of the SIM. It should be stressed that only one computation cycle of the mapping for each point of the initial interval is required. This idea of mapping is close to that used in (Guckenheimer and Wechselberger, 2006) for the analysis of chaotic attractors of the relaxation oscillations in the state space of lower dimensionality.

Not every trajectory which starts from the lower fold of the SIM will reach the initial interval ($R = [\Theta_1, \Theta_2]$) since it can go to an alternative attractor at the upper or the lower branch of the SIM, if the latter exists. Of course, only those points which are mapped into the interval R can carry sustained relaxation oscillations.

To illustrate the construction of the return map we depict all four stages of its construction separately in Fig. 2.55 for the system parameters $A = 0.4, \lambda = 0.2, \sigma = 0$, $A_{1crit} < A < A_{2crit}$. At the first stage (Fig. 2.55a) the R interval defined earlier is mapped to the upper stable branch of the SIM according to (2.294). The second stage consists of the slow drift of the system on the upper stable branch until the upper fold line (N_2) is reached (Fig. 2.55b). The fast jumps from the upper fold N_2 to the lower stable branch of SIM constitute the third stage of the mapping process in accordance with (2.295) (Fig. 2.55c). The fourth stage contains the slow drift of the system until the initial interval R is reached (Fig. 2.55d). Composition of these four stages results in a single one-dimensional map of R into itself (Fig. 2.55e).

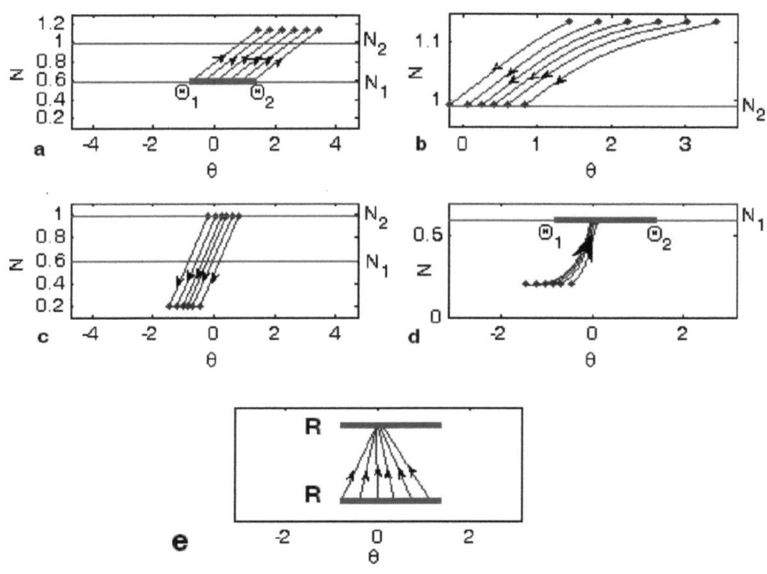

Fig. 2.55 Return map construction: (**a**) Mapping from the R interval to the *upper stable* branch of SIM (N_u) (**b**) Mapping from the line N_u to the *upper fold* N_2 (**c**) Mapping from the N_2 to the *lower stable* branch of SIM (N_d) (**d**) Mapping from N_d to R. (**e**) Complete return map. $A = 0.4, \lambda = 0.2, \sigma = 0, A_{1crit} < A < A_{2crit}$

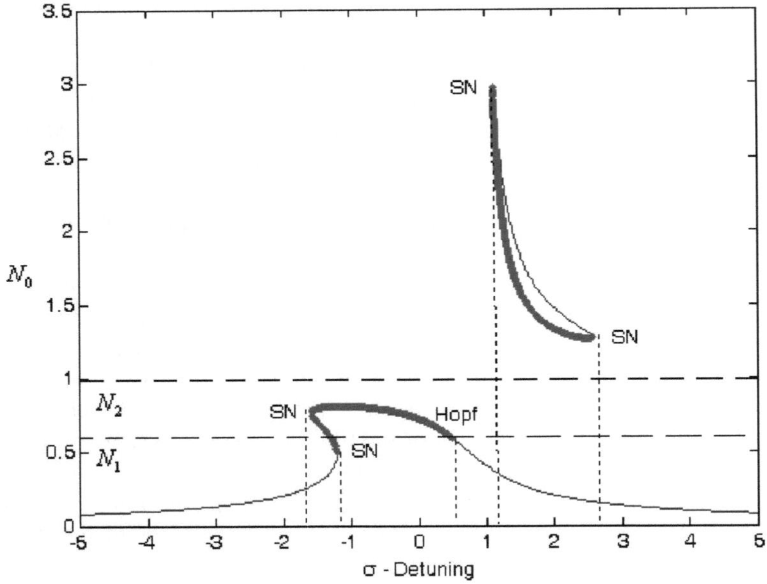

Fig. 2.56 Frequency response diagram. *Bold lines* refer to the unstable regions of the periodic solutions when the *thin lines* refer to the stable regions. *Lower* and *upper folds* of the SIM are marked on the diagram with the *thin dashed lines*. System parameters: $A = 0.4, \lambda = 0.2, \sigma = 0$

The frequency response described by Eq. (2.285) is plotted in Fig. 2.56 for the system parameters: $A = 0.4, \lambda = 0.2, \varepsilon = 0.01$. The fold lines of the SIM are also marked there. As one can expect from a cubic equation for the square of the response amplitude (2.285), for some zones of the parameter values one or three fixed points can exist. These fixed points undergo generic bifurcations (saddle-node and Hopf). More details on the structure and properties of the regular fixed points of the flow are presented in (Starosvetsky and Gendelman, 2008).

Observing the frequency response diagram presented in Fig. 2.56, one can notice that for $A = 0.4, \lambda = 0.2$ there are no stable regular fixed points for zero detuning value. Therefore, every point of R is mapped into R (it has no way out). The map is obviously contracting; therefore one can expect the existence of a stable attractor corresponding to the regime of the relaxation oscillations (or SMR). In this case it is a single-period cycle originating at a point $\theta \approx 0$ (see Fig. 2.55).

By increasing the detuning parameter value to $\sigma = 2$ (the values of the forcing amplitude and the damping parameters are the same, $A = 0.4, \lambda = 0.2, A_{1crit} < A < A_{2crit}$) one can notice the changes the map undergoes (Fig. 2.57).

Revisiting the frequency response diagram in Fig. 2.56 one can notice that there are two stable periodic responses for $\sigma = 2$. One of the attractors is on the lower stable branch and the second one resides on the upper stable branch. We do not illustrate the attractor of the upper stable branch in Fig. 2.57 since it does not affect the trajectories of the return map as it comes from Fig. 2.57b. Namely all the fraction of points mapped from R to the upper stable branch are successfully mapped (in the

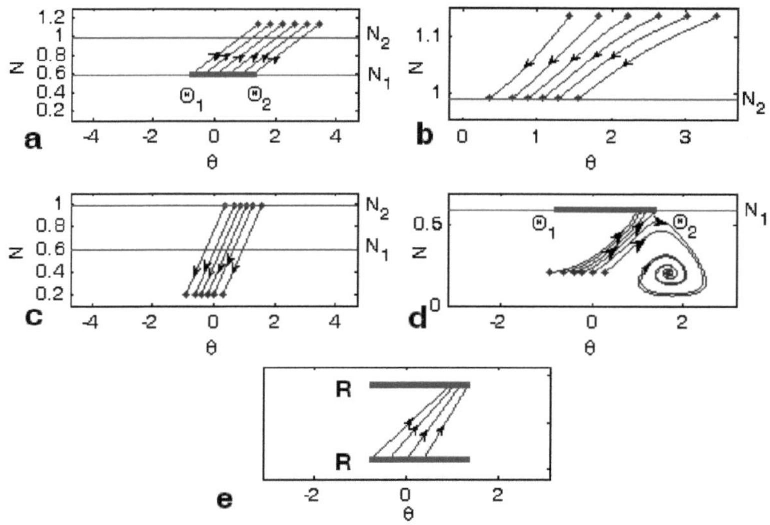

Fig. 2.57 Return map construction (**a**) Mapping from the R interval to the upper stable branch of SIM (N_u) (**b**) Mapping from the line N_u to the *upper fold* N_2 (**c**) Mapping from the N_2 to the lower stable branch of SIM (N_d) (**d**) Mapping from N_d to R. (**e**) Complete return map. $A = 0.4, \lambda = 0.2, \sigma = 2, A_{1crit} < A < A_{2crit}$

second stage) to the fold line N_2 (Fig. 2.57b), this without being attracted to the stable focus. However, by observing the final stage of the mapping (Fig. 2.57d) one can notice how the attractor on the lower stable branch of SIM modifies the return map. Several trajectories are attracted to the stable focus when the remaining fraction returns back to the R interval as it is illustrated in Fig. 2.57d. This means that only a certain subinterval of R is mapped into R.

Looking at the mapping diagram (Fig. 2.57e), one can see that all the mapping lines tend to the right and there is also a region on the basin which does not contain any lines. This region relates to the unaccomplished cycles, namely to the phase trajectories which started from the region and have been attracted to the periodic response attractor before they have reached the basin once more as it is demonstrated in Fig. 2.57d. Thus, the empty regions of the basin in the diagrams will be related to the trajectories which do not reach the basin once more. It is clear from Fig. 2.57e that there is no stable attractor of the SMR and for every initial condition on the basin the system finally (after sufficient number of cycles) leaves the basin. Consequently, the relaxation oscillations will exist in transient response under certain initial conditions, but not in the sustained response. Quite obviously, such a situation requires the presence of an alternative attractor.

While changing the values of detuning σ (keeping the rest of the system parameters fixed) and for each step performing the mapping, one can track the value of σ for which the attractor vanishes. Thus, one obtains a tool to determine the frequency region where the SMR exists. For the following set of system parameters ($A = 0.4, \lambda = 0.2$), the boundaries of the detuning parameter within which the

SMR exists are $\sigma_R = 1.597 > \sigma > \sigma_L = -1.275$. Returning to the initial frequency domain of Eq. (2.265), we can conclude that the SMR exists in O(ε) – vicinity of the exact resonance. This finding confirms earlier results of direct numeric simulations.

Our next goal is to investigate the mechanism of "birth" of the limit cycle related to the SMR when the detuning parameter passes its critical value. In Fig. 2.58, a sequence of maps close to the upper critical value of the detuning parameter ($\sigma = \sigma_R$) is presented for the following system parameters: $A = 0.4, \lambda = 0.2$.

For the critical value ($\sigma = \sigma_R = 1.597$) we can see undistinguishable stable and unstable cycles; they split off as the detuning parameter decreases. This scenario corresponds to a simple fold bifurcation of a 1D map – the creation of stable and unstable fixed points. In the framework of the current problem, it corresponds to the fold of stable and unstable limit cycles in 4D state space of the averaged problem (2.268). This global bifurcation is not related to the behavior of fixed points or homoclinic orbits of the problem (the latter are absent in this generic case) and cannot be addressed by local analysis within the initial framework. Still, presence of the small parameter related to the mass ratio allows us to reduce the global flow to the 1D nonlinear map and thus to demonstrate this bifurcation in terms of a local bifurcation of the map.

Hence, we have depicted the mechanism of creation and annihilation of the stable and unstable cycles near the right boundary (σ_R) of the SMR existence. It is essential to check what happens about the left boundary (σ_L). For this reason we plot the

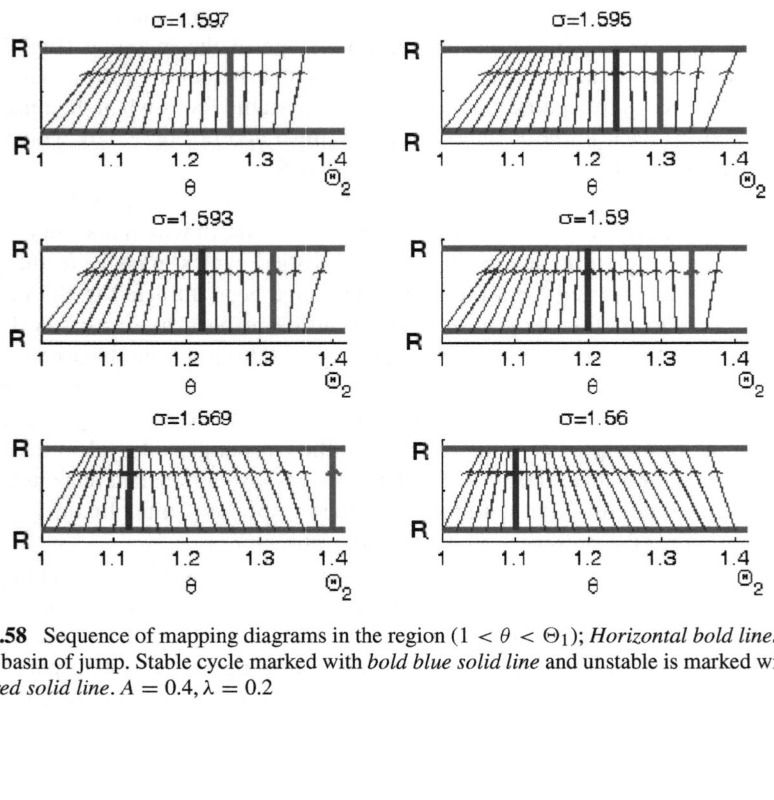

Fig. 2.58 Sequence of mapping diagrams in the region ($1 < \theta < \Theta_1$); *Horizontal bold lines* refer to the basin of jump. Stable cycle marked with *bold blue solid line* and unstable is marked with the *bold red solid line*. $A = 0.4, \lambda = 0.2$

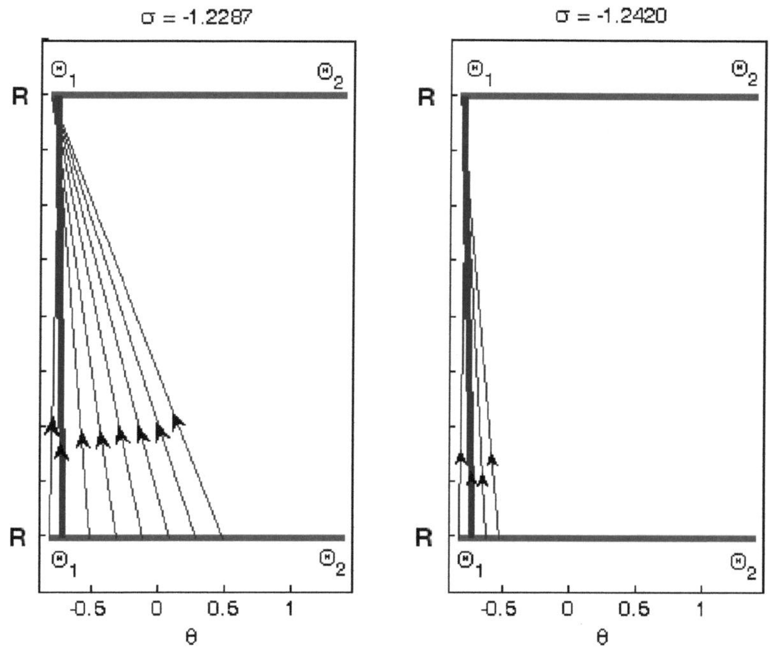

Fig. 2.59 Sequence of the one dimensional mapping diagrams for ($\sigma = -1.2287, \sigma = -1.242$). The *bold blue line* refers to the stable one period cycle $A = 0.4, \lambda = 0.2$

diagrams for gradually decreasing values of the detuning parameter near the left boundary (σ_L). The maps are presented in Fig. 2.59.

It is clear from the diagram that when the frequency detuning σ decreases, two phenomena happen. Firstly, the stable cycle (marked with the bold line on the diagrams (Fig. 2.59)) moves towards the left boundary of the jump basin. Secondly, the major fraction of the mapping trajectories vanishes. The second phenomenon points out to the attraction of the SMR trajectories to the stable periodic regimes (by the mechanism presented in Fig. 2.57). Unlike the saddle node bifurcation of the SMR cycles observed in the vicinity of the right boundary σ_R we see that more and more mapping trajectories are attracted to the stable periodic attractor with decrease of the detuning value. Therefore, the SMR disappears due to the attraction of the entire set of the mapping trajectories to the alternative periodic attractor (the map itself disappears).

In order to provide an additional illustration of the SMR cycle, we have drawn the cycle on the phase portrait plane. These cycles include the fast jumps (from one stable branch of SIM to another) and slow evolution on the SIM (Fig. 2.60).

The behavior of the SMR is described by a 1D nonlinear map. Consequently, generic bifurcations of the 1D maps are also expected to be observed for these limit cycles in 4D state space. One of such generic bifurcations is period doubling, which exists for certain values of parameters. For example, picking a set of system

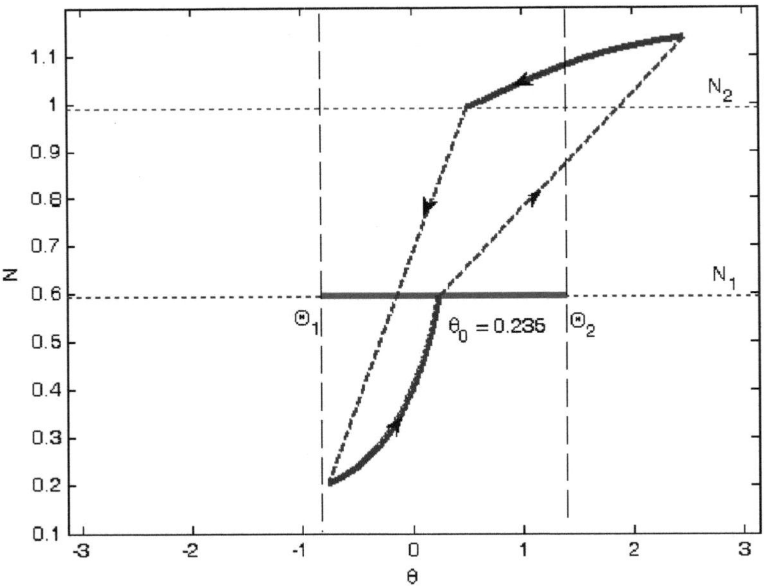

Fig. 2.60 The entire cycle of SMR; *Horizontal bold line* refers to the initial interval $\Theta_1 < \theta < \Theta_2$. *Dashed lines* refer to the fast jumps; *solid line* refers to the slow evolution on SIM, $A = 0.4, \lambda = 0.2, \sigma = 0.5$

parameters $A = 1, \lambda = 0.05, \sigma = 0$ one obtains the double period cycle of the one dimensional mapping (Fig. 2.61).

Additional period doubling bifurcations (e.g. from double period to 4-period cycle) were not observed in the mappings, however, the period doublings are rater ubiquitous. In Fig. 2.62, the zones with double period are depicted on the plane λ-A for zero detuning.

It should be mentioned that the analytic approach developed above is valid for the limit $\varepsilon \to 0$. The approximation for finite values of ε requires the computation of the higher-order expansions for Eq. (2.271) and is a rather cumbersome task. Moreover, the value of such enhancement is questionable, since the correction will be less than the error due to the averaging procedure. Then, we restrict ourselves by comparison of the analytic predictions with numeric simulations of the original system (2.265) and of the averaged system (2.268).

Our next goal is to numerically verify an analytical prediction of the SMR attractor existence described in the previous section. We start from the validation of the one-dimensional mapping procedure developed above by comparing it with the averaged system (2.268). In Figs. 2.63, 2.64 and 2.65, the one-dimensional mapping procedure is compared to the numeric solutions of the averaged system (2.268) for various values of the small parameter ε; the other system parameters are kept fixed ($A = 0.4, \lambda = 0.2, \sigma = 0.5$).

"Fast" parts of the mapping cycle are computed from (2.294, 2.295) and are therefore not related to the actual trajectory (it spins around the slow manifold due

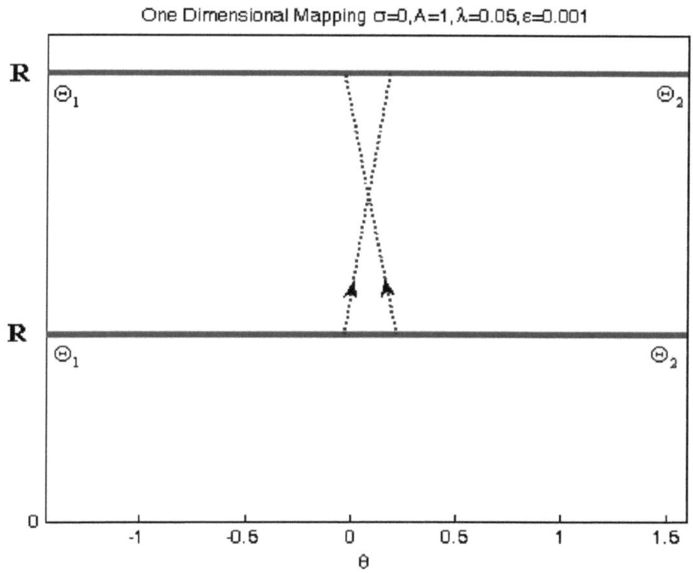

Fig. 2.61 Double period cycle of the one dimensional mapping: $A = 1, \lambda = 0.05, \sigma = 0$

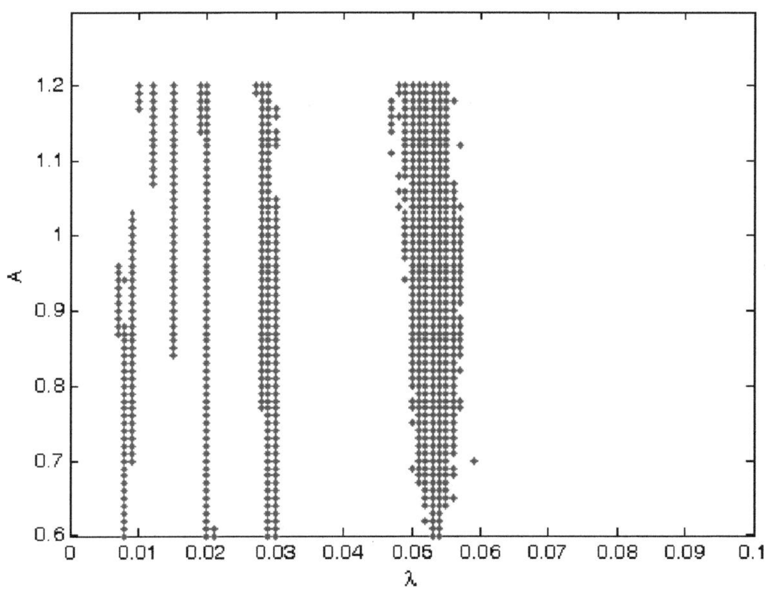

Fig. 2.62 Zones of period doubling ($\sigma = 0$)

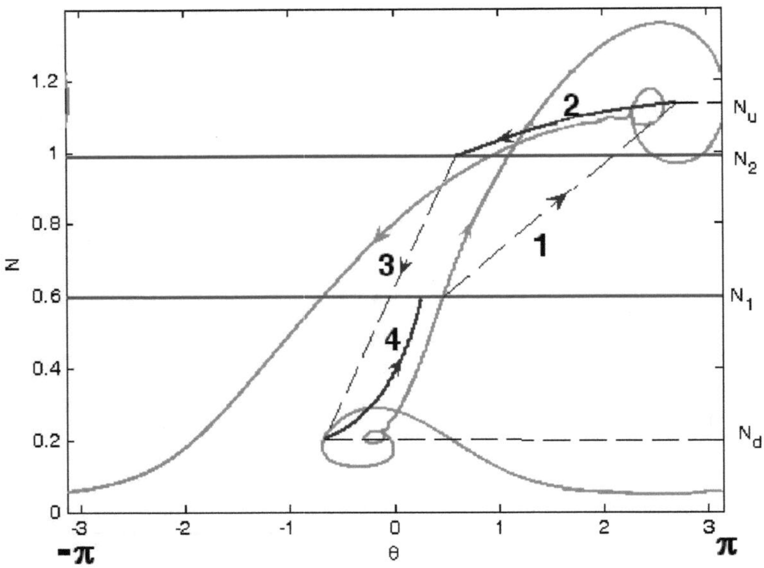

Fig. 2.63 Comparison between prediction of the mapping approach (numbered, online – *blue line*) and numeric simulation of the averaged system (2.268) (not numbered, online – *green line*). The "fast" parts of the mapping trajectory are denoted by *dashed lines*. System parameters: $A = 0.4, \lambda = 0.2, \sigma = 0.5, \varepsilon = 0.01$

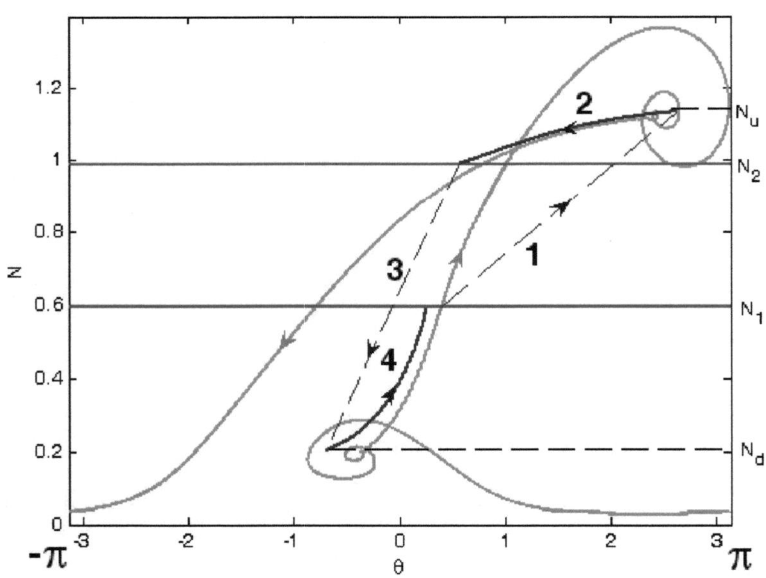

Fig. 2.64 Comparison between prediction of the mapping approach (numbered, online – *blue line*) and numeric simulation of the averaged system (2.268) (not numbered, online – *green line*). The "fast" parts of the mapping trajectory are denoted by *dashed lines*. System parameters: $A = 0.4, \lambda = 0.2, \sigma = 0.5, \varepsilon = 0.005$

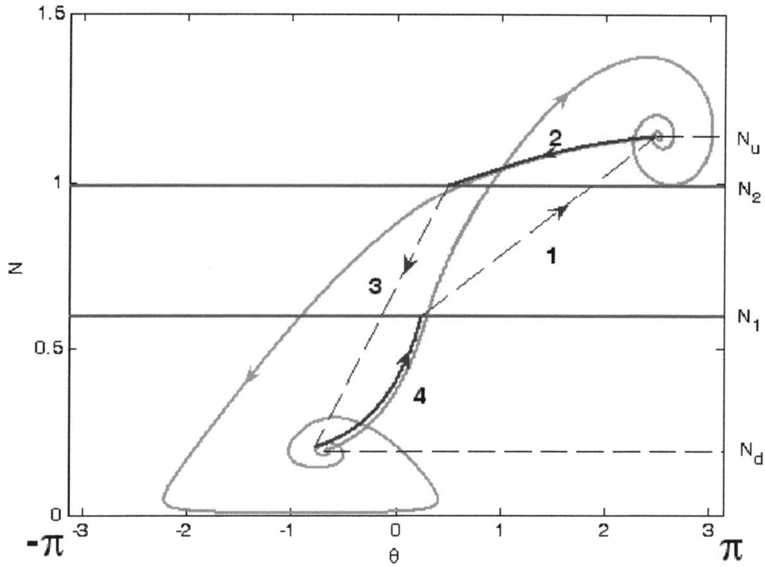

Fig. 2.65 Comparison between prediction of the mapping approach (numbered, online – *blue line*) and numeric simulation of the averaged system (2.268) (not numbered, online – *green line*). The "fast" parts of the mapping trajectory are denoted by *dashed lines*. System parameters: $A = 0.4, \lambda = 0.2, \sigma = 0.5, \varepsilon = 0.001$

to a pair of complex conjugate eigenvalues). Thus, only the "slow" parts of the mapping cycle should be compared with the numeric simulation data. It is clear from the plots that the "true" phase trajectory of the averaged system slightly deviates from the one predicted by the mapping; the deviation grows with growth of ε, as one should expect. For moderate values of ε, the method of mapping can be used at least for qualitative predictions.

Then, we compare the numeric solution of the original system (2.265) with the initial conditions constrained to the SIM (with the same parameters as used for the plot in Fig. 2.55, $A = 0.4, \lambda = 0.2, \sigma = 0.0, \varepsilon = 0.001$) with the analytic predictions (Fig. 2.66).

The frequency detuning interval obtained for the analytically predicted existence of the SMR attractor is $\sigma \in [-1.275, 1.597]$ when the numerical simulation demonstrates that for $\varepsilon = 0.01$ the SMR exists as $\sigma \in [-1.3, 1.3]$. Quite naturally, the accuracy of the analytical prediction increases for smaller ε.

Additional use of one dimensional mapping discussed in the previous section is the ability to predict peculiar transient behavior of the response – the system can exhibit a few cycles of the relaxation oscillations before it is attracted to the stable periodic response. In order to verify it we present the case for which there is no stable SMR attractor. Thus, for the system parameters ($A = 0.3, \lambda = 0.2, \sigma = 0.35$) by picking the initial value the of phase angle ($\theta_0 = 0$) and plotting the sequence of mappings, we get the picture in Fig. 2.67.

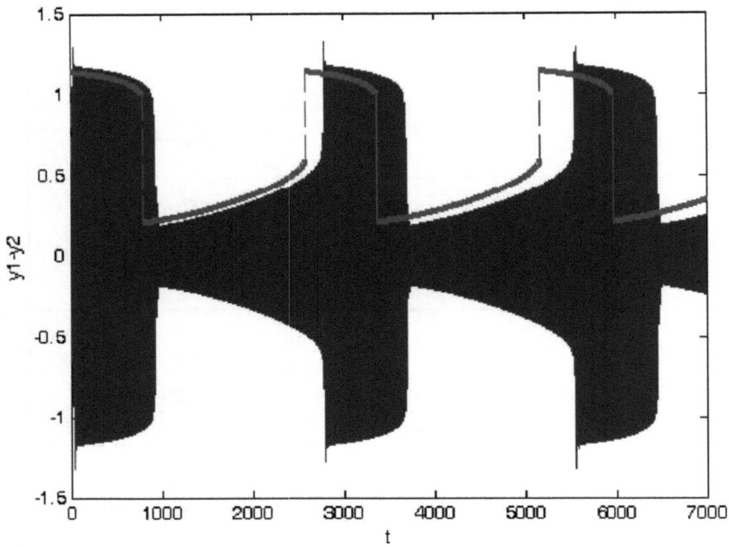

Fig. 2.66 Strongly modulated response compared to prediction of one dimensional mapping (*the modulation envelope*). System parameters: $A = 0.4, \lambda = 0.2, \sigma = 0.0, \varepsilon = 0.001$

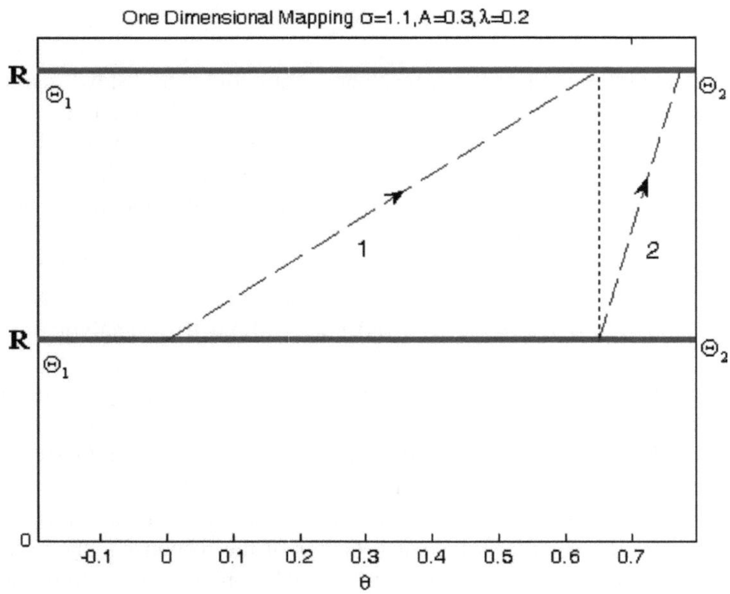

Fig. 2.67 The mapping diagram in the absence of stable SMR, $(A = 0.3, \lambda = 0.2, \sigma = 1.1)$

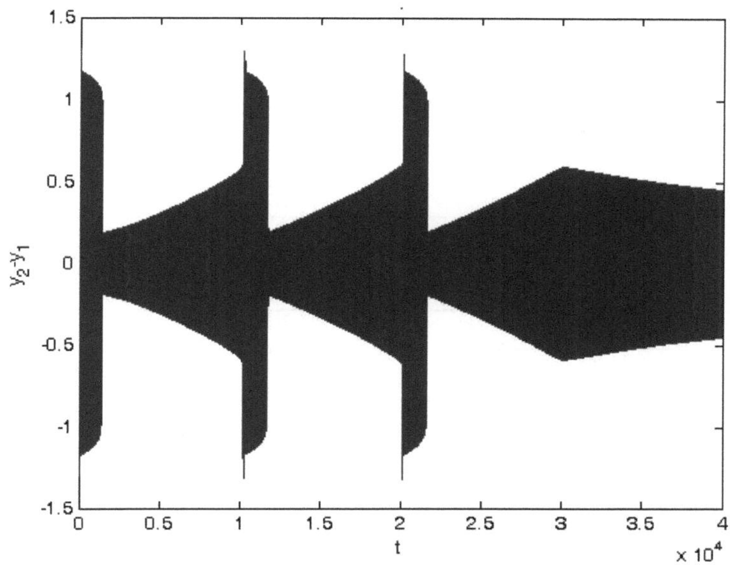

Fig. 2.68 Time series of the original system with initial conditions constrained to the SIM and similar to those of the one dimensional mapping (Fig. 2.62), $A = 0.3, \lambda = 0.2, \sigma = 1.1, \varepsilon = 0.0005$

The system parameters were specially chosen in order to obtain more than one transient cycle on the mapping diagram. In the case presented in Fig. 2.67, the number of mapping cycles is 2. In order to check the prediction, we supply the similar initial conditions to the original system as for the mapping and plot the time series for the initial system (2.265). The result is plotted in Fig. 2.68.

It is clear from Fig. 2.68 that before a system is attracted to the periodic response it exerts two cycles of relaxation type as it was predicted by the one-dimensional map. It should be mentioned that in order to get such a coincidence, the parameter ε is picked to be extremely small. The latter requirement reflects the fact that the transient relaxation oscillations are observed in close vicinity of the bifurcation of stable–unstable limit cycles and the flow structure is thus very sensitive to minor changes of parameters.

The next simulation is related to period doubling, in accordance with the predictions in Fig. 2.62. We performed simulations (Poincare Section, Time Series) in the zone of predicted period doubling ($A = 0.8, \lambda = 0.053$) and in the neighboring zone of one period ($A = 0.7, \lambda = 0.065$) where the small parameter and detuning are $\varepsilon = 0.005, \sigma = 0$. Poincare section and time series presented in Fig. 2.69 refer to the first pair of parameters$A = 0.8, \lambda = 0.053$ in the zone of period doubling. Poincare section and time series presented in Fig. 2.70 refer to the second pair of parameters $A = 0.8, \lambda = 0.065$ in the zone of one period, very close to the period doubling point.

It is interesting to see that the time series clearly reveal the period doubling. As for Poincare sections, they demonstrate that in fact both responses are chaotic-like

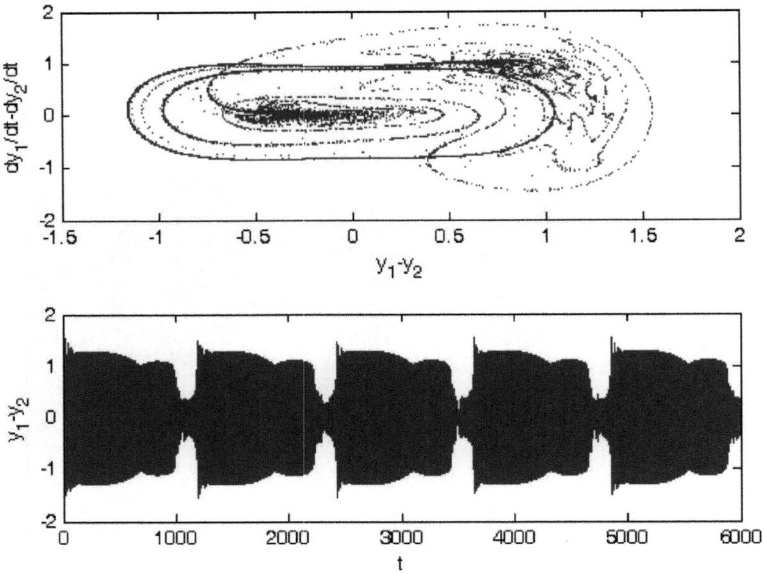

Fig. 2.69 Poincare section, time series response for system parameters in a zone of period doubling. System parameters: $(A = 0.8, \lambda = 0.053, \varepsilon = 0.005, \sigma = 0)$

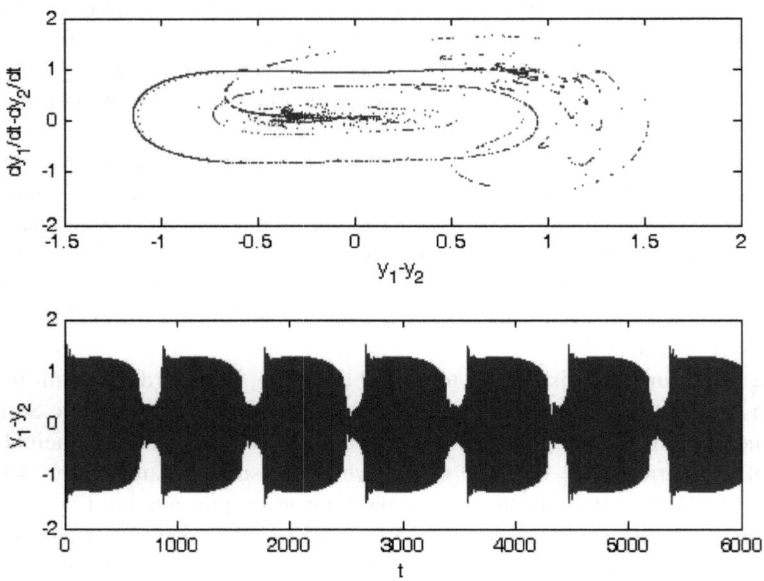

Fig. 2.70 Poincare section, time series response for system parameters in zone of one period. System parameters: $(A = 0.8, \lambda = 0.065, \varepsilon = 0.005, \sigma = 0)$

at time scale O (ε^{-2}). Of course, the treatment presented above says nothing about the behavior of the system at this time scale.

The approach developed is formally valid for $\varepsilon \to 0$ and care is required when using it for the system having a finite mass ratio. Some effects are smashed due to mixing between a fast and a slow time scale; some others, like chaotization, are not captured by the analytic approach. Still, the mapping can be used at least for the qualitative evaluation of possible response regimes. The treatment cannot be considered as mathematically rigorous: the averaging is performed beyond the conditions of the averaging theorem; the mapping includes a numeric part and is valid only for $\varepsilon \to 0$. Anyway, the predictive power of the approach is confirmed by comparison to the results of direct numerical simulations.

In general, the responses similar to the SMR may be present in a variety of systems having essential nonlinearity and mass (or modal mass) asymmetry, if appropriate multiple scale decomposition provides the necessary geometry of invariant manifolds to provide the possibility for the relaxation oscillations. In such cases, the method presented above may be used in order to establish the zone in the space of parameters where the SMR exists, as well as its domain of attraction.

2.7 Coupled Nonlinear Oscillators with Time Delays

2.7.1 Analytic Model

A concept of nonlinear normal mode (NNM) as analytic continuation of well-known normal modes of linear system originates from the work of Lyapunov (1947). As it was mentioned above, the concept has been refined by Rosenberg and other researchers (Rosenberg, 1960, 1962; Kauderer, 1958); they have primarily considered the NNMs as particular solutions of nonlinear dynamical systems, which are characterized by synchronous evolution of all variables. Recently, there was a plenty of activity in the field of the NNMs, including successful analysis of multi-DOF systems, transient motions etc (Vakakis et al., 1996; Shaw and Pierre, 1991, 1993). See also the Addendum at the end of this book.

Account of time delay is a necessary modification of many dynamical models if one takes into account finite speed of the signal propagation, retarded waves etc. Such models are widely studied in relation with cutting and milling processes, neural networks and other systems (Sen and Rand, 2003; Wirkus and Rand, 2002; Atay, 2003). Usually, the nonlinearities in these models are treated by means of asymptotic expansions and averaging; these methods are very powerful but it is well-known that they can miss some effects related to essential nonlinearity. In the dynamical systems without delays, the computation of the NNMs can help to reveal these solutions (Vakakis et al., 1996).

Our present goal is to investigate whether the concept of the NNM may be useful for investigation of an essentially nonlinear dynamical system with time delays. Of course, one can rely on the linear normal modes no more – due to the time delay, the state space has an infinite dimension from the very beginning. Instead, it is proposed to treat the NNMs as synchronous solutions of the system, i.e. to include the modes which are described by closed continuous curves in the configuration space. The study of in-phase and out-of-phase motions in a system of coupled Van-der-Pol oscillators with time delays has been performed in (Sen and Rand, 2003). The phase-locked solutions in the case of relaxation oscillations were studied in (Wirkus and Rand, 2002). We adopt here a somewhat different approach – the dynamical system is chosen in a form which allows exact computation of some modal shapes, with subsequent numeric verification.

The system of two similar homogeneous coupled essentially nonlinear oscillators is postulated to have the form

$$\begin{aligned} u_{1,tt} + Cu_1^m + G(u_1 - u_2(t-\tau))^m = 0 \\ u_{2,tt} + Cu_2^m + G(u_2 - u_1(t-\tau))^m = 0 \end{aligned} \tag{2.296}$$

where $u_i \equiv u_i(t), i = 1, 2$, C is the nonlinear stiffness of the oscillator, G is the coupling strength, τ is the time delay and m is an odd positive number. In every equation, the time delay is introduced only to the coupling term related to the other oscillator; such a choice of model is suggested to describe the retarded reaction of the coupling spring. For $\tau = 0$ system (2.296) is reduced to system (1.2.1) in the book (Vakakis et al., 1996) without the linear part.

After simple rescaling

$$t = \xi\tau, u_i = gy_i, i = 1, 2, g = (\tau^2 C)^{-\frac{1}{m-1}}, k = G/C \tag{2.297}$$

system (2.296) is reduced to a dimensionless form

$$\begin{aligned} \ddot{y}_1 + y_1^m + k(y_1 - y_2(t-1))^m = 0 \\ \ddot{y}_2 + y_2^m + k(y_2 - y_1(t-1))^m = 0 \end{aligned} \tag{2.298}$$

which is the basis for further analysis. The dot denotes the differentiation with respect to ξ.

Periodic synchronous solutions of system (2.298) are searched in a form

$$y_1 = cy_2(\xi - T) \tag{2.299}$$

where c is constant ratio of amplitudes and T – constant phase shift. Substitution of (2.299) to (2.298) yields

$$\begin{aligned} c\ddot{y}_2(\xi - T) + c^m y_2^m(\xi - T) + k(cy_2(\xi - T) - y_2(\xi - 1))^m = 0 \\ \ddot{y}_2 + y_2^m + k(y_2 - cy_2(\xi - 1 - T))^m = 0 \end{aligned} \tag{2.300}$$

Shifting the first equation of System (2.300) by T, one obtains

$$\begin{aligned} \ddot{y}_2 + c^{m-1}y_2^m + kc^{-1}(cy_2 - y_2(\psi + T - 1))^m = 0 \\ \ddot{y}_2 + y_2^m + k(y_2 - cy_2(\xi - 1 - T))^m = 0 \end{aligned} \tag{2.301}$$

In order to have a consistent solution, the first and the second equations of system (2.301) should be equivalent. Periodic solutions are searched for; let us consider that the solution of system (2.301) is periodic with minimal period Δ. Due to the symmetry of initial equations, one can suggest

$$y_2 = y_2(\xi + \Delta), y_2 = -y_2\left(\xi + \frac{\Delta}{2}\right) \tag{2.302}$$

System (2.301) will be consistent, if the following relationships hold:

$$T - 1 = \frac{n\Delta}{2}, T + 1 = \frac{l\Delta}{2}, n, l \in \mathbb{Z} \tag{2.303}$$

From (2.303) it is easy to obtain

$$\Delta = \frac{4}{l-n}, T = (l+n)\frac{\Delta}{4} = \frac{l+n}{l-n}, l, n \in \mathbb{Z} \tag{2.304}$$

From (2.304) it follows that possible values of period are discrete and time shift T is a multiple of $\Delta/4$. The latter conclusion allows a description of the modal shapes on configuration plane y_1–y_2: for l+n even the modal shapes will be straight lines and for l+n odd-ovals. Solutions similar to the latter type of modes were described, for instance, in (Manevich and Manevitch, 2005) for coupled conservative oscillators with weak nonlinearity. There they were referred to as "elliptic" modes. Here this term is not justified, since solutions of essentially nonlinear system (2.296) are very different from sine and cosine functions and therefore the modal curves are not ellipses. Therefore the term "oval modes" is used. Case l = n corresponds to the trivial solution $y_1 = y_2 = 0$.

Let us consider different possibilities for l and n, which will yield different consistency conditions for system (2.301). Due to the symmetry of the system without restriction of generality it is possible to adopt l>n.

A. Both l and n are even.

In this case $\Delta = 2/q, q = 1, 2, 3\ldots$ and T = 0 or $\Delta/2$. The condition of consistency for system (2.301) is written as

$$1 + k(1-c)^m = c^{m-1} + kc^{-1}(c-1)^m \tag{2.305}$$

Equation (2.305) always allows solutions $c = \pm 1$, corresponding to symmetric and antisymmetric modes (due to conditions (2.299) and (2.301), transformation $c \to -c$, T → T+$\Delta/2$ keeps the system unchanged). Besides the symmetric and the antisymmetric modes, at a certain critical value of k, solutions of equation (2.305) bifurcate, giving rise to localized modes. The bifurcation diagram for $m = 3$ is presented in Fig. 2.71, the bifurcation point is $k = 0.25$, $c = -1$.

This figure is similar to Fig. (1.2.2b) in Vakakis et al. (1996), besides the issue of stability which is discussed below. However, one should keep in mind that for

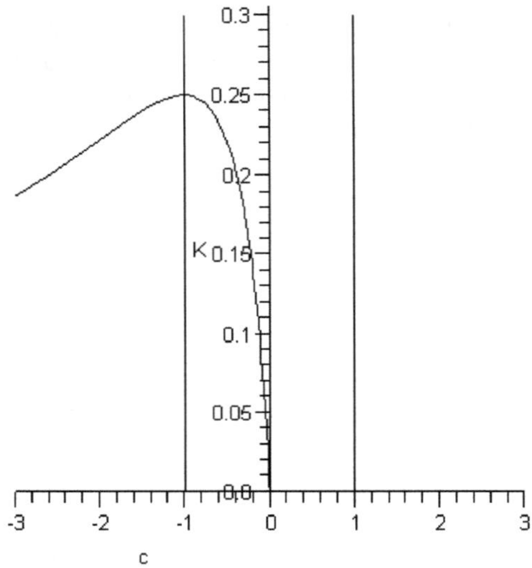

Fig. 2.71 Solutions of Eq. (2.305) – k versus c

the system with time delay both solutions with $T = 0$ and $T = \Delta/2$ are possible, depending on values of l and n. Thus, in accordance with (4) for $c < 0$ one will obtain both physically symmetric and antisymmetric solutions. Contrary to the case of $\tau = 0$, both symmetric and antisymmetric mode can bifurcate and yield the localized solutions.

B. Both l and n are odd.

Also in this case, $\Delta = 2/q, q = 1, 2, 3 \ldots$ and $T = 0$ or $\Delta/2$. The condition of consistency for system (2.301) is written as

$$1 + k(1 + c)^m = c^{m-1} + kc^{-1}(c + 1)^m \tag{2.306}$$

Transformation $c \to -c$ yields Eq. (2.305); therefore, this series of solutions is equivalent to the previous case.

C. 1 *is even, n is odd.*

In this case, $\Delta = 4/(2q + 1), q = 0, 1, 2, 3 \ldots$ and $T = \Delta/4$ or $3\Delta/4$. In other terms, these solutions correspond to the "oval modes" mentioned above. The condition of consistency for system (2.301) is written as

$$1 + k(1 + c)^m = c^{m-1} + kc^{-1}(c - 1)^m \tag{2.307}$$

The structure of solutions for Eq. (2.307) is very different from that of (2.305). First of all, the values $c = \pm 1$ do not satisfy the equation for every value of k, although they may satisfy it for some particular coupling. Therefore, for almost all values of the coupling coefficient the solutions are localized. Moreover, it is possible to demonstrate that these solutions exist only if k is some bounded interval. Reshaping Eq. (2.307), one obtains

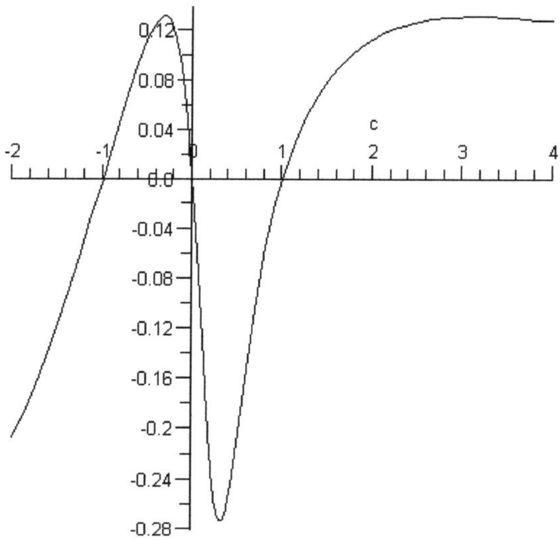

Fig. 2.72 Solutions of Eq. (2.308) – k versus c

$$k = \frac{c^m - c}{c(1+c)^m + (1-c)^m} \tag{2.308}$$

It is easy to demonstrate that the denominator in (2.308) never vanishes for any odd positive m. For the case $m = 3$, the plot of function (2.308) is presented in Fig. 2.72

D. l is odd, n is even.

It is easy to demonstrate that the sign inversion of c brings this case to (2.307, 2.308).

Following this, one can conclude that system (2.301) possesses two families of synchronous periodic solutions. The first family corresponds to the cases (A) and (B) and can be presented by straight lines on the configuration plane. The second one corresponds to the cases (C) and (D); on the configuration plane this is described by ovals.

In order to complete the computation, one should determine the amplitude of y_2 for each case. If two equations in system (2.301) are consistent, both of them are reduced to the form

$$\ddot{y}_2 + Qy_2^m = 0 \tag{2.309}$$

where Q is a constant depending on the selected mode. This equation has a well-known solution (see, e.g., Salenger et al., 1999):

$$y_2(t) = Av\left(\sqrt{\frac{2Q}{m+1}}A^{\frac{m-1}{2}}t + \varphi_0\right) \tag{2.310}$$

where A and ϕ_0 are determined by the initial conditions, and periodic function v(t) with unit amplitude is determined by the quadrature

$$dt = \frac{dv}{\sqrt{1 - v^{m+1}}} \tag{2.311}$$

The minimal period of solution (2.310) is determined as

$$T = 2A^{\frac{1-m}{2}} \sqrt{\frac{2\pi}{Q(m+1)} \frac{\Gamma(1/_{m+1})}{\Gamma(1/2 + 1/_{m+1})}} \tag{2.312}$$

Equations (2.305) and (2.307) allow presenting the coefficient Q in Eq. (2.309) as an explicit function of c for every case. For "straight" modes (cases A and B) this function will read

$$Q_{straight} = \frac{\pm c^m + 1}{\pm c + 1} \tag{2.313}$$

A positive sign corresponds to the case (A) and a negative one to (B)

For "oval" modes (cases C and D), a combination of System (2.301) and Eq. (2.308) yields:

$$Q_{oval} = \frac{\pm c^m (1 \pm c)^m + (1 \mp c)^m}{\pm c(1 \pm c)^m + (1 \mp c)^m} \tag{2.314}$$

The upper sign corresponds to C and the lower – to D).

A combination of the expression for period (2.312) with expression (2.313) or (2.314) for the relevant mode (amplitude ratio c should be selected from the solutions of Eqs. (2.306) or (2.308) respectively with the given value of coupling *k* yields the condition for discrete values of amplitude):

$$A_{straight} = q^{\frac{2}{m-1}} \left(\frac{Q_{straight}(m+1)}{2\pi} \frac{\Gamma^2(1/2 + 1/_{m+1})}{\Gamma^2(1/_{m+1})} \right)^{-\frac{1}{m-1}}, q = 1, 2, \ldots \tag{2.315}$$

$$A_{oval} = (2q+1)^{\frac{2}{m-1}} \left(\frac{2Q_{oval}(m+1)}{\pi} \frac{\Gamma^2(1/2 + 1/_{m+1})}{\Gamma^2(1/_{m+1})} \right)^{-\frac{1}{m-1}}, q = 1, 2, \ldots \tag{2.316}$$

In the theory of nonlinear normal modes, it is common to investigate *asymptotic* and *orbital* stability (Vakakis et al., 1996) – if the perturbation of the initial conditions which determine the given mode is small enough, the distance between the perturbed and the unperturbed orbit will vanish in the former case and remains bounded in the latter. In the case of systems with time delay, the above standard definitions should be changed, since in order to determine the orbit unambiguously one should also define the pre-history of the motion at the delay time. So, the orbit $y_i(t)$, $i = 1,2$ of Eq. (2.296) is a small perturbation of the orbit $y_{0i}(t)$, $i = 1,2$ if

$$\max_{-\tau \le t \le 0} \left\| (y_i(t) - y_{0i}(t), \dot{y}_i(t) - \dot{y}_{0i}(t))^T \right\| = \varepsilon << 1 \tag{2.317}$$

The orbit $y_{0i}(t)$ is asymptotically stable, if for any $y_i(t)$ satisfying (2.317) for ε small enough the distance between the unperturbed and the perturbed orbit vanishes

$$\lim_{t \to \infty} dist(y_i(t), y_{0i}(t)) = 0 \tag{2.318}$$

and is orbitally stable, if for any δ small enough, for which condition (2.317) is satisfied, there exists such ε that

$$\max_{0 \le t < \infty} dist(y_i(t), y_{0i}(t)) = \delta(\varepsilon) \tag{2.319}$$

and δ(ε) vanishes if ε vanishes. In the limit $\tau \to 0$, both definitions are reduced to standard ones for common NNMs.

The system under consideration is not conservative; therefore the solutions may be asymptotically stable. As will be demonstrated below, linear analysis of the stability may be insufficient for this system and a complete nonlinear analysis is hardly possible.

Still, one particular case may be easily investigated analytically. The trivial solution $y_i = 0$, $i = 1,2$ clearly satisfies Eq. (2.298); let us consider its stability. One should introduce the perturbation of the trivial solution as follows:

$$y_1(t) = \delta_1(t), y_2(t) = \delta_2(t), \|\delta_1\| \sim \|\delta_2\| \sim O(\delta) << 1 \tag{2.320}$$

Equations (2.298) are reduced to the form

$$\begin{aligned} \ddot{\delta}_1 + \delta_1^m + k(\delta_1 - \delta_2(t-1))^m = 0 \\ \ddot{\delta}_2 + \delta_2^m + k(\delta_2 - \delta_1(t-1))^m = 0 \end{aligned} \tag{2.321}$$

This system is not easier than System (2.298). Still, the balance of orders of terms yields the conclusion concerning the rate of time evolution of the variables:

$$d/dt \sim O(\delta^{m-1/2}) \tag{2.322}$$

Then, time evolution is relatively slow and one can use a Taylor series expansion for the delay terms:

$$\delta_i(t-1) = \delta_i(t) - \dot{\delta}_i(t) + \frac{1}{2!}\ddot{\delta}_i(t) + \ldots, i = 1,2 \tag{2.323}$$

If one keeps only the first term of (2.323) in both Eq. 2.316), the effect of delay is not taken into account, the system is conservative and the origin is a neutrally stable equilibrium point. That is why the second terms in (2.323) should be also kept. The approximate equations are

$$\begin{aligned} &\ddot{\delta}_1 + \delta_1^m + k(\delta_1 - \delta_2)^m + km\dot{\delta}_2(\delta_1 - \delta_2)^{m-1} = 0 \\ &\ddot{\delta}_2 + \delta_2^m + k(\delta_2 - \delta_1)^m + km\dot{\delta}_1(\delta_2 - \delta_1)^{m-1} = 0 \end{aligned} \tag{2.324}$$

If one introduces function positively defined in the vicinity of the origin and vanishing in the origin itself

$$E = \frac{1}{2}(\dot{\delta}_1^2 + \dot{\delta}_2^2) + \frac{1}{m+1}(\delta_1^{m+1} + \delta_2^{m+1} + k(\delta_1 - \delta_2)^{m+1}) \tag{2.325}$$

then it is easy to conclude from (2.324) that

$$\frac{dE}{dt} = -2\,km\dot{\delta}_1\dot{\delta}_2(\delta_1 - \delta_2)^{m-1} \tag{2.326}$$

For antisymmetric perturbation $\delta_1(t) = -\delta_2(t)$, one obtains dE/dt>0 if the perturbations are not constant; therefore, the origin is unstable. Of course, E is the total energy of the system in the absence of delay.

If one considers the stability of the set of periodic solutions of system (2.298) ($y_{01}(t)$, $y_{02}(t)$), where both functions are not zeros and $y_{01}(t) \neq y_{02}(t \pm 1)$, then the problem of stability may be treated by linear approximation (the particular case $y_{01}(t) = y_{02}(t \pm 1)$ is addressed in the next section). Small perturbations are introduced as follows:

$$y_i = y_{0i} + \delta_i,\ \|\delta_i\| << \|y_i\|\ , i = 1,2 \tag{2.327}$$

In the linear approximation the following homogeneous system is obtained:

$$\begin{aligned} &\ddot{\delta}_1 + my_{10}^{m-1}\delta_1 + km(y_{10} - y_{20}(t-1))^{m-1}(\delta_1 - \delta_2(t-1)) = 0 \\ &\ddot{\delta}_2 + my_{20}^{m-1}\delta_2 + km(y_{20} - y_{10}(t-1))^{m-1}(\delta_2 - \delta_1(t-1)) = 0 \end{aligned} \tag{2.328}$$

If periods of $y_{10}(t)$ and $y_{20}(t)$ are equal, as in the treatment above, or at least commensurate, as it is the case for the phase-locked solution presented in the next section, then the stability of orbits is determined by location of eigenvalues of infinite monodromy matrix of system (2.328) (see, e.g. Saaty, 1981 for details).

Needless to say, even approximate computation of this matrix is a cumbersome task, especially in the case of the strong nonlinearity treated in this paper. Therefore, the stability of particular solutions for various modes described above was investigated numerically.

All numeric simulations presented below were performed for system (2.298) with $m = 3$ (unless otherwise stated explicitly), for different values of the coupling parameter k and different initial functions $y_1(t)$ and $y_2(t)$, t<0.

2.7.2 Numeric Verification – Straight Modes

The first series of simulations illustrates modes which are described by straight lines on the configurational plane. Figure 2.73a, b, demonstrates the displacements y_1 and y_2 respectively for the symmetric mode ($c = 1, T = 0$).

It is easy to see that the symmetric mode is realized with an amplitude close to 15. Substituting $c = 1$ and $m = 3$ to (2.313) and (2.315), one obtains:

$$A_{straight} \approx 3.708q \tag{2.329}$$

Figure (2.73a, b) demonstrate the solution with $q = 4$, the period being equal to $2/q$ =0.5, in complete agreement with the analytic predictions above. Similar results may be obtained for higher amplitudes as well (Fig. 2.74a, b).

It is easy to see that the solution now corresponds to $q = 8$. It should be mentioned that the initial conditions for both simulations above are symmetric, but do

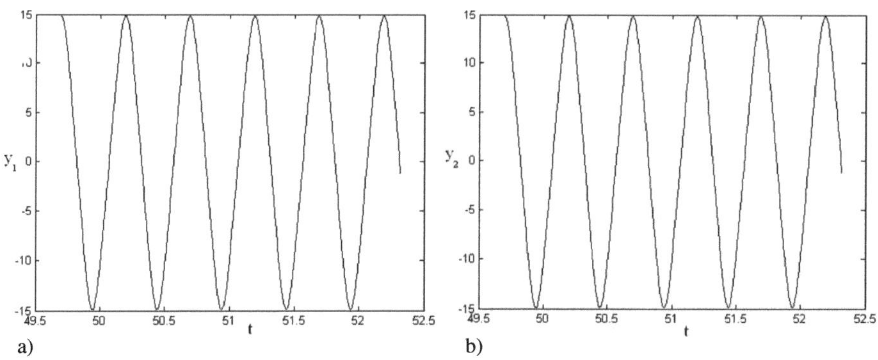

Fig. 2.73 Displacements for the symmetric mode, k = 0.1, $y_1(t) = y_2(t) = 13$, $t \leq 0$; (**a**) $y_1(t)$; (**b**) $y_2(t)$

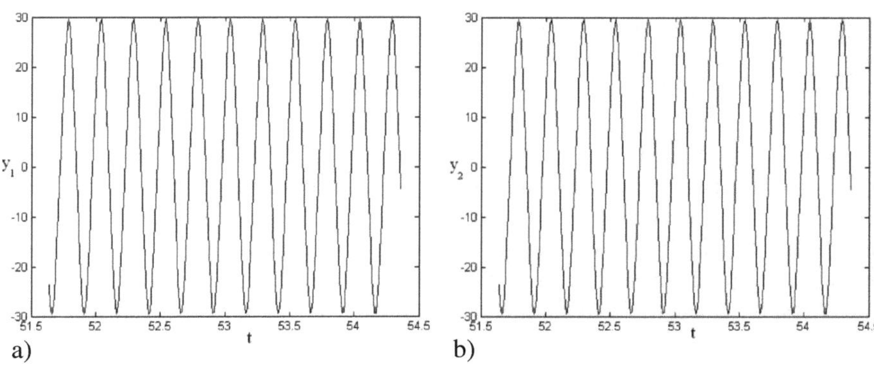

Fig. 2.74 Displacements for the symmetric mode, k = 0.13, $y_1(t) = y_2(t) = 17$, $t \leq 0$; (**a**) $y_1(t)$; (**b**) $y_2(t)$

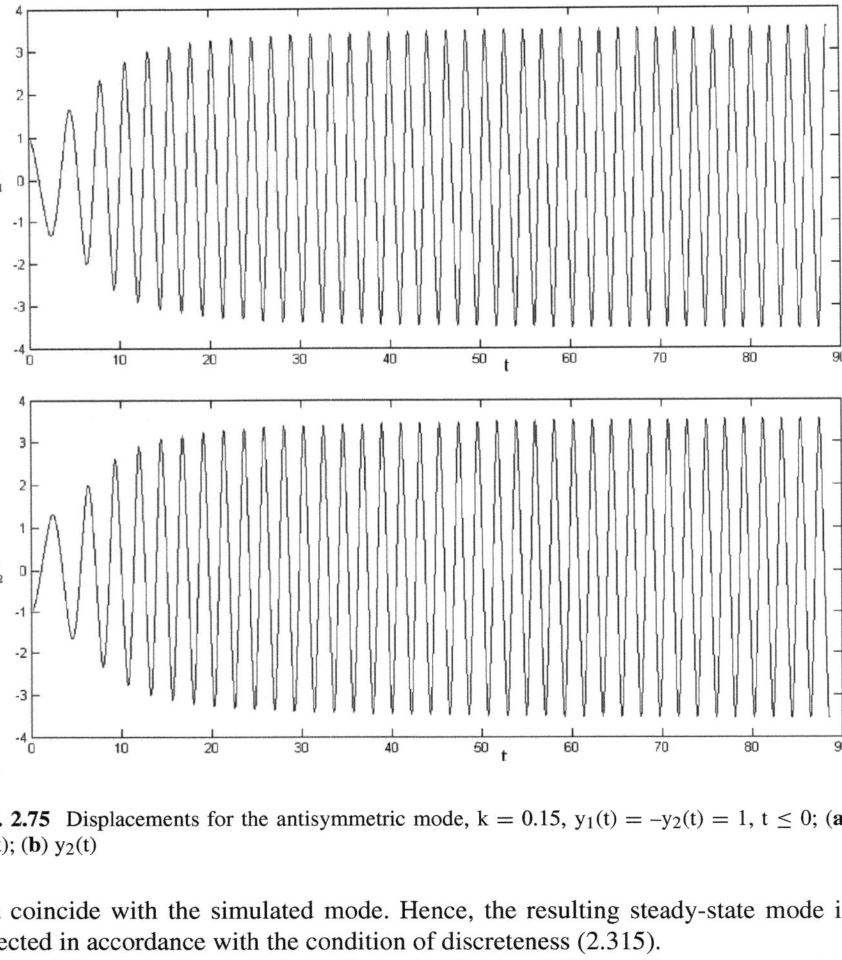

a)

b)

Fig. 2.75 Displacements for the antisymmetric mode, $k = 0.15$, $y_1(t) = -y_2(t) = 1$, $t \leq 0$; (**a**) $y_1(t)$; (**b**) $y_2(t)$

not coincide with the simulated mode. Hence, the resulting steady-state mode is selected in accordance with the condition of discreteness (2.315).

Similarly, the antisymmetric mode can be simulated. The results are presented in Fig. 2.75a, b.

It is easy to see that this regime corresponds to $q = 1$. In fact, here $c = 1$ and $T=\Delta/2$.

Stability of the symmetric and the antisymmetric solutions mentioned above is a rather subtle issue. From one side, direct simulation of the exact antisymmetric response has not indicated any divergence; moreover, from Fig. 2.75 one can see that the mode actually attracts the flow. At least, it seems that the numeric errors do not bring about the divergence. Still, this is not sufficient to conclude on the stability of this response regime. To illustrate the point, the simulation with a relatively small initial deviation from ideal antisymmetry was performed (Fig. 2.76a, b).

In the beginning of the process, the responses are very similar to those presented in Fig. 2.75a, b but finally the flow leaves the antisymmetric mode and diverges – the

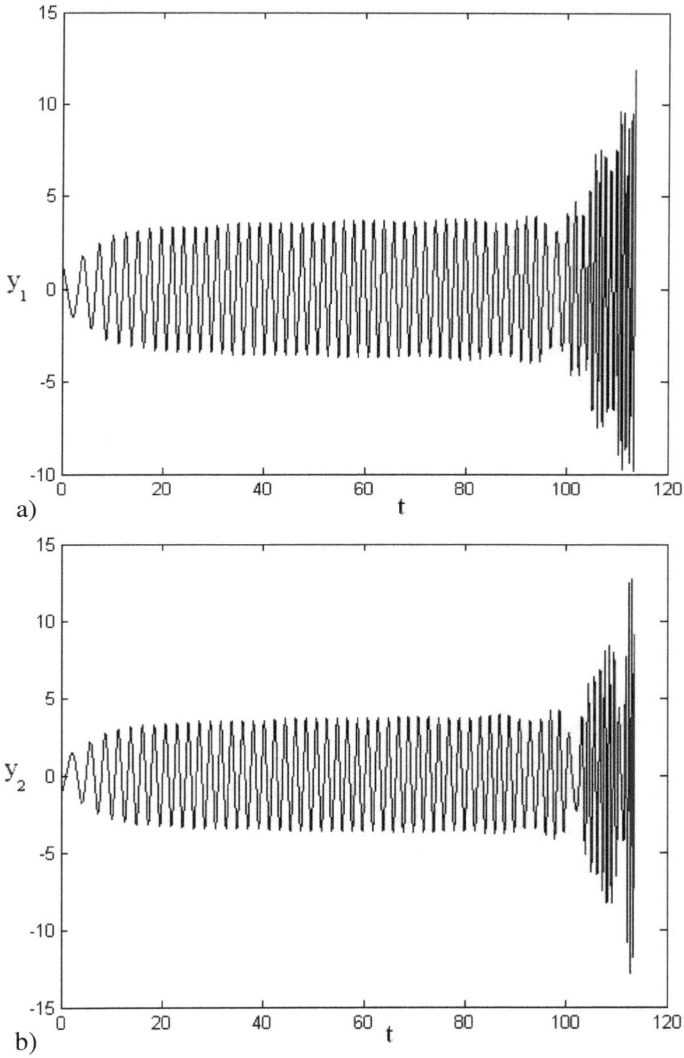

Fig. 2.76 Displacements for the squeezed antisymmetric mode, $k = 0.15$, $y_1(t) = 1.2$, $y_2(t) = -1$, $t \le 0$; (**a**) $y_1(t)$; (**b**) $y_2(t)$

segment of the plot for $t > 100$ may be problematic with respect to numeric accuracy but a trend towards divergence is obvious. Simulations with smaller deviation from the exact antisymmetry also yielded the approximate capture to the antisymmetric mode with subsequent divergence, but it took much more time. The behavior of the symmetric mode is rather similar.

It is possible to explain this phenomenon with a perturbation analysis of system (2.298) by investigation of the vicinity of the exact solution with $c = 1$. For the sake of simplicity, the analysis will be performed only for the symmetric mode with $y_1(t) = y_2(t-1)$:

$$\begin{aligned} y_1(t) &= Y_0(t) + \delta_1(t) \\ y_2(t) &= Y_0(t) + \delta_2(t) \\ Y_0(t) &= Y_0(t-1), \delta_j << 1 \end{aligned} \tag{2.330}$$

Substituting (2.330) with (2.298), one obtains:

$$\begin{aligned} \ddot{\delta}_1 + \{mY_0^{m-1}\delta_1 + \ldots\} + k(\delta_1 - \delta_2(t-1))^m = 0 \\ \ddot{\delta}_2 + \{mY_0^{m-1}\delta_2 + \ldots\} + k(\delta_2 - \delta_1(t-1))^m = 0 \end{aligned} \tag{2.331}$$

The terms in figured parentheses stem from the uncoupled system and cannot bring about the divergence; the only reason for the instability can be related to the last term. It means, first of all, that the stability of the symmetric and the antisymmetric modes cannot be analyzed in linear approximation. Then, if the deviation is extremely small (like in the case of numeric errors), it will not affect the stability of the numeric solution – the corrections yielded by these deviations will not be taken into account by a numeric scheme, since it has a finite accuracy. So, the situation here is a bit paradoxical – the regime is in fact unstable, but the instability cannot be captured in the linear approximation and does not reveal itself via numeric errors.

The situation with the other modes where the linear coupling terms do not disappear due to the symmetry (e.g. "straight" localized modes) is quite different – it seems that they are unstable and cannot be revealed by direct numeric simulation. Quite surprisingly, the only stable attractors revealed for the system under consideration are those synchronous solutions which do not exist at all for the system with zero delay – the "oval" modes.

2.7.3 Numeric Verification – "Oval" Modes and Phase – Locked Solutions

A typical example of the "oval" mode is presented in Fig. 2.77.

It should be mentioned that the initial conditions are rather "far" from the mode obtained; it demonstrates remarkable stability, unlike the "straight" modes. Stable "oval" modes are ubiquitous in the system under consideration for relatively small values of the coupling coefficient; for higher values of k they cease to exist (see Fig. 2.72).

Another interesting phenomenon may be observed for the same value of the coupling but for a somewhat higher amplitude of the initial function (Fig. 2.78).

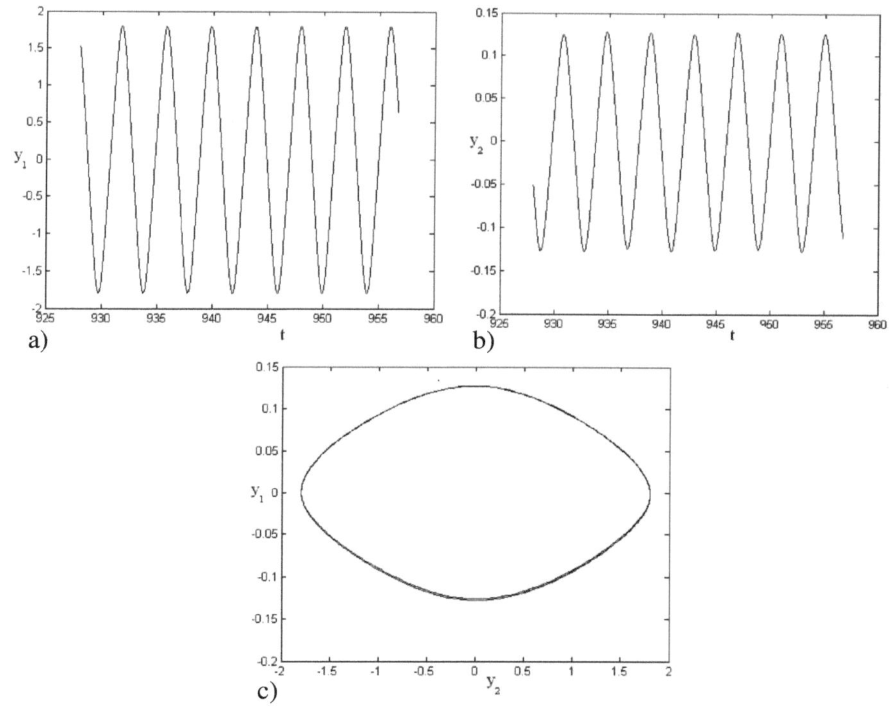

Fig. 2.77 Displacements for the "oval" mode, $k = 0.06$, $y_1(t) = 2$, $y_2(t) = 0$, $t \leq 0$; (**a**) $y_1(t)$; (**b**) $y_2(t)$; (**c**) configuration plane: y_1 versus y_2

One obtains a stable phase-locked solution with a period ratio of 2:3. Such phase-locked solutions are also rather ubiquitous for the system; of course, the synchronous solutions described above are a particular case of the phase locking with a period ratio of 1:1.

The last simulation treats the case $m = 7$. It is easy to obtain the "oval" mode for this case too (Fig. 2.79).

One can see that the modal shape is very different from an ellipse; this difference is a manifestation of the extreme nonlinearity of the system.

The investigation presented above demonstrates that the concept of a nonlinear normal mode makes sense for the essentially nonlinear system with time delay. For such a system the NNM cannot be treated as an analytic continuation of the linear normal mode; instead, one should refer to the definition of "synchronous motion". In this case, the NNM turns to be a particular case of phase locking. The NNMs determined in this way may serve as stable attractors of the dynamical flow; such situation is not possible in a conservative system.

The special structure of system (2.296) allows an exact investigation of the synchronous solutions. For other systems this will not be the case. Vast literature devoted to the NNMs describes many approximate methods of their computation. It

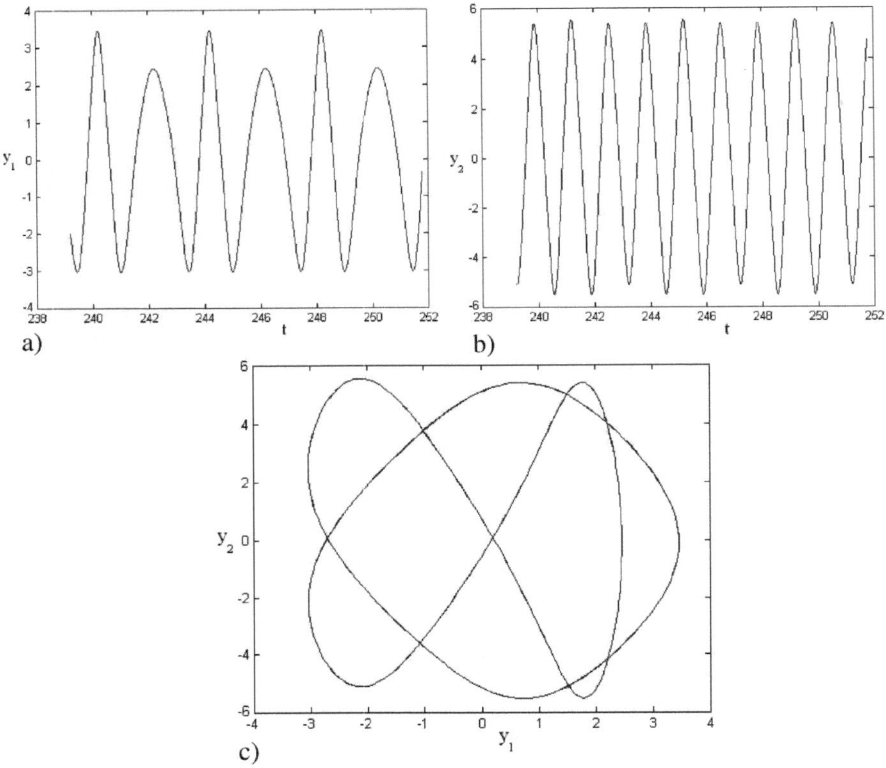

Fig. 2.78 Displacements for the phase – locked mode, $k = 0.06$, $y_1(t) = 3$, $y_2(t) = 0$, $t \leq 0$; (**a**) $y_1(t)$; (**b**) $y_2(t)$; (**c**) configuration plane: y_1 versus y_2

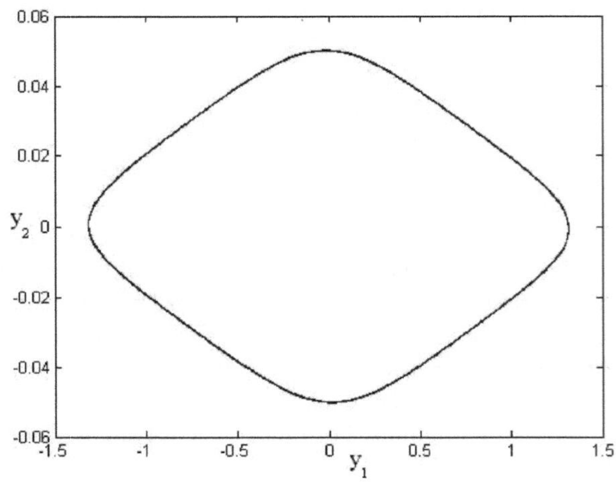

Fig. 2.79 Displacements for the "oval" mode, $m = 7$, $k = 0.03$, $y_1(t) = 1$, $y_2(t) = 0$, $t \leq 0$; configuration plane: y_1 versus y_2

would be desirable to develop the methods to treat the NNMs in the systems with time delays. It would allow one to find the solutions which are not available by standard averaging or asymptotic schemes.

2.8 Low-DOF Discrete Nonlinear Systems

The problem of energy localization and energy transfer in discrete nonlinear systems has a long history in physics, chemistry and the mechanical sciences. The discovery of intrinsic localized modes (ILM) in an infinite oscillatory chain by Sievers and Takeno (1988) gave rise to a very wide area of studies of periodic discrete systems. The intrinsic localized modes have been detected and studied experimentally in such different systems as interacting Josephson junction ladders (Binder et al., 2000), weakly coupled nonlinear optical waveguides (Eisenberg et al., 1998), lattice vibrations in crystals (Swanson et al., 1999), antiferromagnetic structures (Shwarz et al., 1999), micromechanical cantilever arrays (Sato et al., 2003, 2004), Bose–Einstein condensates loaded on optical lattices (Eiermann et al., 2004), and layered high-Tc superconductors (Machida and Koyama, 2004).

There are several universal models that take the regularities inherent to wide class of nonlinear systems into account. The Fermi-Pasta-Ulam (FPU) chain is one of these models because its potential energy contains the terms of third and fourth order together with parabolic ones. Beginning from pioneer work (Fermi et al., 1955), predominant attention in the study of FPU chains was paid to large systems, which are integrable only within the continuum limit. However, it turns out that even a discrete FPU model with symmetric potential (β-FPU) can be integrable if the number of particles is equal to 3 and the periodicity conditions are satisfied (Feng, 2006).

We consider a more general situation when the potential contains an asymmetric part (αβ-FPU) and the number of particles N in the periodic chain is small enough but can be more than 3 ($N = 3,4,\ldots$). The common peculiarity of the systems under consideration is the presence of doubly degenerate frequencies in the oscillatory spectrum with wide gaps between the pairs. In such cases, the localization of energy in one of the degenerate eigenmodes may be considered (in terms of normal coordinates) as the immobile ILM. The resonant energy exchange between two degenerate modes in this situation corresponds to the ILM mobility in the periodic FPU chain. When the number of particles increases, resonance relations between different frequencies turn out to be possible and the present consideration is not valid.

Let us consider a common FPU-system, the Hamiltonian which can be written as follows:

$$H_0 = \sum_j^N \frac{1}{2}p_j^2 + \frac{1}{2}(q_{j+1} - q_j)^2 + \alpha\frac{1}{3}(q_{j+1} - q_j)^3 + \frac{\beta}{4}(q_{j+1} - q_j)^4 \qquad (2.332)$$

where q_j and p_j are the coordinates and conjugate moments, respectively, and N is the number of particles. As it is well known, the system (2.332) with periodic

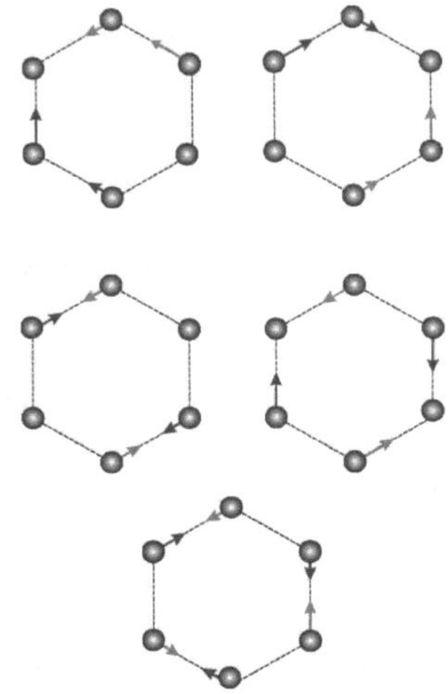

Fig. 2.80 Normal modes of 6-particles FPU chain. *Red* and *blue arrows* correspond to positive and negative displacements, respectively. *Left column* contains respective eigenvalues

boundary conditions has N eigenvalues ω_j in the harmonic limit, the majority of them are doubly degenerate (e.g., see Fig. 2.80 for $N = 6$).

If N is an even number, then the highest eigenvalue ω_{N-1} is not degenerate. The equations of motion in the terms of linear normal modes

$$\begin{aligned} \xi_j &= \sum_{k=1}^{N} a_{jk} q_k \\ \eta_j &= \sum_{k=1}^{N} a_{jk} p_k \end{aligned} \tag{2.333}$$

where a_{jk} are components of the transitional matrix, can be written as

$$\begin{aligned} &\frac{d\xi_j}{dt} - \eta_j = 0 \\ &\frac{d\eta_j}{dt} + \omega_j^2 \xi_j + F_j(\{\xi\}) = 0 \end{aligned} \tag{2.334}$$

Here, the functions F_j contain the nonlinear terms depending on ξ_j and on the interaction between different modes.

Introducing the complex combinations of ξ_j and η_j:

$$\Psi_j = \frac{1}{\sqrt{2}}(\eta_j + i\omega_j \xi_j) \quad \bar{\Psi}_j = \frac{1}{\sqrt{2}}(\eta_j - i\omega_j \xi_j) \tag{2.335}$$

one can write the coupled nonlinear dispersionless Schrödinger equations:

$$i\frac{d\Psi_j}{dt} + \omega_j \Psi_j + F_j(\{\Psi, \bar{\Psi}\}) = 0 \tag{2.336}$$

In the case of a small-amplitude motion we can represent the functions Ψ as:

$$\Psi_j = e^{i\omega_j t}(\varepsilon\chi_j + \varepsilon^2\chi_{j,2} + \varepsilon^3\chi_{j,3} + \ldots) \tag{2.337}$$

We have extracted the mono-frequency motion in Eq. (2.6) by the factor $\exp(i\omega_j t)$ to focus attention on the nonlinear inter-mode coupling effects. To analyze them, we have to introduce the time hierarchy:

$$\tau_0 = t, \quad \tau_1 = \varepsilon t, \quad \tau_2 = \varepsilon^2 t$$

In such an expansion we can see that the amplitudes of main-order approximation χ_j are independent of the "fast time" τ_0. Indeed, for order zero of the small parameter ε we get:

$$\varepsilon^0 : \frac{\partial\chi_j}{\partial\tau_0} = 0$$

It is easy to show that the equations in the next order of the small parameter ε have the following form:

$$\varepsilon^1 : \frac{\partial\chi_j}{\partial\tau_1} = 0, \quad \chi_{j,1} = f(\{\chi_j, \bar{\chi}_j\}) \tag{2.338}$$

where f is some quadratic form of main-order amplitudes χ_j. The last relation in (2.338) follows from the condition of the absence of secular terms. Then, one can remove the first-order amplitudes $\chi_{j,1}$ from the equations of second order of the small parameter.

Let us consider the system with four particles, as an example. In this case the eigenfrequencies are:

$$\omega_0 = 0; \quad \omega_1 = \omega_2 = \sqrt{2}; \quad \omega_3 = 2$$

The matrix of transition to the normal modes is

$$A = \begin{pmatrix} -\frac{1}{2} & \frac{1}{2} & -\frac{1}{2} & \frac{1}{2} \\ 0 & -\frac{1}{\sqrt{2}} & 0 & \frac{1}{\sqrt{2}} \\ -\frac{1}{\sqrt{2}} & 0 & \frac{1}{\sqrt{2}} & 0 \\ \frac{1}{2} & \frac{1}{2} & \frac{1}{2} & \frac{1}{2} \end{pmatrix} \tag{2.339}$$

The reduced system with an immobile centre of mass has three degrees of freedom. The Hamiltonian in terms of linear normal coordinates is

$$H_0 = \sum_j^4 \tfrac{1}{2}\eta_j^2 + 2\zeta_1^2 + \zeta_2^2 + \zeta_3^2 + 3\beta\zeta_1^2(\zeta_2^2 + \zeta_3^2) + \tfrac{3}{2}\zeta_2^2\zeta_3^2 + 4\alpha\zeta_1\zeta_2\zeta_3 + \\ +\beta(\zeta_1^4 + \zeta_2^4 + \tfrac{1}{4}\zeta_3^4) \tag{2.340}$$

After some manipulations, using Eqs. (2.335 and 2.338), we can obtain the following modulation equations:

$$\begin{aligned}
&i\frac{d\chi_1}{d\tau_2}+\kappa_0|\chi_3|^2\chi_1+\kappa_1|\chi_1|^2\chi_1+\kappa_2|\chi_2|^2\chi_1+\kappa_3\chi_2^2\bar{\chi}_1=0\\
&i\frac{d\chi_2}{d\tau_2}+\kappa_0|\chi_3|^2\chi_2+\kappa_1|\chi_2|^2\chi_2+\kappa_2|\chi_1|^2\chi_2+\kappa_3\chi_1^2\bar{\chi}_2=0\\
&i\frac{d\chi_3}{d\tau_2}+\sigma_1|\chi_3|^2\chi_3+\sigma_2(|\chi_1|^2+|\chi_2|^2)\chi_3=0
\end{aligned} \tag{2.341}$$

In Eq. (2.341), the envelope functions χ_1 and χ_2 correspond to degenerate normal modes with eigenvalues $\omega_1=\omega_2$ and function χ_3 corresponds to eigenvalue ω_3. The coefficients in the Eq. (2.341) have a rather simple structure

$$\begin{aligned}
&\sigma_1=3\beta;\quad \sigma_2=\tfrac{3\beta-2\alpha^2}{4\sqrt{2}}\\
&\kappa_0=\tfrac{3\beta-2\alpha^2}{8};\quad \kappa_1=\tfrac{3\beta}{8};\quad \kappa_2=\tfrac{3\beta}{16};\quad \kappa_3=\tfrac{3\beta-4\alpha^2}{16}
\end{aligned} \tag{3.342}$$

In the invariant subspace $\chi_3=0$, Eq. (2.341) are reduced to a more simple system:

$$\begin{aligned}
&i\frac{d\chi_1}{d\tau_2}+\kappa_1|\chi_1|^2\chi_1+\kappa_2|\chi_2|^2\chi_1+\kappa_3\chi_2^2\bar{\chi}_1=0\\
&i\frac{d\chi_2}{d\tau_2}+\kappa_1|\chi_2|^2\chi_2+\kappa_2|\chi_1|^2\chi_2+\kappa_3\chi_1^2\bar{\chi}_2=0
\end{aligned} \tag{2.343}$$

The corresponding Hamiltonian has the form:

$$H_\chi=\frac{\kappa_1}{2}(|\chi_1|^4+|\chi_2|^4)+\kappa_2|\chi_1|^2|\chi_2|^2+\frac{\kappa_3}{2}(\bar{\chi}_1^2\chi_2^2+\chi_1^2\bar{\chi}_2^2) \tag{2.344}$$

It is easy to show that Eq. (2.343) have an additional integral of motion

$$N_1=|\chi_1|^2+|\chi_2|^2 \tag{2.345}$$

Taking into account the Eq. (2.345), we can rewrite expression (2.344) as

$$H_\chi=\frac{\kappa_1}{2}N_1^2+(\kappa_2-\kappa_1)|\chi_1|^2|\chi_2|^2+\frac{\kappa_3}{2}(\bar{\chi}_1^2\chi_2^2+\chi_1^2\bar{\chi}_2^2) \tag{2.346}$$

Let us note that despite a certain similarity of Hamilton function (2.344) to that for the well-known DST model (Eilbeck et al., 1985), the system under consideration has essential peculiarities which will be discussed below.

We derived Eq. (2.343) for partial case $N=4$, however, it can be shown that their structure is similar for any pair of resonant modes in the systems with an arbitrary number of particles.

As this takes place, there are three different cases for the coefficients relationship:

(i) $\kappa_1 - \kappa_2 - \kappa_3 = 0$
(ii) $\kappa_1 - \kappa_2 + \kappa_3 = 0$
(iii) $\kappa_1 - \kappa_2 \pm \kappa_3 \neq 0$

(i) In this case, the value $iG_1^- = (\chi_1\bar{\chi}_2 - \bar{\chi}_1\chi_2)$ is the integral of motion, as it is easy to derive from Eq. (2.341). It allows us to linearize Eq. (2.343):

$$i\frac{dx_1}{d\tau_2} + k_1N_1x_1 - ik_3G_1^-x_2 = 0$$
$$i\frac{idx_2}{d\tau_2} + k_1N_1x_2 - ik_3G_1^-x_1 = 0 \quad (2.347)$$

The partial solution of Eq. (2.347) is

$$\chi_1 = \sqrt{\tfrac{N_1}{2}}\exp\left[i\left(\Omega\tau_2 + \tfrac{\delta_0}{2}\right)\right]$$
$$\chi_2 = \sqrt{\tfrac{N_1}{2}}\exp\left[i\left(\Omega\tau_2 - \tfrac{\delta_0}{2}\right)\right] \quad (2.348)$$

with

$$\Omega = \kappa_1N_1 - \kappa_3G_1^-, \quad \delta_0 = {}^{\pi}/_{2}$$

and $G_1^- = -N_1$ or

$$\Omega = \kappa_1N_1 + \kappa_3G_1^-, \quad \delta_0 = -{}^{\pi}/_{2}$$

and $G_1^- = N_1$
This solution corresponds to the elliptic mode (Manevich and Manevitch, 2005). The general solution can be presented in the form:

$$\chi_1 = \sqrt{N_1}e^{i\kappa_1N_1\tau_2}\cos\theta e^{i\delta/2}$$
$$\chi_2 = \sqrt{N_1}e^{i\kappa_1N_1\tau_2}\sin\theta e^{-i\delta/2} \quad (2.349)$$

The Eq. (2.347) have the following form in the terms of variables θ and δ:

$$\sin 2\theta\left[\frac{d\theta}{d\tau_2} + \kappa_3G_1^-\cos\delta\right] = 0$$
$$\sin 2\theta\left[\frac{d\delta}{d\tau_2} - 2\kappa_3G_1^-ctg2\theta\sin\delta\right] = 0 \quad (2.350)$$

It is easy to show that the parameter G_1^- is connected with the variables (δ, θ) by this relation:

$$\sin 2\theta\sin\delta = C = G_1^-/N_1$$

The behavior of the system can be analyzed efficiently on the reduced phase plane (δ, θ) (see Fig. 2.81). There are two characteristic phase trajectories. The

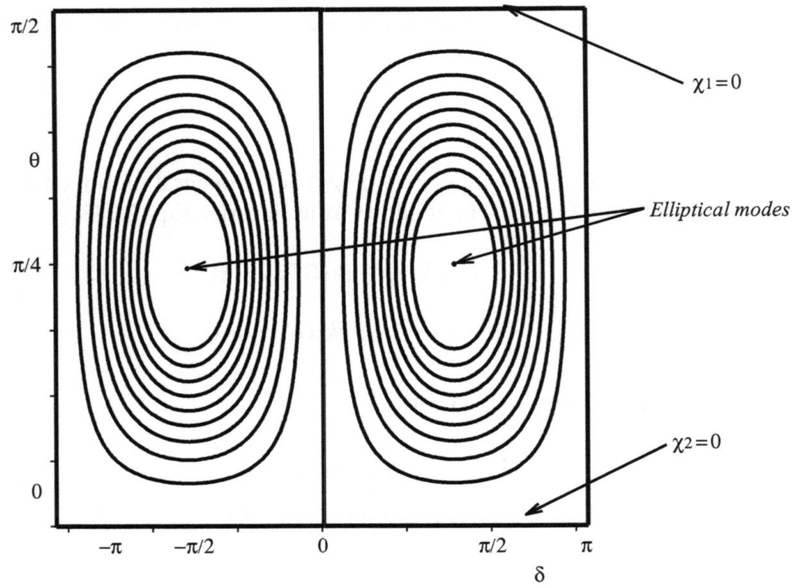

Fig. 2.81 Phase plane of Eq. (2.351) in the terms (δ, θ). The stationary point ($\delta=\pm\pi/2$, $\theta=\pi/4$) corresponds to elliptical mode ($G_1^- = N_1$). The *rectangular* trajectory corresponds to $G_1^- = 0$. Both presented quadrants are equivalent ones because of periodicity conditions

first of them is the central point ($G_1^-/N_1=1$, $\delta=\pi/2$, $\theta=\pi/4$) corresponding to elliptical normal modes (2.348). The trajectories close to this point correspond to weak intermode energy exchange. The second trajectory ($G_1^-/N_1=0$) is the limiting phase trajectory (LPT) describing a complete energy transfer. Let us note that the rate of energy exchange rises with increasing G_1^- and is minimal if $G_1^-/N_1=0$. So, both intensity and rate of energy transfer are controlled by the parameter G_1^-/N_1. The most intensive energy transfer is attained for the trajectories close to LPT (for LPT itself the rate is equal to zero, this is strongly different from the case of two linearly coupled nonlinear oscillators (Manevitch, 2007), where complete interparticle energy exchange occurs with a finite rate).

Certain results for the computer simulation of the high frequency part of normal modes (Fig. 2.80, modes with $\omega_{3,4} = \sqrt{3}$) in the β–FPU chain with 6 particles are shown in Fig. 2.82.

(ii) In this case, $G_1^+ = (\bar{\chi}_1\chi_2+\chi_1\bar{\chi}_2)$ is the integral of motion. The linearized form of Eq. (2.344) is:

$$i\frac{d\chi_1}{d\tau_2} + \kappa_1 N_1 \chi_1 + \kappa_3 G_1^+ \chi_2 = 0$$
$$i\frac{d\chi_2}{d\tau_2} + \kappa_1 N_1 \chi_2 + \kappa_3 G_1^+ \chi_1 = 0 \qquad (2.351)$$

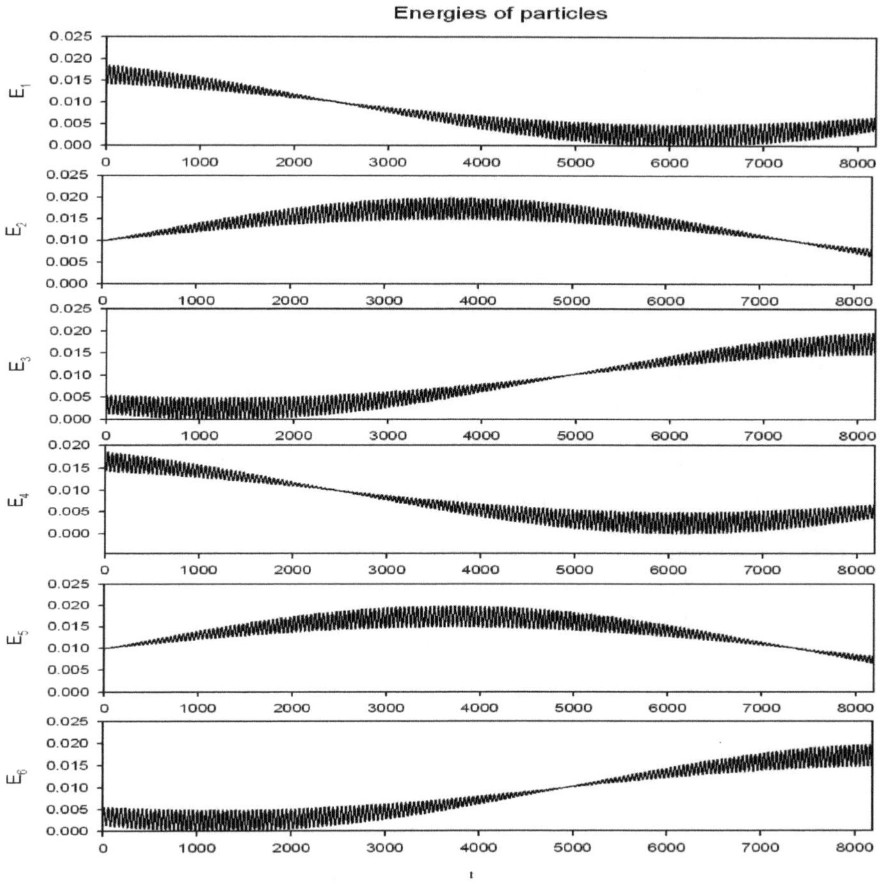

Fig. 2.82 Full energy exchange between high-frequency resonant modes leads to excitation mobility through the 6-particles FPU-chain. The initial excitation corresponds to $G_1^- \ll N_1$, which leads to a very long energy transfer

The stationary solutions have the following form:

$$\begin{aligned} \chi_1 &= \sqrt{\frac{N_1}{2}} \exp\left[i\left(\Omega\tau_2 + \frac{\delta_0}{2}\right)\right] \\ \chi_2 &= \sqrt{\frac{N_1}{2}} \exp\left[i\left(\Omega\tau_2 - \frac{\delta_0}{2}\right)\right] \end{aligned} \tag{2.352}$$

where $\Omega = \kappa_1 N_1 + \kappa_3 G_1^+, \delta_0 = 0, G_1^+ = N_1$ or otherwise $\Omega = \kappa_1 N_1 - \kappa_3 G_1^+, \delta_0 = \pi, G_1^+ = -N_1$.

Equations (2.352) in the terms of variables (δ, θ) are written as

$$\begin{aligned}
&\sin 2\theta \left[\frac{d\theta}{d\tau_2} + \kappa_3 G_1^+ \sin\delta \right] = 0 \\
&\sin 2\theta \left[\frac{d\delta}{d\tau_2} + 2\kappa_3 G_1^+ ctg 2\theta \cos\delta \right] = 0
\end{aligned} \tag{2.353}$$

The last equations determine a relation between the angle variables in the form:

$$\sin 2\theta \cos\delta = C = G_1^+/N$$

In this case, the central point $G_1{}^+/N_1 = 1$ corresponds to a "supernormal mode" which is the combination of both resonant modes with identical weights. As for LPT ($G_1{}^+/N_1 = 0$), it describes again a complete intermode transfer with zero rate, so that actual slow energy transfer may be observed for trajectories close to LPT. The data of a computer simulation study confirm the analytical results (see Fig. 2.83).

(iii) This condition leads to a more complicated description than (i–ii). In such a case, Eq. (2.343) cannot be linearized, but they are written in terms of angular variables as above:

$$\begin{aligned}
&\sin 2\theta \left[\frac{d\theta}{d\tau_2} + \frac{1}{2}\kappa_3 N_1 \sin 2\theta \sin 2\delta \right] = 0 \\
&\sin 2\theta \left\{ \frac{d\delta}{d\tau_2} - N_1 \cos 2\theta \left[(\kappa_1 - \kappa_2) - \kappa_3 \cos 2\delta \right] \right\} = 0
\end{aligned} \tag{2.354}$$

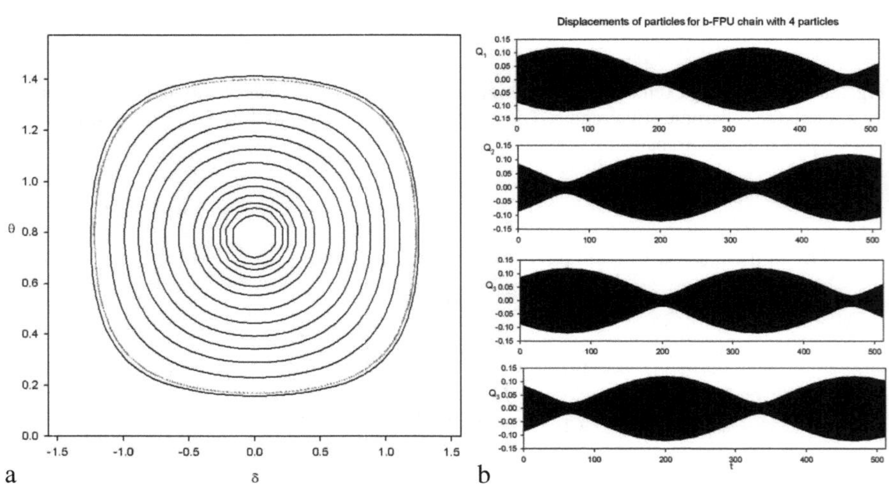

Fig. 2.83 (**a**) Phase plane for the case (ii). The stationary point (δ=0, θ=π/4) corresponds to the "supernormal" mode (*red points* – computer simulation results), (**b**) displacements of particles

The relation between the variables (δ,θ) following from Eq. (2.354) is:

$$\sin 2\theta \sqrt{\left|\frac{\kappa_1 - \kappa_2}{\kappa_3} - \cos 2\delta\right|} = C$$

Here, the phase plane has a structure depending on the relationship between the constants κ_1, κ_2, κ_3, which are controlled by the parameter α of the potential asymmetry. The case $\alpha = 0$ leads to relationship (ii) with the supernormal mode as a stationary point (Fig. 2.83a). Any trajectory close to LPT corresponds to a full energy exchange between modes χ_1 and χ_2 with large period. Otherwise, occurrence of any asymmetry of the potential function leads to the creation of a separatrix.

The singular (saddle) point for small values $0<\kappa_3<\kappa_1-\kappa_2$ (small α values) corresponds to unstable elliptical mode ($\delta=\pi/2$, $\theta=\pi/4$). As a result we get a phase plane as shown in Fig. 2.84a. The separatrix passing through the point ($\delta = \pi/2$, $\theta = \pi/4$) separates the domain of closed trajectories with partial energy exchange

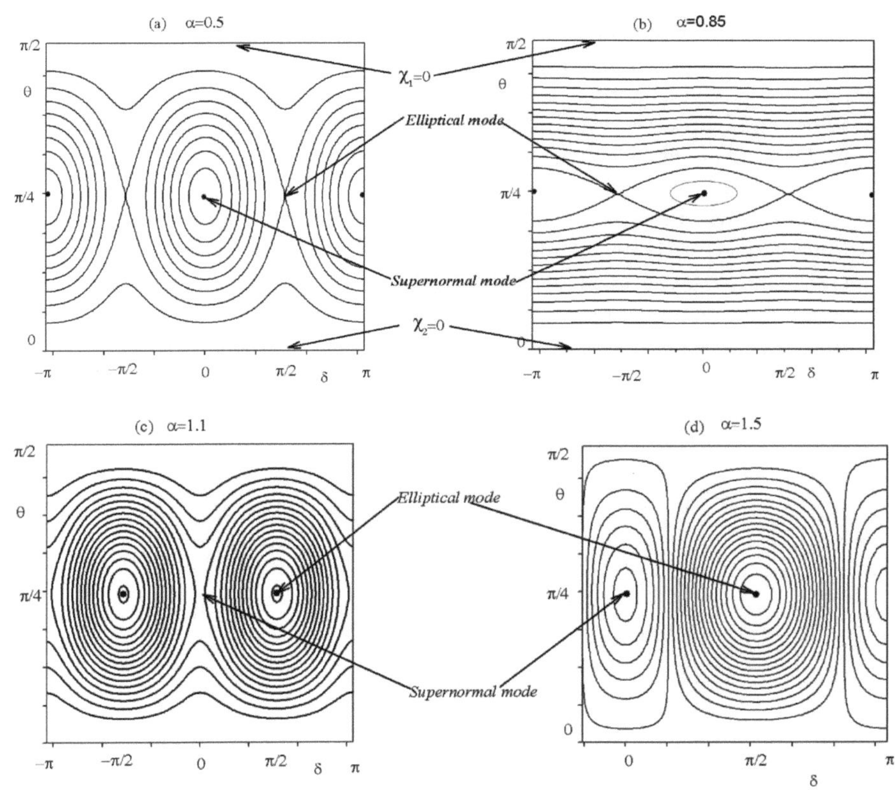

Fig. 2.84 Transformation of the phase plane of Eq. (2.354) with a variation of potential's asymmetry parameter α (see text)

from the transit-time trajectories for which an energy exchange is practically absent. The domain of closed trajectories decreases while the parameter κ_3 tends to zero at $\alpha=0.866$ (Fig. 2.84b). The two branches of the separatrix are reduced to straight lines and the phase plane is filled with transit-time trajectories, which are mutually parallel ones. The supernormal mode ($\delta = 0$, $\theta = \pi/4$) appears as the saddle point within the interval $-(\kappa_1-\kappa_2) < \kappa_3 < 0$ for $0.866 < \alpha < 1.2248$ (Fig. 2.84c). Additionally, there are two stationary points ($\delta = \pi/2$, $\theta = \pi/4$) and ($\delta = 0$, $\theta = \pi/4$) if $\kappa_3 < -(\kappa_1-\kappa_2)$ ($\alpha > 1.2248$, Fig. 2.84d). In such a case, the trajectories circling over both supernormal and elliptic stationary points and close to the LPT correspond to a complete energy exchange between the resonant normal modes. In this way, the dynamics of the particles drastically change with a variation of parameter κ_3, which, in turn, is determined by the asymmetry of potential.

To check the main analytical results we have performed the computer simulations of dynamics of several FPU-chains with various number of particles ($N = 3,\ldots,8$). Some results of these simulations in the particular case $N = 4$ are shown in Figs. 2.85, 2.86 and 2.87.

The nonlinear excitations discussed up to this point are similar to discrete q-breathers of an extended discrete chain (Flach et al., 2006) in the case of small FPU-systems. A small nonlinearity manifests itself in formation of q-breathers, which correspond to the preservation of energy in the nonlinear normal mode (or in the combination of several normal modes belonging to a certain integral manifold). This behavior can be observed even in the case when the interaction with other integral manifolds is taken into account.

If the number of particles grows, a new opportunity arises. Namely, the resonant interaction between the modes having close linear frequencies becomes possible. Two qualitatively different scenarios, related to various possible structures of the integral manifolds, can be singled out. For small energies a simple superposition of the normal vibrations occurs; it is qualitatively similar to linear approximation. However, above a certain energy threshold the resonant interaction mentioned above

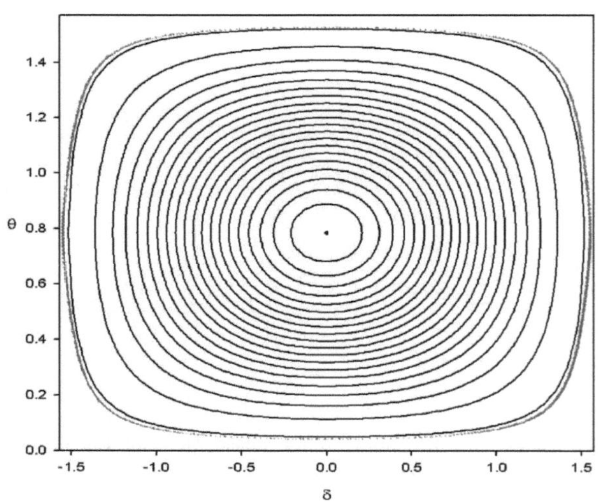

Fig. 2.85 The separatrix curve creation in the FPU chain with 4 particles at small parameter of asymmetry of potential function $\alpha=0.1$. *Red points* correspond to computer simulation data

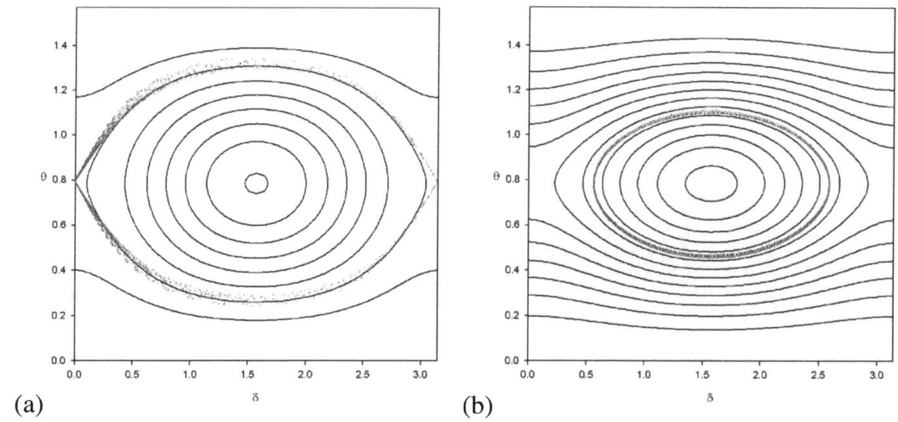

Fig. 2.86 Two phase trajectories for different initial conditions in the αβ-FPU chain with 4 particles. Parameter of asymmetry α=1.1. (**a**) Starting point is corresponding to supernormal mode (separatrix), (**b**) starting point is in the area of partial energy exchange. Scattered, online – *red* correspond to computer simulation data

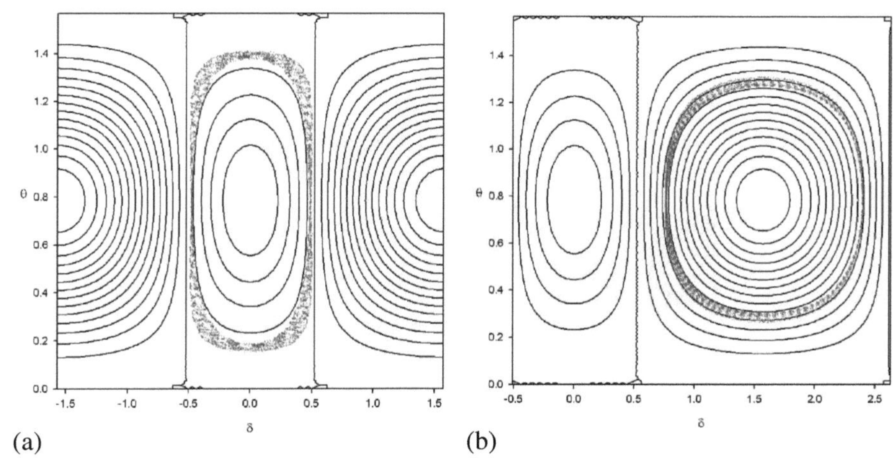

Fig. 2.87 Two phase trajectories for different starting points in the αβ-FPU chain (4 particles) with large asymmetry α=1.5. Scattered, online – *red* correspond to computer simulation data

occurs and it leads to some localization of the excitation which can be moved along the chain. This change in the behavior of the system can be interpreted as a transition from q-breathers to conventional ones typical for a system with large number of the particles and, consequently, with many almost resonant relations between different linear frequencies.

Two types of the behavior discussed above are illustrated in the system comprising 8 particles. For the case of very small energies one can observe the solution close to superposition of given normal modes (Fig. 2.88). But with growth of amplitudes the transition occurs to localized breather-like excitation (Figs. 2.89 and 2.90).

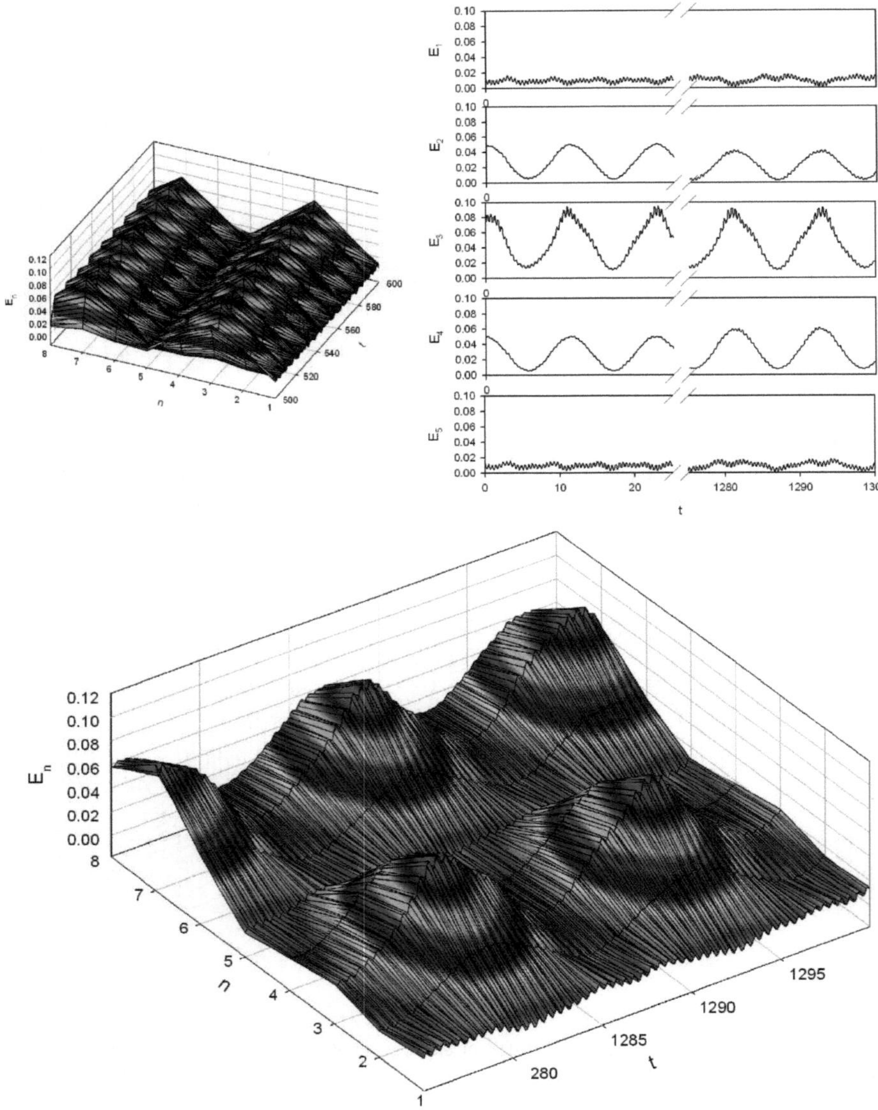

Fig. 2.88 Energy distribution in the αβ-FPU system of 8 particles. The *left picture* is a 3D surface of the energy. *Right pictures* exhibit the energy of different particles. The initial condition corresponds to excitation of non-resonant mode in combination with one of the resonant modes. Total occupation number n is 0.1, n is particle number, and the time t is measured in the periods of the high-frequency resonant mode

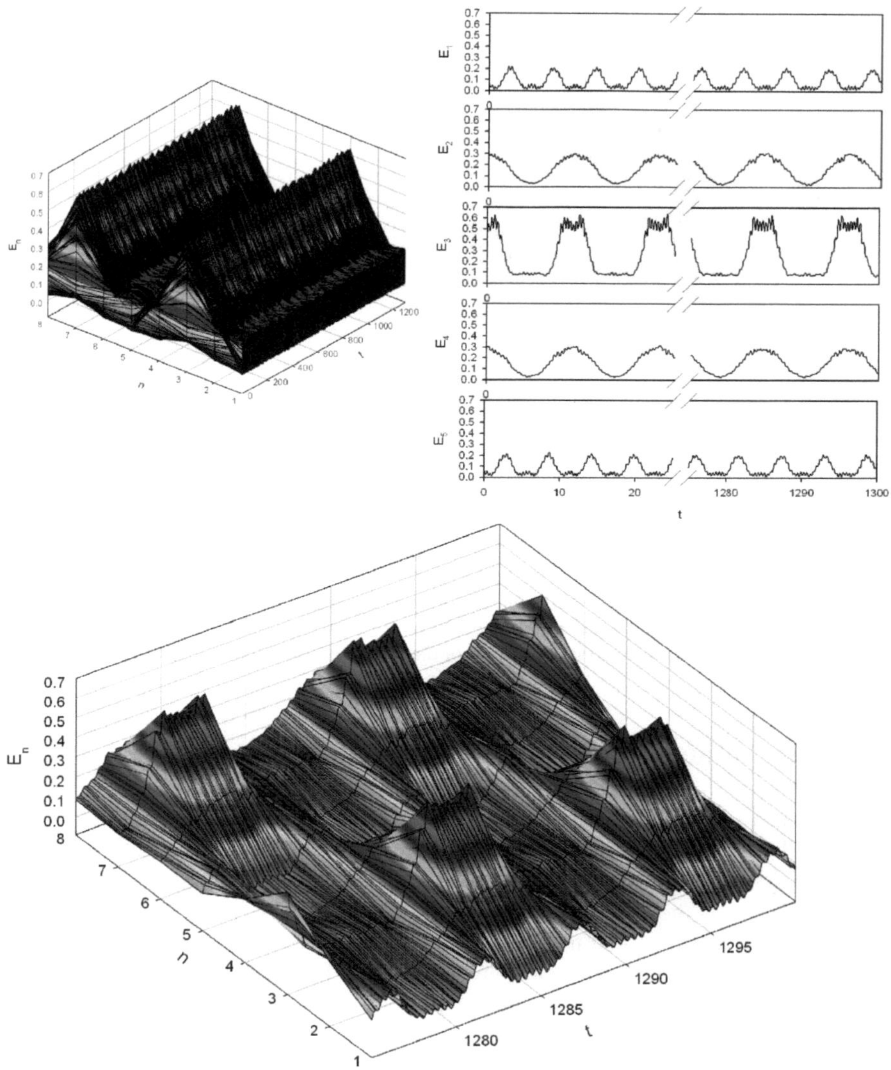

Fig. 2.89 The same as in Fig. 2.88 with the occupation number n is equal 0.5

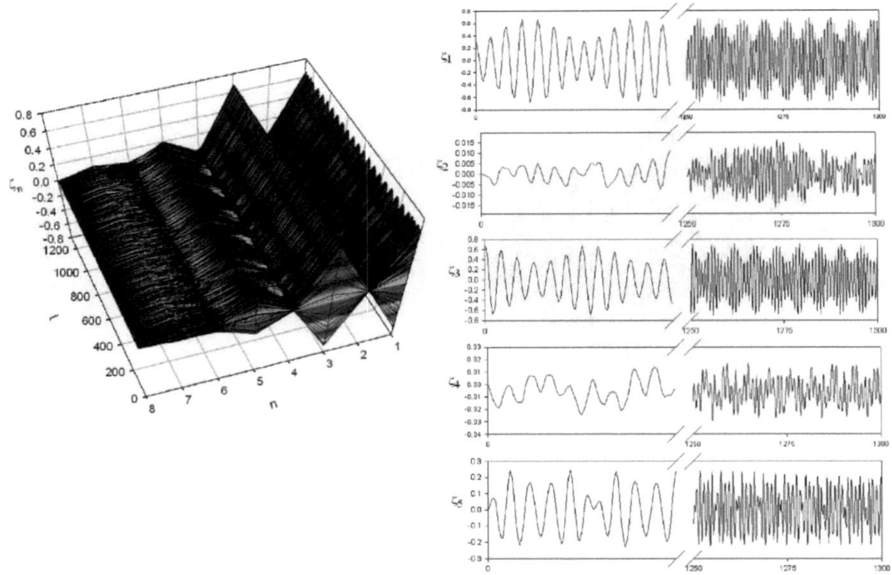

Fig. 2.90 3D-surface of normal modes in the case corresponding to Fig. 2.89 (the *left half* of the picture). In spite of the initial condition consists only two modes, some other ones are excited during the modeling time. Note that the scales of diagrams are different in the *right pictures*

2.9 Concluding Remarks

If one tries to summarize how to arrive at the tractable models for discrete system with moderate number of degrees of freedom, one should first of all set the exactly solvable systems aside. Some examples of such systems, linear and nonlinear, are presented above, the others are beyond the scope of this book. Exactly solvable models might qualify to be tractable. However, even simple examples of common single or coupled linear oscillators demonstrate that by formulating relevant asymptotic approximation one can gain an essentially deeper understanding of the system behavior in different limit cases, even if more general model is exactly solvable. Moreover, these ideas of approximation often can be efficiently used for more complicated systems, where the exact solutions are not available. If some asymptotic limits are similar, one can speak of "generic behavior" for wide groups of dynamical systems in a certain regime – no matter whether the systems belonging to these groups are exactly solvable or not.

Many problems in this chapter were solved by means of different variations of perturbation and averaging techniques. We do not analyze further mathematical details of these methods here – the interested reader may refer to standard textbooks on these topics (Arnold, 1978; Arnold et al., 2006; Nayfeh, 2000). As it was demonstrated above, the main difficulty in obtaining the tractable model is not the application of particular mathematical technique of the perturbation analysis, but an efficient choice of basic paradigms (for instance, the LPTs alongside with the NNMs) and of appropriate small parameters. The same problem can have different small parameters, and very different tractable models are obtained this way.

The other important hint for obtaining a tractable model is a reduction of dimension. For instance, in Sect. 2.8, the general number of normal modes in the system may be relatively large, far beyond the possibility of an exact solution. Still, in some interesting dynamical regimes only few of these modes really participate and interact. So, in terms of these participating modes, one can obtain the tractable system with a low number of degrees of freedom.

Sometimes no analysis is available beyond the computing of some particular solutions, for instance the nonlinear normal modes. In the absence of superposition, such particular solutions do not allow a complete description of the system dynamics. Still, they allow insight into particular regimes of behavior and sometimes even turn out to be attractors of the dynamic flow, as in Sect. 2.7. In addition to that, the approaches initially developed for computation of these periodic orbits, such as saw – tooth time transforms, are useful for approximate solutions of general initial value problems in rather complicated nonlinear systems (Sects. 2.3, 2.5).

If only some particular solutions are available, it is not clear whether one can qualify such a model as tractable. We think that the answer is positive, at least partially. If some solutions can be computed exactly or approximately, then one has some independent information to interpret and verify the data of a full-scale computer simulation.

References

Akhmeriev, N.N., Ankiewicz, A.: Solitons: Nonlinear Pulses and Beams. Chapman and Hall, London (1992)

Andrianov, I.V.: Asymptotic solutions for nonlinear systems with high degree of nonlinearity. PMM J. App. Math. Mech. **57**, 941–943 (1993)

Arnold, V.I.: Mathematical Methods of Classical Mechanics. Springer, New York, NY (1978)

Arnold, V.I., Afrajmovich, V.S., Il'yashenko, Yu.S., Shil'nikov, L.P.: Dynamical Systems V. Encyclopedia of Mathematical Sciences. Springer, Berlin (1994)

Arnold, V.I., Kozlov, V.V., Neishtadt, A.I., Khukhro, E.: Mathematical Aspects of Classical and Chelestial Mechanics. Springer, Berlin (2006)

Atay, F.M.: Distributed delays facilitate amplitude death of coupled oscillators. Phys. Rev. Lett. **91**, 094101 (2003)

Azeez, M.A.F., Vakakis, A.F., Manevich, L.I.: Exact solutions of the problem of the vibro-impact oscillations of a discrete system with two degrees of freedom. J. App. Math. Mech. **63**, 527–530 (1999)

Babitsky, V.I., Veprik, A.M.: Universal bumpered vibration isolator for severe environment. J. Sound Vib. **218**, 269–292 (1998)

Berinde, V.: Iterative Approximation of Fixed Points. Springer, New York, NY (2007)

Binder, P., Abraimov, D., Ustinov, A.V., Flach, S., Zolotaryuk, Y.: Observation of breathers in Josephson ladders. Phys. Rev. Lett. **84**, 745–748 (2000)

Blekhman, I.I.: Vibrational Mechanics: Nonlinear Dynamic Effects, General Approach, Applications. World Scientific, Singapore (2000)

Bluman, G.W., Kumei, S.: Symmetries and Differential Equations. Springer, New York, NY (1989)

Burns, T.J., Jones, C.K.R.T.: Mechanism for capture into resonance. Physica D. **69**, 85–106 (2003)
Courant, R., Hilbert, D.: Methods of Mathematical Physics, vol. 2, pp. 136–147. Wiley, New York, NY (1962)
Den Hartog, J.P.: Mechanical Vibrations. McGraw-Hill, New York, NY (1956)
Eiermann, B., Anker, T., Albiez, M., Taglieber, M., Marzlin, K.P., Oberthaler, M.K.: Bright Bose - Einstein gap solitons of atoms with repulsive interaction. Phys. Rev. Lett., **92**, 230401 (2004)
Eilbeck, J.C., Lomdahl, P.S., Scott, A.C.: The discrete self – trapping equation. Physica D. **16**, 318–338 (1985)
Eisenberg, H.S., Silberberg, Y., Morandotti, R., Boyd, A.R., Aitchison, J.S.: Discrete spatial optical solitons in waveguide arrays. Phys. Rev. Lett. **81**, 3383 (1998)
Feng, B.-F.: An integrable three particle system related to intrinsic localized modes in Fermi-Pasta-Ulam-beta chain. J. Phys. Soc. Jpn. **75**, 014401 (2006)
Fermi, E., Pasta, J., Ulam, S.: Los Alamos Science Laboratory Report No. LA-1940 unpublished; (Reprinted in Collected Papers of Enrico Fermi, E. Segre (ed.). University of Chicago Press, Chicago, 1965), vol. 2, p. 978 (1955)
Flach, S., Ivanenchenko, M.V., Kanakov, O.V.: q-breathers in Fermi-Pasta-Ulam chains: existence, localization, and stability. Phys. Rev. E. **73**, 036618 (2006)
Gendelman, O.: Transition of energy to a nonlinear localized mode in a highly asymmetric system of two oscillators. Nonlinear Dyn. **25**, 237–253 (2001)
Gendelman, O.V.: Bifurcations of nonlinear normal modes of linear oscillator with strongly nonlinear damped attachment. Nonlinear Dyn. **37**, 115–128 (2004)
Gendelman, O.V.: Modeling of inelastic impacts with the help of smooth functions. Chaos Solitons Fractals. **28**, 522–526 (2006)
Gendelman, O.V., Gorlov, D.V., Manevitch, L.I., Musienko, A.I.: Dynamics of coupled linear and essentially nonlinear oscillators with substantially different masses. J. Sound Vib. **286**, 1–19 (2005)
Gendelman, O., Manevitch, L.I., Vakakis, A.F., M'Closkey, R.: Energy pumping in nonlinear mechanical oscillators I: dynamics of the underlying hamiltonian systems. ASME J. App. Mech. **68**, 34–41 (2001)
Gendelman, O.V., Meimukhin, D.: Response regimes of integrable damped strongly nonlinear oscillator under impact periodic forcing. Chaos Solitons Fractals. **32**(2), 405–414 (2007)
Gendeman, O.V., Starosvetsky, Y., Feldman, M.: Attractors of harmonically forced linear oscillator with attached nonlinear energy sink I. Description of response regimes. Nonlinear Dyn. **51**, 31–46 (2008)
Goldreich, P., Peale, S.: Spin-orbit coupling in the solar system. Astron. J. **71**, 425–438 (1966)
Gourdon, E., Lamarque, C.H.: Energy pumping with various nonlinear structures: numerical evidences. Nonlinear Dyn. **40**, 281–307 (2005)
Grosberg, Y.A., Khokhlov, A.R.: Statistical Physics of Macromolecules. Nauka, Moscow (1989) [in Russian]
Guckenheimer, J., Hoffman, K., Weckesser, W.: Bifurcations of relaxation oscillations near folded saddles. Int. J. Bifurcat. Chaos. **15**, 3411–3421 (2005)
Guckenheimer, J., Holmes, P.: Nonlinear Oscillations, Dynamical Systems, and Bifurcations of Vector Fields. Springer, New York, NY (2002)
Guckenheimer, J., Wechselberger, M.: Lai-Sang Young: chaotic attractors of relaxation oscillators. Nonlinearity. **19**, 701–720 (2006)
Jackson, E.A.: Perspectives of Nonlinear Dynamics, vol. 1. Cambridge University Press, Cambridge (1991)
Kauderer, H.: Nichtlineare Mechanik. Springer, Berlin (1958)
Kevorkian, J., Cole, J.D.: Multiple Scale and Singular Perturbation Methods. Springer, Berlin/NewYork (1996)
Khasnutdinova, K.R., Pelinovsky, D.E.: On the exchange of energy in coupled Klein-Gordon equations. Wave Motion. **38**, 1–10 (2003)

Korn, G.A., Korn, T.M.: Mathematical Handbook for Scientists and Engineers, 2nd edn. Dover Publications, New York, NY (2000)

Kosevitch, A.M., Kovalyov, A.S.: Introduction to Nonlinear Dynamics. Naukova Dumka, Kiev (1989) [in Russian]

Landa, P.S.: Nonlinear Oscillations and Waves in Dynamical Systems. Springer, Berlin (1996)

Landau, L.D., Lifshits, E.M.: Mechanics. Butterworth-Heinemann, Boston, MA (1976)

Lin, W.A., Reichl, L.E.: External field induced chaos in an infinite square well potential. Physica D. **19**, 145–152 (1986)

Lyapunov, A.: The General Problem of the Stability of Motion. Princeton University Press, Princeton, NJ (1947)

Machida, M., Koyama, T.: Localized rotating-modes in capacitively coupled intrinsic Josephson junctions: systematic study of branching structure and collective dynamical instability. Phys. Rev. B. **70**, 024523 (2004)

Manevich, A.I., Manevitch, L.I.: The Mechanics of Nonlinear Systems with Internal Resonances. Imperial College Press, London (2005)

Manevitch, L.I.: Complex representation of dynamics of coupled nonlinear oscillators. In: Uvarova, L., Arinstein, A., Latyshev, A. (eds.) Mathematical Models of Non-Linear Excitations, Transfer Dynamics and Control in Condensed Systems and Other Media, pp. 269–300. Kluwer Academic Publishers, Dordrecht (1999)

Manevitch, L.I.: The description of localized normal modes in a chain of nonlinear coupled oscillators using complex variables. Nonlinear Dyn. **25**, 95–109 (2001)

Manevitch, L.I.: New approach to beating phenomenon in coupled nonlinear oscillatory chains. In: Avrejcewicz, J. (ed.) Dynamical Systems:Theory and Applications, vol. 1, pp. 119–137. Lodz, Poland (2005)

Manevitch, L.I.: New approach to beating phenomenon in coupled nonlinear oscillatory chains. Arch. Appl. Mech. **77**, 301–312 (2007)

Manevitch, L.I., Azeez, M.A.F., Vakakis, A.F.: Exact solutions for a discrete systems undergoing free vibro – impact oscillations. In: Babitsky, V.I. (ed.) Dynamics of Vibro – Impact Systems, Proceedings of the Euromech Colloquium 15–18 September 1998. Springer, New York, NY (1998)

Manevitch, L.I., Gourdon, E., Lamarque, C.-H.: Towards the design of an optimal energetic sink in a strongly inhomogeneous two-degree-of-freedom system. ASME J. App. Mech. **74**, 1078–1086 (2007)

Manevitch, L.I., Mikhlin, Yu.V., Pilipchuk, V.N.: The Method of Normal Vibrations for Essentially Nonlinear Systems. Nauka, Moscow (1989) [in Russian]

Meirovitch, L.: Principles and Techniques of Vibrations. Prentice Hall, Upper Saddle River, NJ (2000)

Melnikov, V.K.: On the stability of the center for time periodic perturbations. Trans. Moscow Math. Soc. **12**, 1–57 (1963)

Nayfeh, A.H.: Perturbation Methods. Wiley, New York, NY (2000)

Nayfeh, A.H., Mook, D.T.: Nonlinear Oscillations. Wiley, New York, NY (1995)

Neishtadt, A.I.: Passage through a separatrix in a resonance problem with slowly varying parameter. J. Appl. Math. Mech. **39**, 594–605 (1975)

Pikovsky, A., Rosenblum, M., Kurtz, J.: Synchronization: Universal Concept in Nonlinear Sciences. Cambridge University Press, Cambridge (2003)

Pilipchuk, V.N.: A transformation of vibrating systems based on a non-smooth periodic pair of functions. Dokl. Akad. Nauk UkrSSR (Ukrainian Acad. Sci. Rep.). **A**(4), 37–40 (1988) [in Russian]

Pilipchuk, V.N.: Application of special non-smooth temporal transformations to linear and non-linear systems under discontinuous and impulsive excitation. Nonlinear Dyn. **18**, 203–234 (1999a)

Pilipchuk, V.N.: Strongly nonlinear vibrations of damped oscillators with two non-smooth limits. J. Sound Vib. **302**, 398–402 (1999b)

Pilipchuk, V.N.: Impact modes in discrete vibrating systems with rigid barriers. Int. J. Non Linear Mech. **36**, 999–1012 (2001)

Quinn, D.D., Rand, R.H., Bridge, J.: The dynamics of resonance capture. Nonlinear Dyn. **8**, 1–20 (1995)

Rand, R.H.: The dynamics of resonance capture. In: Guran, A. (ed.) Proceedings of the First International Symposium on Impact and Friction of Solids, Structures and Intelligent Machines, pp. 91–94. World Scientific, Ottawa, ON (1998)

Rand, R.H.: Lecture Notes on Nonlinear Vibrations. The Internet-First University Press, Cornell University. http://ecommons.library.cornell.edu/handle/1813/62 (2009). Accessed 7 Aug 2009

Rand, R.H., Quinn, D.D.: Resonant capture in a system of two coupled homoclinic oscillators. J. Vibr. Control. **1**, 41–56 (1995)

Rosenberg, R.M.: Normal modes in nonlinear dual-mode systems. J. App. Mech. **27**, 263–268 (1960)

Rosenberg, R.M.: The normal modes of nonlinear n-degree-of-freedom systems. J. App. Mech. **29**, 7–14 (1962)

Rosenberg, R.M.: On nonlinear vibrations of systems with many degrees of freedom. Adv. Appl. Mech. **9**, 155–242 (1966)

Saaty, T.L.: Modern Nonlinear Equations. Dover Publications Inc., New York, NY (1981)

Salenger, G., Vakakis, A.F., Gendelman, O.V., Andrianov, I.V., Manevitch, L.I.: Transitions from strongly- to weekly-nonlinear motions of damped nonlinear oscillators. Nonlinear Dyn. **20**, 99–114 (1999)

Sato, M., Hubbard, B.E., Sievers, A.J., Ilic, B., Craighead, H.G.: Optical manipulation of intrinsic localized vibrational energy in cantilever arrays. Europhys. Lett. **66**, 318 (2004)

Sato, M., Hubbard, B.E., Sievers, A.J., Ilic, B., Czaplewski, D.A., Craighead, H.G.: Observation of locked intrinsic localized vibrational modes in a micromechanical oscillator array. Phys. Rev. Lett. **90**, 044102 (2003)

Schwarz, U.T., English, L.Q., Sievers, A.J.: Experimental generation and observation of intrinsic localized spin wave modes in an antiferromagnet. Phys. Rev. Lett. **83**, 223 (1999)

Scott, A.S., Lomdahl, P.S., Eilbeck, J.C.: Between the local-mode and normal-mode limits. Chem. Phys. Lett. **113**, 29–36 (1985)

Sen, A.K., Rand, R.H.: A numerical investigation of the dynamics of a system of two time-delay coupled relaxation oscillators. Commun. Pure App. Anal. **2**, 567–577 (2003)

Shaw, S.W., Pierre, C.: Nonlinear normal modes and invariant manifolds. J. Sound Vib. **150**, 170–173 (1991)

Shaw, S.W., Pierre, C.: Normal modes for nonlinear vibratory systems. J. Sound Vib. **164**, 85–124 (1993)

Sievers, A.J., Takeno, S.: Intrinsic localized modes in anharmonic crystals. Phys. Rev. Lett. **61**, 970–973 (1988)

Sokolov, I.J., Babitsky, V.I., Halliwell, N.A.: Hand-held percussion machines with low emission of hazardous vibration. J. Sound Vib. **306**, 59–73 (2007)

Starosvetsky, Y., Gendelman, O.V.: Attractors of harmonically forced linear oscillator with attached nonlinear energy sink II: Optimization of a nonlinear vibration absorber. Nonlinear Dyn. **51**, 47–57 (2008)

Swanson, B.I., Brozik, J.A., Love, S.P., Strouse, G.F., Shreve, A.P., Bishop, A.R., Wang, W.Z., Salkola, M.I.: Observation of intrinsically localized modes in a discrete low-dimensional material. Phys. Rev. Lett. **82**, 3288 (1999)

Uzunov, I.M., Muschall, R., Gölles, M., Kivshar, Y.S., Malomed, B.A., Lederer, F.: Pulse switching in nonlinear fiber directional couplers. Phys. Rev. E. **51**, 2527–2537 (1995)

Vakakis, A.F., Gendelman, O.: Energy pumping in nonlinear mechanical oscillators II: resonance capture. ASME J. App. Mech. **68**, 42–48 (2008)

Vakakis, A.F., Gendelman, O.V., Bergman, L.A., McFarland, D.M., Kerchen, G., Lee, Y.-S.: Nonlinear Targeted Energy Transfer in Mechanical and Structural Systems. Springer, Berlin (2008)

Vakakis, A.F., Manevitch, L.I., Mikhlin, Yu.V., Pilipchuk, V.N., Zevin, A.A.: Normal Modes and Localization in Nonlinear Systems. Wiley, New York, NY (1996)

Wirkus, S., Rand, R.H.: The Dynamics of two coupled van der pol oscillators with delay coupling. Nonlinear Dyn. **30**, 205–221 (2002)

Zhuravlev, V.F., Klimov, D.M.: Applied Methods in Vibration Theory. Nauka, Moscow (1988)

Chapter 3
Infinite Discrete Systems

Discrete systems comprising an infinite number of particles are rather challenging for a formulation of tractable models. The simplifications related to a low number of the degrees of freedom or normal modes cannot normally be implemented directly. Transition to a continuum and to a description in terms of partial differential equations may be involved and normally requires a number of assumptions; each of them should be carefully checked. In particular, for different regimes of motion of the discrete system one can obtain rather different continuous approximations. On the other side, such systems are extremely important both in mechanics (for modeling arrays of structural elements etc.) and in physics. For the study of the relationship between macroscopic properties and the microscopic structure in classical physics, the models of this sort are the best possible.

3.1 Dynamics of Infinite Nonlinear Chains

The lowest possible level of complexity corresponds to an infinite single nonlinear chain of particles. As it was mentioned above, this step is necessary for the transition from microscopic or discrete objects (such as individual molecules or lumped – mass models of vibrating systems) to macroscopic objects characterized by a thermodynamically large number of degrees of freedom. The main challenge here is to relate the common macroscopic continuum description of the bulk system to its discrete structure. This section deals with this problem and demonstrates how the discrete structure of the model reveals itself in different continuum limits.

3.1.1 Long-Wavelength Approximation. Equation of Supersonic Extension Solitons in an Infinite FPU Chain

Let us begin from the simplest approach to long wavelength dynamics of an infinite FPU chain with an asymmetric potential of a gradient type (Manevitch and Smirnov, 2008)

L.I. Manevitch, O.V. Gendelman, *Tractable Models of Solid Mechanics*, Foundations of Engineering Mechanics, DOI 10.1007/978-3-642-15372-3_3,

$$V(U_j) = \sum_{j=-\infty}^{\infty} \left[\frac{c_1}{2}(U_{j+1} - U_j)^2 + \frac{c_2}{3}(U_{j+1} - U_j)^3 + \frac{c_3}{4}(U_{j+1} - U_j)^4 \right] \quad (3.1)$$

Here, U_j are the displacements of the particles ($-\infty < j < +\infty$). Such simplified consideration is justified because the real structure of chain is not important when dealing with long-wavelength approximation. The values of parameters determining the solution of the problem can be found in more realistic models that take the structure of a particular chain into account.

When transformed to dimensionless variables and parameters, the system of classical equations of motion takes the form

$$\begin{aligned} &\frac{d^2u_j}{dt^2} + (2u_j - u_{j+1} - u_{j-1})\{1 + \varepsilon\alpha_1(u_{j+1} - u_{j-1}) + \\ &+ \varepsilon^2\alpha_2[(u_{j+1} - u_{j-1})^2 + (u_j - u_{j+1})(u_j - u_{j-1})]\} = 0, \end{aligned} \quad (3.2)$$

where $\tau = \sqrt{\frac{c_1}{m}}t$, $u_j = \frac{U_j}{\varepsilon r_0}$, $\alpha_1 = \frac{c_2 r_0}{c_1}$, $\alpha_2 = \frac{c_2 r_0^2}{c_1}$, m – mass of the particle, c_i are the first coefficients of power expansion of potential (3.1), and r_0 is the distance between the particles. Small parameter ϵ reflects relative smallness of displacements with respect to interparticle distance.

Measuring the distance between the atoms in units of $\varepsilon^{-1} r_0$ and introducing a continuous space coordinate ζ, we can reduce the system above in the long-wavelength approximation to a Korteweg–de Vries equation. Expanding the differences in the equation of motion in a Taylor series:

$$u_{j\pm1} = u_j \pm \varepsilon\frac{\partial u}{\partial\zeta} - \frac{1}{2}\varepsilon^2\frac{\partial^2 u}{\partial\zeta^2} \pm \frac{1}{6}\varepsilon^3\frac{\partial^3 u}{\partial\zeta^3} + \frac{1}{24}\varepsilon^4\frac{\partial^4 u}{\partial\zeta^4} + \ldots$$

we obtain one nonlinear partial differential equation:

$$\frac{\partial^2 u}{\partial\tau^2} - \varepsilon^2\frac{\partial^2 u}{\partial\zeta^2}\left[1 + \varepsilon^2\alpha_1\frac{\partial u}{\partial\zeta} + \varepsilon^4\alpha_2\left(\frac{\partial u}{\partial\zeta}\right)^2\right] - \frac{\varepsilon^4}{12}\frac{\partial^4 u}{\partial\zeta^4} + \ldots = 0,$$

where $u = u(\zeta,\tau)$ and "..." denotes higher order terms in the parameter ε.

It is natural to take into account the smallness of ε by changing the space and time variables: $\xi = \zeta - \varepsilon\tau$ and $\tau_1 = \varepsilon^3\tau$.

Substituting these expressions into the equation of motion, we obtain

$$\frac{\partial^2 u}{\partial\tau_1\partial\xi} + \frac{1}{2}\alpha_1\frac{\partial^2 u}{\partial\xi^2}\frac{\partial u}{\partial\xi} + \frac{\partial^4 u}{\partial\xi^4} + \ldots = 0$$

The new space coordinate ξ is measured from the front of the linear (sound) wave, whereas the new time variable is slow compared to τ.

Introducing the notation $\frac{\partial u}{\partial\xi} = w$ and keeping only the principal approximation, we obtain the Korteweg–deVries equation in the following form:

$$\frac{\partial w}{\partial \tau_1} + \frac{1}{2}\alpha_1 w \frac{\partial w}{\partial \xi} + \frac{1}{24}\frac{\partial^3 w}{\partial \xi^3} = 0 \tag{3.3}$$

In this equation, all terms have the same order of smallness with respect to ε. This equation describes the nonlinear dynamics of a one-dimensional crystal in long-wave approximation in the case of the asymmetric anharmonism. Along with periodic solutions, the Korteweg–deVries equation has localized soliton solutions (solitons and multisoliton waves). In particular, the soliton is a localized wave of compression or extension (depending on the sign of α_1), described by the equation

$$w = A/\cosh^2 \frac{\xi - v\tau}{D}. \tag{3.4}$$

The amplitude A and the soliton localization region size D are related to the velocity or, more specifically, to the difference between the soliton velocity v and the sound velocity v_0 as $A \sim (v^2 - v_0^2)$ and $D \sim (v^2 - v_0^2)^{-1/2}$.

3.1.2 Zigzag Chain and Long-Wave Solitons

Let us consider now the more realistic model of a zigzag chain, taking into account its spatial structure (Manevitch and Savin, 1997, 2005; Manevitch and Smirnov, 1992, 2008). For this model system, every particle in the corner of a zigzag is considered as a "united atom" consisting of carbon and two hydrogen atoms interacting due to the presence of the angular potential (if only the planar dynamics are considered and the valence bonds are supposed to be absolutely rigid) Of course, edges of the trans-zigzag correspond to valence bonds. Thus, the Hamiltonian of a polymer chain in a *trans*-zigzag conformation (if only the planar dynamics are considered and the valence bonds are supposed to be absolutely rigid) without consideration for the interchain interaction reads

$$H = \sum \left[\frac{1}{2} M \left(\frac{\partial u_n}{\partial t} \right)^2 + \frac{1}{2} M \left(\frac{\partial v_n}{\partial t} \right)^2 + \frac{1}{2} K_2 (\theta_n - \theta_0)^2 \right], \tag{3.5}$$

where u_n and v_n are displacements of the CH_2 group in the direction parallel to the chain axis and in a perpendicular direction, θ_n are angles between neighbor C–C bonds, and θ_0 is their equilibrium value. The condition of rigidity of the valence bonds in the continuum approximation takes the form

$$\Delta\rho = \rho_0 \sin^2 \frac{\theta_0}{2} u_x - 2\cos\frac{\theta_0}{2} v +$$
$$+ \frac{1}{2\rho_0} \left(\rho_0 \cos\frac{\theta_0}{2} \sin\frac{\theta_0}{2} u_x + 2 \sin\frac{\theta_0}{2} v \right)^2 + \ldots = 0$$

Here, ρ_0 is the length of the non-deformed bond.

This condition suggests that the transverse displacement can be expressed through the longitudinal displacement as

$$v = \frac{1}{2}\frac{\sin^2 \theta_0/2}{\cos \theta_0/2}\rho_0 u_x - \frac{1}{4}\frac{\sin^2 \theta_0/2}{\cos^3 \theta_0/2}\rho_0 u_x^2.$$

The change in the valence angle in the principal anharmonic approximation can be presented as a nonlinear function of the derivatives of the transverse displacement:

$$\Delta\theta = a_1 u_x + a_2 u_x^2 + a_3 u_{xx} + \ldots$$

Substituting this expression into the Hamiltonian, we obtain the following equation of motion:

$$\frac{\partial^2 u}{\partial t^2} - c_0 \frac{\partial^2 u}{\partial x^2} - p_1 \frac{\partial u}{\partial x}\frac{\partial^2 u}{\partial x^2} - p_2 \frac{\partial^4 u}{\partial x^4} - p_3 \frac{\partial^4 u}{\partial x^2 \partial t^2} = 0, \tag{3.6}$$

where

$$p_1 = 6a_1 a_2 K_2/M = 12\sin^4 \frac{\theta_0}{2} K_2/M \cos^2 \frac{\theta_0}{2}$$
$$p_2 = 2a_1 a_2 K_2/M = 2\sin^4 \frac{\theta_0}{2}\left(2 + \sin^2 \frac{\theta_0}{2}\right) K_2 \rho_0^2/6M\cos^2 \frac{\theta_0}{2}$$
$$p_3 = \rho_0^2 \sin^4 \frac{\theta_0}{2}/4\cos^2 \frac{\theta_0}{2}$$

Here, $u(x, t)$ is the longitudinal displacement of the chain, and c_0 is the sound velocity in the harmonic approximation.

As in the case of the simplest polymer chain, the obtained nonlinear differential equation in partial derivatives can be asymptotically reduced to the Korteweg–de Vries equation.

After introduction of the new variable $\xi = x - ct$ (where c is the velocity of propagation of a stationary wave), the analysis of stationary excitations leads to an ordinary nonlinear differential equation for the function $w(\xi) = u_\xi(\xi)$

$$(p_2^2 + p_3 c^3) w_\xi^2 + \frac{1}{3} p_1 w^3 + (c_0^2 - c^2) w^2 = 0 \tag{3.7}$$

(the subscript at the variable denotes differentiation with respect to ξ). This equation has a soliton-type spatially localized solution

$$w = A/\cosh^2(\xi/L) \tag{3.8}$$

or

$$u(x,t) = AL\tanh\left(\frac{x - ct}{L} + x_1\right) \tag{3.9}$$

where $A = 3(c^2 - c_0^2)/p_1$ is the amplitude of the soliton, $L = 2\sqrt{(p_2 + p_3 c^2)/(c^2 - c_0^2)}$ is the soliton width, and $c > c_0$ is the velocity of the soliton.

This solution corresponds to the soliton of the Korteweg–de Vries equation and describes the supersonic propagation of an extension or compression wave in the long-wave approximation. The type of the wave is determined by the sign of the amplitude A, which, in turn, is determined by the sign of the coefficient p_1 of the nonlinear term in Eqs. (3.6) and (3.7). In the case of infinitely rigid bonds, the sign of the coefficient p_1 is positive and, therefore, only extension solitons propagate over the chain ($A > 0$), the propagation velocity being limited only by the conditions of validity of the long-wave (continuum) approximation. Indeed, as follows from solution (3.9), the amplitude of the soliton is equal to zero at $c = c_0$, while the soliton width tends to infinity at $c \to c_0$ and decreases monotonically at $c > c_0$ with increasing c, which means that, at a certain value of c, the continuum approximation becomes invalid.

A more precise analysis of the nonlinear dynamics of a polymer chain performed on the basis of a numerical variational method (Savin et al., 1999) makes it possible to take the finite stiffness of the valence bonds into account and to use an improved (periodic) form of the valence-angle potential. In this case, the periodic character of the potential leads to restriction of the region of existence of solitons even in the long-wave approximation. The type of the soliton depends on the parameter characterizing the ratio of the valence angle stiffness to the valence bond stiffness $k_2/(k_1\rho_0^2)$. This ratio was estimated to be equal to 0.019 or 0.078 (Savin et al., 1999; Manevitch, 2001). It reflects the relative contributions of the physical anharmonicity of the valence-bond potential and the geometric anharmonicity associated with the nonlinear dependence of the valence angle on the longitudinal displacement. At the ratio of 0.019, geometric anharmonicity dominates ($A > 0$), and, therefore, the elementary excitations in a polymer chain are extension solitons. At the ratio of 0.078, the physical anharmonicity dominates ($A < 0$), indicating that compression solitons can propagate over the chain (Savin et al., 1999) (Fig. 3.1).

If geometric anharmonicity dominates, in addition to the above-mentioned extension soliton with a finite velocity spectrum, supersonic extension solitons with unique characteristic widths and propagation velocities exist. However, such solitons have considerably higher energies and amplitudes corresponding to the transition to a new ground state. Figure 3.1 presents typical profiles of a small-amplitude extension soliton at the initial moment and upon traveling over distance equivalent to 10^5 chain monomers, which practically coincide. Figure 3.2 presents the regions of existence of extension and compression solitons, whereas Fig. 3.3 illustrates an elastic collision of these two solitons.

3.1.3 Envelope Solitons

In the short-wavelength limit in the framework of the simplest approximation of a one-dimensional chain, the situation differs markedly from that typical of the long-wavelength approximation (Manevitch, 2001). The system of equations of motion

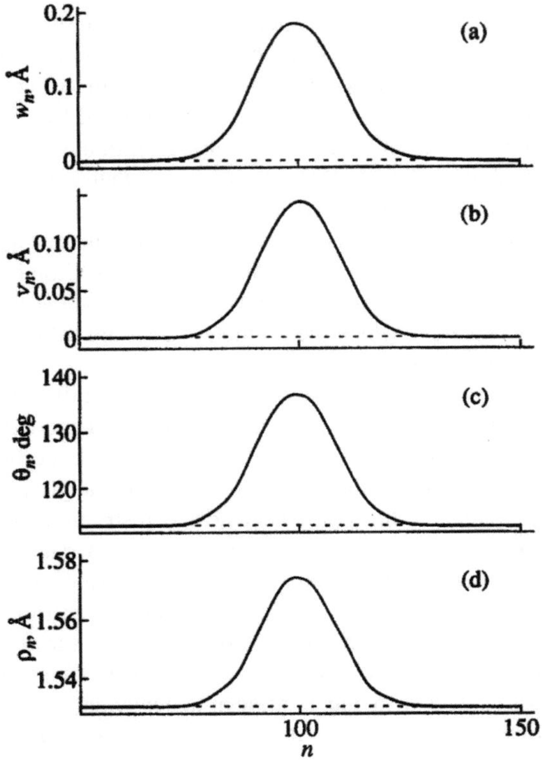

Fig. 3.1 Profiles of supersonic extension solitons with respect to the components (**a**) $w_n = u_{n+1} - u_n$, (**b**) v_n, (**c**) θ_n, and (**d**) ρ_n at the initial moment of time ($t = 0$) and at $t = 160.682$ ps after it begins to travel over a chain composed of 105 united atoms. The dimensionless velocity is $s = 0.94$ (c=7,940.21 m/s)

(3.2) has an exact solution in the form of a nonlinear standing wave of minimum length:

$$u_j = w_1(\tau), \quad u_{j+1} = -w_2(\tau),$$
$$u_{j+2} = w_1(\tau), \quad u_{j+3} = -w_2(\tau), \ldots .$$
$$(-\infty < j < +\infty)$$

To analyze these short-wavelength modes, we switch to new variables:

$$u_j = w_{j,1}(\tau), \quad u_{j+1} = -w_{j+1,2}(\tau),$$
$$u_{j+2} = w_{j+2,1}(\tau), \quad u_{j+3} = -w_{j+3,2}(\tau), \ldots$$

Then, the equations of motion take the form

$$\begin{cases} \dfrac{d^2 w_{j,1}}{d\tau^2} + (2w_{j,1} + w_{j-1,2} + w_{j,2})\{1 + \varepsilon\alpha_1(-w_{j,2} + w_{j-1,2}) + \\ +\varepsilon^2\alpha_2[(w_{j,2} - w_{j-1,2})^2 + (w_{j,1} + w_{j,2})(w_{j,1} + w_{j-1,2})]\} = 0 \\ -\dfrac{d^2 w_{j,2}}{d\tau^2} - (2w_{j,2} + w_{j,1} + w_{j+1,1})\{1 + \varepsilon\alpha_1(-w_{j+1,1} + w_{j,1}) + \\ +\varepsilon^2\alpha_2[(w_{j+1,1} - w_{j,1})^2 + (w_{j,2} + w_{j+1,1})(w_{j,2} + w_{j,1})]\} = 0 \end{cases} \tag{3.10}$$

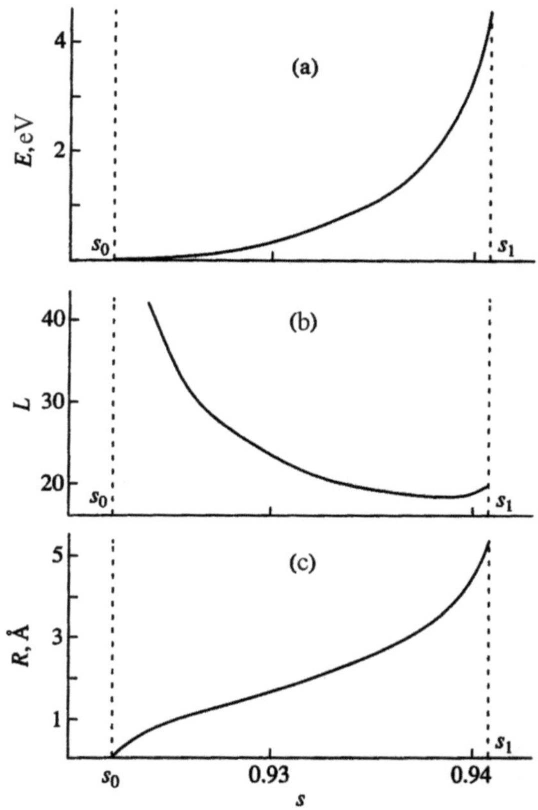

Fig. 3.2 (**a**) Energy E, (**b**) width L, and (**c**) amplitude R of the soliton as functions of the dimensionless velocity s

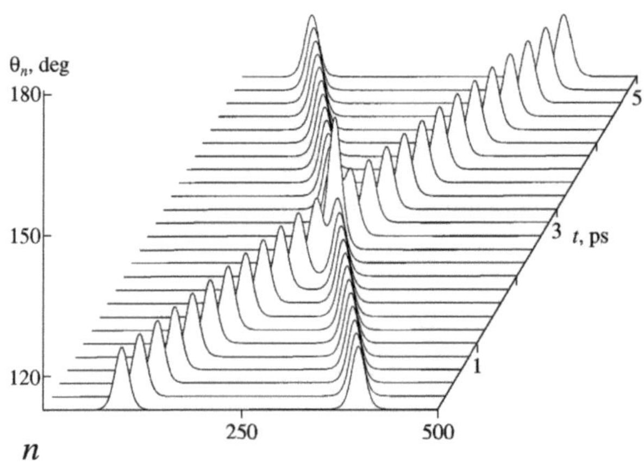

Fig. 3.3 Elastic collision of two extension solitons at a dimensionless velocity of $s = 0.935$

Let us introduce two functions of two variables that describe the dynamics of a chain of atoms within the continuum limit under the assumption that the *modulations* of the nonlinear mode of minimum length are characterized by a size substantially higher than the interatomic distance, which equals ε in the accepted units of measure. Thus,

$$w_{j\pm 1,1} = w_1(\zeta,\tau) \pm \varepsilon \frac{\partial w_1}{\partial \zeta} + \frac{1}{2}\varepsilon^2 \frac{\partial^2 w_1}{\partial \zeta^2} + ...,$$

where $j\varepsilon = \zeta$.

Based on this expression, we obtain the following continuum equations of motion:

$$\begin{cases} \dfrac{\partial^2 w_1}{\partial \tau^2} + 2(w_1 + w_2)\left\{1 - \varepsilon^2 \left[2\alpha_1 \dfrac{\partial w_2}{\partial \zeta} - \alpha_2 (w_1 + w_2)^2\right]\right\} + \varepsilon^2 \dfrac{\partial w_2}{\partial \zeta^2} + ... = 0 \\ \dfrac{\partial^2 w_2}{\partial \tau^2} + 2(w_1 + w_2)\left\{1 - \varepsilon^2 \left[2\alpha_1 \dfrac{\partial w_1}{\partial \zeta} + \alpha_2 (w_1 + w_2)^2\right]\right\} - \varepsilon^2 \dfrac{\partial w_2}{\partial \zeta^2} + ... = 0 \end{cases}$$

where "· · · " denotes higher order terms in the parameter ε.

Combining these equations, we obtain the relatively simple system

$$\begin{aligned} &\frac{\partial^2 \tilde{W}_1}{\partial \tau^2} + 4\tilde{W}_1 + 4\alpha_1 \varepsilon_1 \tilde{W}_1 \frac{\partial \tilde{W}_2}{\partial \zeta} + 4\alpha_2 \varepsilon_1 \tilde{W}_1^3 + \varepsilon_1 \frac{\partial^2 \tilde{W}_1}{\partial \zeta^2} + ... = 0 \\ &\frac{\partial^2 \tilde{W}_2}{\partial \tau^2} - 2\alpha_1 \varepsilon_1 \frac{\partial (\tilde{W}_1^2)}{\partial \zeta} - \varepsilon_1 \frac{\partial^2 \tilde{W}_2}{\partial \zeta^2} + ... = 0 \end{aligned} \quad . \tag{3.11}$$

Here, $\tilde{W}_1 = w_1 + w_2$, $\tilde{W}_2 = w_1 - w_2$, and $\varepsilon_1 = \varepsilon^2$.

Note that the different terms of these equations are characterized by different orders of smallness in the parameter ε, and, hence, the possible asymptotic reduction is not complete yet. In contrast to the long-wave approximation, the equations of motion in the short-wavelength approximation contain a nongradient term. As a result, the interaction between the atoms, which is described by the gradient terms, is comparatively weak. Thus, it is reasonable to use a complex presentation of the variables for the displacement $\tilde{W}_1$ only, because the nongradient term contains this variable.

Using the substitution $\tau_0 = 2\tau$, we introduce the complex conjugate functions

$$\psi(\zeta,\tau) = \frac{\partial \tilde{W}_1}{\partial \tau_0} + i\tilde{W}_1, \quad \psi^*(\zeta,\tau) = \frac{\partial \tilde{W}_1}{\partial \tau_0} - i\tilde{W}_1$$

Considering only one of the two conjugated equations with respect to φ and φ*, we have

$$\begin{aligned} &\frac{\partial \varphi}{\partial \tau_0} - \frac{i\varepsilon_1}{2}\alpha_1(\varphi - \varphi^* e^{-i2\tau_0})\frac{\partial W}{\partial \zeta} + \frac{i\varepsilon_1}{8}\alpha_2 c(\varphi e^{i\tau_0} - \varphi^* e^{-i\tau_0})^3 e^{-i\tau_0} - \\ &-\frac{i}{8}\varepsilon_1 \left(\frac{\partial^2 \varphi}{\partial \zeta^2} - \frac{\partial^2 \varphi^*}{\partial \zeta^2} e^{-2i\tau_0}\right) = 0, \\ &\frac{\partial^2 W}{\partial \tau_0^2} + \frac{\varepsilon_1}{2}\alpha_2 \frac{\partial}{\partial \zeta}(\varphi e^{-i\tau_0} - \varphi^* e^{-i\tau_0})^2 - \frac{1}{4}\varepsilon_1 \frac{\partial^2 W}{\partial \zeta^2} + ... = 0 \end{aligned} \tag{3.12}$$

where

$$W = \tilde{W}_2, \quad \varphi = \psi e^{-i\tau_0}$$

Introducing the slow characteristic times (along with the fast time)

$$\tau_1 = \varepsilon_1 \tau_0, \quad \tau_2 = \varepsilon_1^2 \tau_0 \dots,$$

and using power expansions of φ, φ^*, and W in the small parameter ε_1

$$\begin{aligned} \varphi &= \varphi_0 + \varepsilon_1 \varphi_1 + \varepsilon_1^2 \varphi_2 + ..., \\ W &= W_0 + \varepsilon_1 W_1 + \varepsilon_1^2 W_2 + ..., \end{aligned}$$

we obtain:

$$\begin{aligned} &\frac{\partial}{\partial \tau_0}(\varphi_0 + \varepsilon_1 \varphi_1 + \varepsilon_1^2 \varphi_2 + ...) + \varepsilon_1 \frac{\partial}{\partial \tau_1}(\varphi_0 + \varepsilon_1 \varphi_1 + \varepsilon_1^2 \varphi_2 + ...) - \\ &-\frac{1}{2} i \varepsilon_1 \alpha_1 [(\varphi_0 + \varepsilon_1 \varphi_1 + \varepsilon_1^2 \varphi_2 + ...) - (\varphi_0^* + \varepsilon_1 \varphi_1^* + \varepsilon_1^2 \varphi_2^* + ...) e^{-2i\tau_0}] \times \\ &\times \left(\frac{\partial W_0}{\partial \zeta} + \varepsilon_1 \frac{\partial W_1}{\partial \zeta} + ... \right) + \frac{1}{8} i \varepsilon_1 \alpha_2 [(\varphi_0 + \varepsilon_1 \varphi_1 + \varepsilon_1^2 \varphi_2 + ...) e^{i\tau_0} - \\ &-(\varphi_0^* + \varepsilon_1 \varphi_1^* + \varepsilon_1^2 \varphi_2^* + ...) e^{-i\tau_0}]^3 e^{-i\tau_0} + \\ &+\frac{1}{8} i \varepsilon_1 \frac{\partial^2}{\partial \zeta^2}(\varphi_0 - \varphi_0^* e^{-2i\tau_0} + \varepsilon_1 \varphi_1 - \varepsilon_1 \varphi_1^* e^{-2i\tau_0} + ...) = 0 \\ &\frac{\partial^2}{\partial \tau_0^2}(W_0 + \varepsilon_1 W_1 + ...) + 2\varepsilon_1 \frac{\partial^2}{\partial \tau_0 \partial \tau_1}(W_0 + \varepsilon_1 W_1 + ...) + \\ &+\frac{1}{8} \alpha_2 \varepsilon_1 \frac{\partial}{\partial \zeta}[(\varphi_0 + \varepsilon_1 \varphi_1 + \varepsilon_1^2 \varphi_2 + ...) e^{i\tau_0} - (\varphi_0^* + \varepsilon_1 \varphi_1^* + \varepsilon_1^2 \varphi_2^* + ...) e^{-i\tau_0}]^2 - \\ &-\frac{1}{4} \varepsilon_1 \frac{\partial^2}{\partial \zeta^2}(W_0 + \varepsilon_1 W_1 + ...) = 0 \end{aligned}$$

Next, setting the coefficients at each power of ε_1 to zero,

$$\varepsilon_1^0 : \quad \frac{\partial \varphi_0}{\partial \tau_0} = 0, \quad \frac{\partial^2 W_0}{\partial \tau_0^2} = 0.$$

As a result, φ_0 can be presented as $\varphi_0 = \varphi_0(\zeta, \tau_1, \tau_2, \dots)$, $W_0 = W_0(\zeta, \tau_1, \tau_2, \dots)$ (taking into account the fact that the equations above must not contain secular terms):

$$\varepsilon_1^1 : \quad \begin{cases} \dfrac{\partial \varphi_1}{\partial \tau_0} + \dfrac{\partial \varphi_1}{\partial \tau_1} - \dfrac{i}{2} \alpha_1 (\varphi_0 - \varphi_0 * e^{-2i\tau_0}) \dfrac{\partial W_0}{\partial \zeta} + \\ + \dfrac{i}{8} \alpha_2 (\varphi_0^3 e^{2i\tau_0} - 3 |\varphi_0|^2 \varphi_0 + 3 |\varphi_0|^2 \varphi_0 * e^{-2i\tau_0} - \varphi_0 * e^{-4i\tau_0}) - \\ - \dfrac{i}{8} \dfrac{\partial^2}{\partial \zeta^2} (\varphi_0 - \varphi_0 * e^{-i\tau_0}) = 0 \\ \dfrac{\partial^2 W_1}{\partial \tau_0^2} + 2 \dfrac{\partial^2 W_0}{\partial \tau_0 \partial \tau_1} + \dfrac{1}{8} \alpha_1 \dfrac{\partial^2}{\partial \zeta^2} (-2 |\varphi_0|^2 + \varphi_0^2 e^{2i\tau_0} + \varphi_0 *^2 e^{-2i\tau_0}) - \dfrac{1}{4} \dfrac{\partial^2 W_0}{\partial \zeta^2} = 0 \end{cases}$$

The secular terms should vanish; then, we obtain the following equations:

$$\frac{\partial \phi_0}{\partial \tau_1} + \frac{i}{2}\alpha_1 \phi_0 \frac{\partial W_0}{\partial \zeta} - \frac{3i}{8}\alpha_2 |\phi_0|^2 \phi_0 - \frac{i}{8}\frac{\partial^2 \phi_0}{\partial \zeta^2} = 0,$$

$$\alpha_1 \frac{\partial |\phi_0|^2}{\partial \zeta} + \frac{\partial^2 W_0}{\partial \zeta^2} = 0$$

The second equation yields

$$\frac{\partial W_0}{\partial \zeta} = -\alpha_1 |\varphi_0|^2.$$

The substitution of this expression into the first equation yields a nonlinear PDE describing the dynamics of the one-dimensional crystal in the principal short-wave approximation:

$$\frac{\partial \varphi_0}{\partial \tau_1} - i\alpha |\varphi_0|^2 \varphi_0 - \frac{i}{8}\frac{\partial^2 \varphi_0}{\partial \zeta^2} = 0 \tag{3.13}$$

where $\alpha=(3\alpha_2-4\alpha_1)/8$. This equation is the nonlinear Schrödinger equation. At $\alpha > 0$, this equation has localized solutions in the form of envelope solitons:

$$\varphi_0(\zeta, \tau_1) = \left(\frac{2S\beta}{\alpha}\right)^{1/2} e^{i(k\zeta - \omega\tau_1)} \sec h\left[S^{1/2}(\zeta - v\tau_1)\right].$$

Here $k = \frac{v}{2\beta}$ $\omega = \frac{v^2}{4\beta^2} - S$ and $\beta = \frac{1}{8}$.

In this case, the amplitude S and the velocity of the soliton v are independent parameters.

3.1.4 Optical Breathers in a Zigzag Chain

Now we would like to demonstrate a more realistic approach to the short-wavelength planar dynamics of a zigzag chain, taking the connection between longitudinal and transversal oscillations into account (Manevitch and Smirnov, 2007, 2008). The geometry of the system under consideration requires using two- or three-dimensional models; we would like to underline the role of geometric nonlinearity in the dynamics of such complex systems. The geometric nonlinearity is significant in very different zigzag systems-macromolecular chain, carbon monolayer, anisotropic crystal or engineering structure. Contrary to simple one-dimensional models, even a zigzag chain with equal masses demonstrates many peculiarities such as the optical branch in the dispersion curve.

Let us consider a zigzag chain with first- and second-neighbor interactions along the chain. The Hamilton function of the system is

$$H = \sum \left\{ M\frac{\dot{r}_{ij}{}^2}{2} + U_1(r_{j+1i} - r_{ji}) + U_2(r_{j+2i} - r_{ji}) + U_3(r_{ji+1} - r_{ji}) \right\}, \tag{3.14}$$

where r_{ji} is a coordinate of the j-th atom in the i-th chain. The potential functions U_1, U_2, U_3 are assumed to be quadratic in coordinates r,

$$U_n(r, r') = \frac{K_n}{2}\left(|r - r'| - |r_0 - r'_0|\right)^2,$$

however, this leads to nonlinear expressions for force factors because of a nonlinear dependence of the deformations on displacements. Here, r, r' and r_0, r_0' are current and equilibrium coordinates of particles, respectively. If the constant $K_2 << K_1$, a longitudinal motion results predominantly in the folding of zigzags. In this context, one has to take a longitudinal as well as a transversal motion of atoms into account. In the opposite case, when $K_2 >> K_1$, one can define two subchains and describe their dynamics almost independently both for longitudinal and transversal degrees of freedom. It is an appropriate field for a one-dimensional approach. An influence of neighbor chains may be taken into account by on-site potential or a direct interchain interaction with mobile neighbor zigzags (Fig. 3.4).

To consider the motion of particles on the plane, it is convenient to introduce the longitudinal and transverse displacements with respect to zigzag axis:

$$r_{ji} = \{x_{ji,0} + w_{ji}, y_{ji,0} + u_{ji}\}$$

with the equilibrium coordinates

$$x_{ji,0} = jl\sin(\theta),\ y_{ji,0} = i(L + l\cos(\theta)) + (-1)^{j+i+1} l\cos(\theta)/2$$

where l is the equilibrium distance between the nearest atoms in the zigzag and $L+l\cos(\theta)$ is the distances between the axes of neighbor zigzags, 2θ is the angle between nearest bounds. For the dimensionless set of variables and parameters (we keep their denotations to reduce the number of symbols)

$$w_{ji} -> w_{ji}/l, \quad u_{ji} -> (-1)^{j+i+1} u_{ji}/l$$
$$K_1 -> K_1 l^2, \quad K_2 -> K_2 l^2, \quad K_3 -> K_3 l^2,$$

we write the potentials in the forms:
for the nearest neighbours

$$U_1 = \frac{K_1}{2}\left[\sqrt{(\sin\theta + w_{j+1i} - w_{ji})^2 + (\cos\theta - u_{j+1i} - u_{ji})^2} - 1\right]^2,$$

for the second ones

$$U_2 = \frac{K_2}{2}\left[\sqrt{(2\sin\theta + w_{j+2i} - w_{ji})^2 + (u_{j+2i} - u_{ji})^2} - 2\sin\theta\right]^2,$$

and for interchain interaction

$$U_3 = \frac{K_3}{2}\left[\sqrt{(w_{ji+1} - w_{ji})^2 + (1 + u_{ji+1} - u_{ji})^2} - 1\right]^2.$$

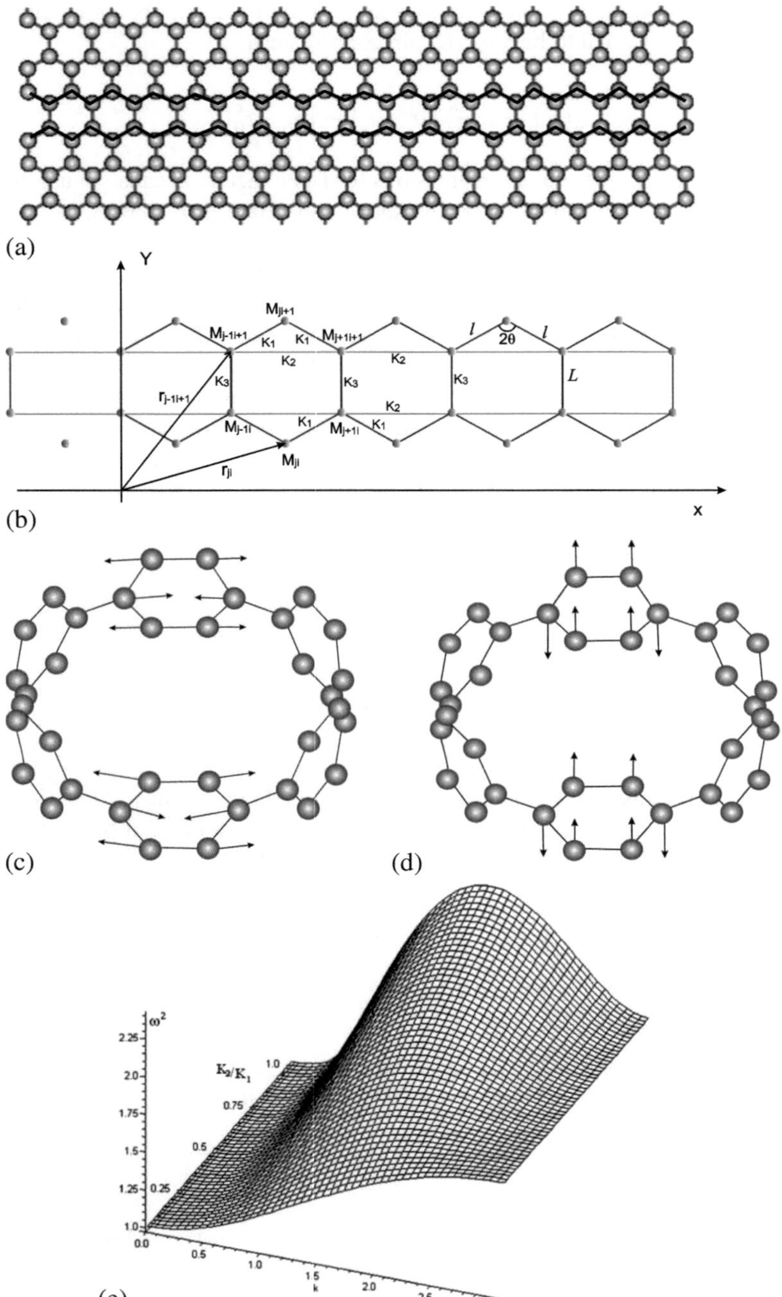

Fig. 3.4 (continued)

Let us briefly discuss the potential U_3. The bonds with rigidities K_1 and K_3 are similar for a carbon sheet (graphene) or carbon nanotube. If the nearest chains perform pure "out of-phase" motion with respect to the selected chain, it is possible to suppose $K_1 = K_3$. In the opposite case of "in-phase" motion we have to accept $K_3 = 0$ because the bonds between nearest chains are not deformed. So, the choice of K_3 is caused by the system symmetry.

The linearized equations of motion can be written as follows:

$$\begin{aligned}
\frac{d^2 w_{ji}}{dt^2} &= K_1\{\sin^2\theta(w_{j+1i} - 2w_{ji} + w_{j-1i}) + \sin\theta\cos\theta(u_{j+1i} - u_{j-1i})\}+ \\
&\quad + K_2(w_{j+2i} - 2w_{ji} + w_{j-2i}) \\
\frac{d^2 u_{ji}}{dt^2} &= -K_1\{\cos^2\theta(u_{j+1i} + 2u_{ji} + u_{j-1i}) - \sin\theta\cos\theta(w_{j+1i} - w_{j-1i})\} \\
&\quad - K_3\{(u_{ji+1} - u_{ji})\lambda_1 - (u_{ji} - u_{j-1i})\lambda_2\}
\end{aligned} \tag{3.15}$$

where the constants λ_1 and λ_2 take into account the difference between odd and even atoms in the chains (e.g., $\lambda_1 = 0$ and $\lambda_2 = 1$ for odd atoms and $\lambda_1 = 1$ $\lambda_2 = 0$ for even ones).

The dispersion relation for the system (3.15) is

$$\begin{aligned}
\omega^2 &= \frac{1}{2}(\omega_0^2 \pm \sqrt{\omega_0^2 - \delta^2}) \\
\omega_0^2 &= 2K_1[1 + \cos 2\theta \cos k] + 2K_2(1 - \cos 2k) + K_3) \\
\delta^2 &= 8K_1K_2\cos^2\theta \sin^2 k(1 + \cos k) + \\
&\quad + 2K_1K_3\sin^2\theta(1 - \cos k) + 4K_2K_3\sin^2 k
\end{aligned} \tag{3.16}$$

where k is a dimensionless wave number ($k = 0 \ldots \pi$).

So, due to the geometric configuration of zigzag, there are two branches of the dispersion curve – the acoustic branch and the optical one. The acoustic branch describes a motion of two subchains of zigzag on the same plane (the left edge $k = 0$ corresponds to a longitudinal displacement of zigzag as a whole and the right edge $k = \pi$ – to a transversal one). The optical branch corresponds to relative motion of the subchains of zigzag (their phases are opposite). It is easy to see that the left edge of the optical branch corresponds to an anti-phase motion of the nearest atoms in the

Fig. 3.4 Zigzag configuration in the graphene mono layer (**a**). Geometry of zigzad chain with geometrical nonlinearity (**b**), K_1, K_2, K_3 – rigidities of first neighbour, second neighbour bond, and interchain bond, l – interatomic distance, and 2q – the angle between interatomic bonds. Transversal (**c**) and longitudinal (**d**) oscillations of zigzag atoms in the carbon nanotubes with armchair configurations. (**e**) – Tranformation of dispersion relation for zigzag chain with change of rigidity ratio direction perpendicular to the zigzag axis and the vector of displacements $\{w_{ji}=0, u_{ji}=(-1)^{j}\}$ is the corresponding solution of the equations of motion. The right edge of the optical branch corresponds to the anti-phase motion of the nearest atoms in the direction along the zigzag axis and the vector $\{w_{ji}=(-1)^{j}, u_{ji}=0\}$ corresponds to the normal mode describing the longitudinal motion. Such motions for the carbon nanotube in armchair configuration are schematically shown in Fig. 3.4c, d. Further we will be interested in the solution corresponding to the optical branch only. Let us consider the dispersion law more accurately (Fig. 3.4e). As we can see, a change of the K_2/K_1 relation leads to drastic changes in the behaviour of the dispersion curve. The frequency $w_{opt}(k)$ is the monotonic function of wave number k for small K_2/K_1 ratio, but it has an extremum point when K_2/K_1 is large. The last circumstance is very important for nonlinear dynamics near the right edge of the spectrum. The presence of interchain interaction changes the gap magnitude, but does not change the curve profile.

To consider the nonlinear dynamics of the particles, we introduce the continuum variables:

$$
\begin{aligned}
h_{n+m} &= \varepsilon[\cos(km)(H+\varepsilon m H_x+\varepsilon^2 m^2 H_{xx}/2+\ldots)+\\
&\quad+\sin(km)(\tilde{H}+\varepsilon m\tilde{H}_x+\varepsilon^2 m^2\tilde{H}_{xx}/2)+\ldots]\\
h_{n+1+m} &= \varepsilon[-\sin(km)(H+\varepsilon m H_x+\varepsilon^2 m^2 H_{xx}/2+\ldots)+\\
&\quad+\cos(km)(\tilde{H}+\varepsilon m\tilde{H}_x+\varepsilon^2 m^2\tilde{H}_{xx}/2)+\ldots]
\end{aligned}
\tag{3.17}
$$

Here h denotes w or u and ε is a small parameter, that reflects a smallness of both the amplitudes relative to interparticle distances and of those with respect to the characteristic wavelength of modulation of k-th normal mode.

After substitution of Eq. (3.17) into the equation of motion we obtain (up to the third order by ε)

$$
\begin{aligned}
&\varepsilon W_{tt}+\varepsilon XW+\varepsilon Z\tilde{U}-\varepsilon^2 X'\tilde{W}_x+\varepsilon^2 Z'U_x-\\
&-\frac{\varepsilon^3}{2}X''W_{xx}-\frac{\varepsilon^3}{2}Z''\tilde{U}_{xx}+\varepsilon F(W,\tilde{W},U,\tilde{U})=0\\
&\varepsilon\tilde{W}_{tt}+\varepsilon X\tilde{W}-\varepsilon ZU+\varepsilon^2 X'W_x+\varepsilon^2 Z'\tilde{U}_x-\\
&-\frac{\varepsilon^3}{2}X''\tilde{W}_{xx}+\frac{\varepsilon^3}{2}Z''U_{xx}+\varepsilon\tilde{F}(W,\tilde{W},U,\tilde{U})=0\\
&\varepsilon U_{tt}+\varepsilon YU-\varepsilon Z\tilde{W}-\varepsilon^2 Y'\tilde{U}_x-\varepsilon^2 Z'W_x-\\
&-\frac{\varepsilon^3}{2}Y''U_{xx}+\frac{\varepsilon^3}{2}Z''\tilde{W}_{xx}+\varepsilon G(W,\tilde{W},U,\tilde{U})=0\\
&\varepsilon\tilde{U}_{tt}+\varepsilon Y\tilde{U}+\varepsilon ZW+\varepsilon^2 Y'U_x-\varepsilon^2 Z'\tilde{W}_x-\\
&-\frac{\varepsilon^3}{2}Y''\tilde{U}_{xx}-\frac{\varepsilon^3}{2}Z''W_{xx}+\varepsilon\tilde{G}(W,\tilde{W},U,\tilde{U})=0
\end{aligned}
,\tag{3.18}
$$

where

$$X = 2[K_1 \sin^2\theta(1-\cos k) + K_2(1-\cos 2k)]$$
$$Y = 2K_1\cos^2\theta(1+\cos k) + K_3$$
$$Z = K_1 \sin 2\theta \sin k$$

Subscript indexes and the apostrophe indicate the differentiation with respect to the corresponding variables and the wave number k, and $F, \tilde{F}, G, \tilde{G}$ are the nonlinear functions of variables $W, \tilde{W}, U, \tilde{U}$ and their derivatives.

To simplify the further analysis, we exclude one pair of variables. Near the left edge of the dispersion curve the transversal motion is predominant. In the first approximation one can get:

$$W = -\frac{Z'}{\omega^2 - X}\tilde{U}$$
$$\tilde{W} = \frac{Z'}{\omega^2 - X}U \tag{3.19}$$

where ω is a "nonlinear" frequency that is different from the "linear" one.

The next step of the reduction procedure is the substitution of Eq. (3.19) into gradient and nonlinear terms of Eq. (3.17). As a result, we get the relations:

$$W = -\frac{1}{X-\omega^2}\Big[Z\tilde{U} - \varepsilon\Big(\frac{ZX'}{X-\omega^2} - Z'\Big)U_x +$$
$$+\frac{\varepsilon^2}{2}\Big(X''\frac{Z}{X-\omega^2} - Z''\Big)\tilde{U}_{xx} + F(U,\tilde{U})\Big]$$

$$\tilde{W} = \frac{1}{X-\omega^2}\Big[ZU - \varepsilon\Big(\frac{ZX'}{X-\omega^2} - Z'\Big)\tilde{U}_x + \tag{3.20}$$
$$+\frac{\varepsilon^2}{2}(X''\frac{Z}{X-\omega^2} - Z'')U_{xx} + \tilde{F}(U,\tilde{U})\Big]$$

One should mention that

$$\omega^2 U \cong \omega_0^2 U - \varepsilon\nu\frac{d\tilde{U}}{dx} + \frac{\varepsilon^2}{2}\mu\frac{d^2U}{dx^2}$$
$$\omega^2\tilde{U} \cong \omega_0^2\tilde{U} + \varepsilon\nu\frac{dU}{dx} + \frac{\varepsilon^2}{2}\mu\frac{d^2\tilde{U}}{dx^2} \tag{3.21}$$

where $\nu = {}^{d\omega_0}/_{dk}$, and $\mu = {}^{d^2\omega_0}/_{dk^2}$.

Substituting (3.20) into (3.18), one obtains:

$$
\begin{aligned}
&U_{tt}+\omega_0^2U+\varepsilon\nu\tilde{U}_x-\frac{\varepsilon^2}{2}\mu U_{xx}+\varepsilon(r_{11}U^2+r_{12}U\tilde{U}+r_{13}\tilde{U}^2)+\\
&\varepsilon^2[q_{12}U\tilde{U}_x+q_{14}\tilde{U}\tilde{U}_x+p_{11}U^3+p_{13}U\tilde{U}^2]=0\\
&\tilde{U}_{tt}+\omega_0^2\tilde{U}-\varepsilon\nu U_x-\frac{\varepsilon^2}{2}\mu\tilde{U}_{xx}+\varepsilon(r_{21}\tilde{U}^2+r_{22}U\tilde{U}+r_{23}\tilde{U}^2)+\\
&\varepsilon^2[q_{21}UU_x+q_{23}\tilde{U}U_x+p_{22}U^2\tilde{U}+p_{24}\tilde{U}^3]=0
\end{aligned}
\quad . \tag{3.22}
$$

where coefficients r, p, and q are the functions of the parameters of considered system.

Now we can introduce the complex variables $\Phi=\tilde{U}_\tau+i\tilde{U}$and $\Psi=U_\tau+iU$ ($\tau{=}\omega_0 t$ – dimensionless time). Then

$$
\begin{aligned}
&i\Psi_\tau+\Psi+\frac{\varepsilon\nu}{2}(\Phi_x-\Phi_x^*)-\frac{\varepsilon^2\mu}{4}(\Psi_{xx}-\Psi_{xx}^*)-\\
&-i\frac{\varepsilon}{4}[r_{11}(\Psi-\Psi^*)^2+r_{12}(\Psi-\Psi^*)(\Phi-\Phi^*)+r_{13}(\Phi-\Phi^*)^2]-\\
&-i\frac{\varepsilon^2}{4}[q_{12}(\Psi-\Psi^*)(\Phi_x-\Phi_x^*)+q_{14}(\Phi-\Phi^*)(\Phi_x-\Phi_x^*)]-\\
&-\frac{\varepsilon^2}{8}[p_{11}(\Psi-\Psi^*)^3+p_{13}(\Psi-\Psi^*)(\Phi-\Phi^*)^2]=0\\
&i\Phi_\tau+\Phi-\frac{\varepsilon\nu}{2}(\Psi_x-\Psi_x^*)-\frac{\varepsilon^2\mu}{4}(\Phi_{xx}-\Phi_{xx}^*)-\\
&-i\frac{\varepsilon}{4}[r_{21}(\Psi-\Psi^*)^2+r_{22}(\Psi-\Psi^*)(\Phi-\Phi^*)+r_{23}(\Phi-\Phi^*)^2]-\\
&-i\frac{\varepsilon^2}{4}[q_{21}(\Psi-\Psi^*)(\Psi_x-\Psi_x^*)+q_{23}(\Psi_x-\Psi_x^*)(\Phi-\Phi^*)]-\\
&-\frac{\varepsilon^2}{8}[p_{22}(\Psi-\Psi^*)^2(\Phi-\Phi^*)+p_{24}(\Phi-\Phi^*)^3]=0
\end{aligned}
\tag{3.23}
$$

where Φ^*, Ψ^* are complex conjugate functions corresponding to Φ and ψ, respectively.
To reduce the number of symbols used, we denote

$$r_{ij}/\omega_0^2\to r_{ij}, q_{ij}/\omega_0^2\to q_{ij}, p_{ij}/\omega_0^2\to p_{ij}.$$

Using the multiple-scale approximation

$$\Phi=\phi e^{-i\omega\tau},\ \phi=\phi_0+\varepsilon\phi_1+\varepsilon^2\phi_2+\cdots\ \Psi=\psi e^{-i\omega\tau},\ \psi=\psi_0+\varepsilon\psi_1+\varepsilon^2\psi_2+\cdots$$

and introducing the slow times $\tau_0{=}\tau$, $\tau_1{=}\varepsilon t$, $\tau_2{=}\varepsilon^2\tau$,... we obtain the next equations for various orders of ε:

$$\varepsilon^0:\ i\partial_{\tau_0}\varphi_0=0,\ i\partial_{\tau_0}\psi_0=0,$$

so, ϕ_0 and ψ_0 do not depend on t_0.

In the order ε^1, the condition of absence of secular terms leads to the equations:

$$\begin{aligned} &i\partial_{\tau_1}\phi_0 + i\partial_{\tau_0}\phi_1 + \frac{\nu}{2}(\partial_x\psi_0 - \partial_x\psi_0^* e^{-2i\tau_0}) = 0 \\ &i\partial_{\tau_1}\psi_0 + i\partial_{\tau_0}\psi_1 - \frac{\nu}{2}(\partial_x\phi_0 - \partial_x\phi_0^* e^{-2i\tau_0}) = 0 \end{aligned} \qquad (3.24)$$

It is easy to show that the pair of equations

$$\begin{aligned} &i\partial_{\tau_1}\psi_0 - \frac{\nu}{2}\partial_x\phi_0 = 0 \\ &i\partial_{\tau_1}\phi_0 + \frac{\nu}{2}\partial_x\psi_0 = 0 \end{aligned}$$

are equivalent to wave-type equations

$$\begin{aligned} &\partial_{\tau_1}^2\psi_0 - \left(\frac{\nu}{2}\right)^2\partial_x^2\psi_0 = 0 \\ &\partial_{\tau_1}^2\varphi_0 - \left(\frac{\nu}{2}\right)^2\partial_x^2\varphi_0 = 0 \end{aligned}$$

These equations are satisfied if ψ_0 and ϕ_0 depend on "wave" coordinate $\xi = x \pm \frac{\nu}{2}\tau_1$. Integrating the rest of relations (3.24) with τ_0, we obtain

$$\begin{aligned} \phi_1 &= \frac{\nu}{2}\partial_\xi\psi_0^* e^{-2i\tau_0} \\ \psi_1 &= -\frac{\nu}{2}\partial_\xi\phi_0^* e^{-2i\tau_0} \end{aligned} \qquad (3.25)$$

In the order ε^2 one obtains:

$$\begin{aligned} &i\partial_{\tau_2}\psi_0 + i\partial_{\tau_0}\psi_2 + i\partial_{\tau_1}\psi_1 + \frac{\nu}{2}\partial_x\varphi_1 - \frac{\mu}{4}\partial_x^2\psi_0 + \frac{3}{8}p_{11}|\psi_0|^2\psi_0 + \\ &+\frac{1}{4}p_{12}|\psi_0|^2\varphi_0 - \frac{1}{8}p_{13}\psi_0^2\bar{\varphi}_0 + \frac{1}{4}p_{13}|\varphi_0|^2\psi_0 + \frac{3}{8}p_{14}|\varphi_0|^2\varphi_0 = 0 \\ &i\partial_{\tau_2}\varphi_0 + i\partial_{\tau_0}\varphi_2 + i\partial_{\tau_1}\varphi_1 - \frac{\nu}{2}\partial_x\psi_1 - \frac{\mu}{4}\partial_x^2\varphi_0 + \frac{3}{8}p_{21}|\psi_0|^2\psi_0 + \\ &+\frac{1}{4}p_{22}|\psi_0|^2\varphi_0 - \frac{1}{8}p_{23}\psi_0^2\bar{\varphi}_0 + \frac{1}{4}p_{23}|\varphi_0|^2\psi_0 + \frac{3}{8}p_{24}|\varphi_0|^2\varphi_0 = 0 \end{aligned}$$

After substituting relations (3.25) into these equations with following integration with respect to "fast" times τ_0, τ_1, the requirement of the absence of secular terms leads to the equations:

$$
\begin{aligned}
&i\partial_{\tau_2}\psi_0 - \frac{1}{4}\left(\mu - \frac{\nu^2}{2}\right)\partial_\xi^2\psi_0 + \frac{3}{8}\alpha\,|\psi_0|^2\,\psi_0 - \frac{1}{4}\beta\,|\phi_0|^2\,\psi_0 = 0 \\
&i\partial_{\tau_2}\phi_0 - \frac{1}{4}\left(\mu - \frac{\nu^2}{2}\right)\partial_x^2\phi_0 + \frac{3}{8}\alpha\,|\phi_0|^2\,\phi_0 - \frac{1}{4}\beta\,|\psi_0|^2\,\phi_0 = 0
\end{aligned}
\quad . \qquad (3.26)
$$

In the common case, parameters a and β in the equations (3.26) depend on wave number k and parameters of the chain in rather complicated manner. Some simplification is possible for special cases only (see below).

System (3.26) allows three relationships between variables ϕ_0 and ψ_0:

(i) $\phi_0 = \kappa\psi_0, \kappa = \pm 1$
(ii) $\psi_0 = 0$;
(iii) $\phi_0 = 0$

One can find for the case (i):

$$
i\partial_{\tau_2}\psi_0 - \frac{1}{4}\left(\mu - {}^{\nu^2}\!/_2\right)\partial_\xi^2\psi_0 + \frac{1}{4}\left(\frac{3}{2}\alpha - \beta\right)|\psi_0|^2\,\psi_0 = 0 \qquad (3.27)
$$

and for the cases (ii) and (iii):

$$
i\partial_{\tau_2}\psi_0 - \frac{1}{4}\left(\mu - {}^{\nu^2}\!/_2\right)\partial_\xi^2\psi_0 + \frac{3}{8}\alpha\,|\psi_0|^2\,\psi_0 = 0 \qquad (3.28)
$$

Equations (3.27) and (3.28) are the nonlinear Scrödinger equations with localized solutions:

$$
\begin{aligned}
&\psi_0(\tau_2, \xi) = \frac{4A}{\sqrt{\gamma}}\exp[i(B\xi - (A^2 - B^2)\tau_2 + C_1)] \times \\
&\times \sec h[\frac{2A\xi}{\sqrt{\mu - {}^{\nu^2}\!/_2}} + 2AB\tau_2 + C_2]
\end{aligned}
\qquad (3.29)
$$

where A, B, C_1, C_2 are free parameters and $\gamma = |3\alpha - 2\beta|$ for the case (i) and $\gamma = 3\alpha$ for the cases (ii) and (iii). Let us note, that the solution (3.29) can exist, if the dispersive and nonlinear terms in Eq 3.17 have the same signs. As it can be shown in the system under consideration this condition corresponds the wave number k<1.0 for the partial case K_1=1.0, K_2=0.05, K_3=1.0 (Fig. 3.5).

If dealing with a localized mode near the right edge of the dispersion curve, one can see from the shape of the dispersion curve at large value K_2, that curvature of ω(k) is positive near k = π. So, one can expect the existence of localized solutions.

Let us consider the limit case k = π to demonstrate some peculiarities of the corresponding procedure. The expansion of the modulating function (3.17) is:

$$
h_{n+m} = \varepsilon(-1)^m\left(H + \varepsilon m H_x + \frac{\varepsilon^2 m^2}{2}H_{xx}\right)
$$

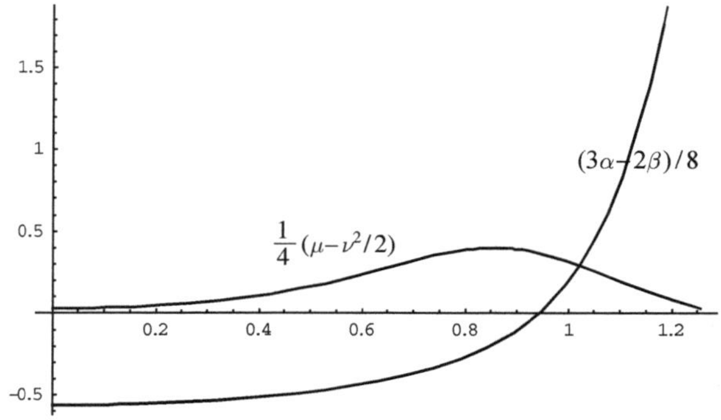

Fig. 3.5 Coefficients of Eq. (3.27) versus the wave number k for strong coupled zigzag chains (K_3=1.0)

After substituting this expansion into a discrete equation of motion, we get:

$$\varepsilon\partial_t^2 W + \varepsilon XW + \varepsilon^2 Z'\partial_x U - \frac{\varepsilon^3}{2}X''\partial_x^2 W + \varepsilon^3 F(W,U) = 0$$
$$\varepsilon\partial_t^2 U + \varepsilon YU - \varepsilon^2 Z'\partial_x W - \frac{\varepsilon^3}{2}Y''\partial_x^2 U + \varepsilon^3 G(W,U) = 0$$

A longitudinal motion is the dominant one near the right edge of the dispersion curve and the next relation between amplitudes W and U is valid:

$$U = \varepsilon\frac{Z'}{Y-\omega^2}\partial_x W.$$

In general, the following procedure of reduction is similar to the one presented above. Thus, for the complex variable

$$\Psi = \dot{W} + iW,\ \ \Psi = (\psi_0 + \varepsilon\psi_1 + \varepsilon^2\psi_2 + ...)e^{i\tau}$$

we can get the following equation:

$$\begin{aligned} &-i\partial_{\tau_2}\psi_0 + \frac{K_1\cos 2\theta + 4K_2}{2\omega_0^2}\partial_x^2\psi_0 + \\ &+3\frac{K_1\cos^2\theta}{\omega_0^2}(4\sin^2\theta - \cos^2\theta)\left|\psi_0\right|^2\psi_0 = 0 \end{aligned} \quad (3.30)$$

Equation (3.30) has a localized solution in the form (3.29) under the condition that the signs of nonlinear and dispersion items are the same. Thus we get the next condition for it:

$$4K_2/K_1 > -\cos 2\theta.$$

Such an approach to the consideration of planar oscillations of a zigzag chain can be used for a number of physical and mechanical problems. Recently, the study of optical oscillations localization has been made for PE macromolecules (Savin and Manevitch, 2003; Manevitch et al., 2006, 2008). In the presence of a physical nonlinearity of angle interaction, the localized solution (envelope soliton) has been found near the minima of the dispersion curve. The results of the computer simulation of optical breather both in isolated zigzag chain and in carbon monolayer and PE macromolecules will be discussed below.

3.1.5 Torsional Solitons

We discussed the nonlinear dynamics of a polymer chain in crystalline PE because most of the studies were carried out with this simplest polymer. However, the nonlinear dynamics of more complex polymers has been recently analyzed, first of all, polytetrafluoroethylene (PTFE) (Savin and Manevitch, 2001, 2003).

In contrast to PE, the ground state of crystalline PTFE has a nonplanar conformation – a three-dimensional helix of type 1*13/6. By applying a numerical variational method, soliton-like excitations of two types were revealed in the PTFE helical chain (Savin and Manevitch, 2000).

Localized excitations of the first type correspond to propagation of the torsional solitary waves along the PTFE helical chain. Figure 3.6 shows typical solitons. Solutions of the second type describe the propagation of a solitary longitudinal compression wave. As can be seen, the deformation of the helix in the region of localization of the first-type solitary wave is largely due to the compression of the conformation angles, whereas second-type waves compress valence angles and bonds (without noticeable variations in conformational angles). The interaction of solitary waves in PTFE chain is inelastic, so that the amount of energy dissipated can be substantial. It was shown that the localized excitation of end conformational angles of the chain leads to the formation of torsional solitons and soliton-like high-frequency packets (envelope solitons), whereas the excitation of end valence angles results in the formation of a longitudinal compression soliton, torsional soliton, and envelope soliton.

Soliton-like torsional excitations can also propagate in a poly(aniline) chain, which has a repeating unit consisting of a nitrogen atom and a phenyl ring. Taking only the energy gain due to the delocalization of the p_x electrons of phenyl rings into account, we found that an idealized structure has a *trans*-zigzag shape. However, this planar configuration is energetically unfavorable, because it is characterized by a large lattice potential, which arises due to steric repulsion. Thus, the ground state of a polyaniline chain has a configuration in which neighbor phenyl rings are rotated in the opposite directions with respect to each other by equal angles of about 50°. In this case, the dihedral angles vary according to the equation $\psi_n = (-1)^n \varphi_n$. This structure provides for the existence of two degenerate ground states corresponding to two different phases (rotations of odd and even rings change places) and the possibility of a transition between them according to the soliton mechanism. In this case, a soliton (kink) is a perturbation of the chain that transfers the chain from one ground state at $x \to -\infty$ to the other at $x \to \infty$. Because the state ψ is physically

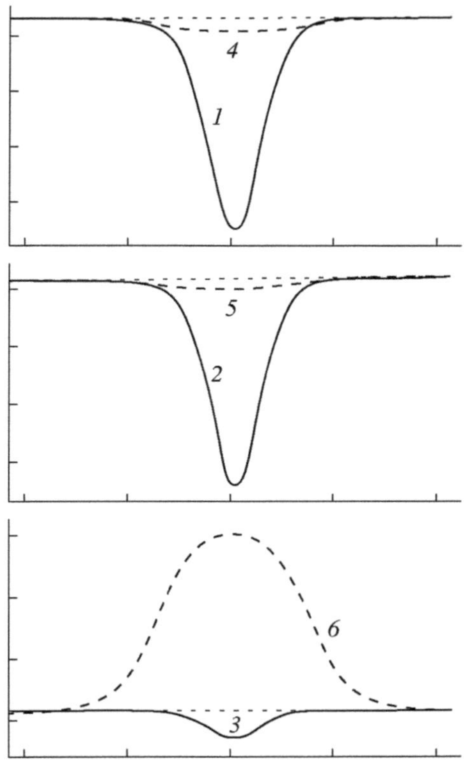

Fig. 3.6 Variations of the valence bond lengths ρ_n, angles θ_n, and conformation angles δ_n in the region of localization of *(1–3)* compression solitons traveling at velocity $s = 1.02$ and *(4–6)* torsion solitons at $s = 0.82$

indistinguishable from the state $\psi + n\pi$, in (MacKenzie et al., 1991), the transition from $-\psi_0 + \pi$ to ψ_0 is considered in addition to the transitions from $-\psi$ to ψ. These two different transitions correspond to the two types of kinks.

The effective potential energy of the lattice can be presented in the form

$$V = \sum_l [V_{1,1}(\sin\psi_l - \sin\psi_{l-1})^2 - V_{2,0}\sin^2\psi_l + V_{4,0}\sin^4\psi_l + const].$$

The corresponding kinetic energy is given by

$$H_{kin} = \sum_l \frac{1}{2} I \left[\frac{d\psi_l}{dt}\right]^2$$

In the continuum limit, the Hamiltonian of the system reads

$$H = \int \left\{ \frac{1}{2} I \left(\frac{\partial\psi}{\partial t}\right)^2 + \left[V_{1,1}\left(\frac{d\sin\psi}{dx}\right)^2 - V_{2,0}\sin^2\psi + V_{4,0}\sin^4\psi + const \right] \right\} dx$$

where x is a dimensionless space variable introduced by normalization to a lattice constant of $a = 4.9$ Å.

The values of the three parameters in this expression can be determined by quantum-mechanical calculations (MacKenzie et al., 1991): $V_{1,1} \sim 0.2$, $V_{2,0} \sim 2$, and $V_{4,0} \sim 1.45$ eV. The energy barriers corresponding to internal rotation angles of 0 and $\pi/2$ are equal to 0.68 and 0.14 eV per ring.

In the static case, it is unnecessary to consider the kinetic energy. In addition, a significant simplification can be achieved by the introduction of the substitution $\sigma = \sin\psi$ (MacKenzie et al., 1991):

$$H = 2V_{1,1} \int \left[\frac{1}{2} \left(\frac{d\sigma}{dx} \right)^2 + \lambda(\sigma^2 - \nu^2)^2 \right] dx,$$

where $\lambda = 2V_{4,0}/V_{1,1} = 14.5$ and $n^2 = V_{2,0}/2V_{4,0} = 0.69$; in this case, $\psi_0 = \arcsin\nu = 56°$, in close agreement with the available experimental data. The continuum Hamiltonian gives the following equation of equilibrium:

$$\frac{d^2\sigma}{dx^2} + \lambda\sigma(\sigma^2 - \nu^2) = 0.$$

The system under study has two basic states $\pm\nu$, while the kink controlling the transition between them can be described as $\sigma_1(x) = \nu\tanh(x/\xi)$, where the characteristic dimension of the kink is $\xi = \sqrt{2V_{1,1}/V_{2,0}}$. The energy of the soliton was estimated $E_1 = \frac{4\nu_2}{3} V_{1,1}\lambda^{1/2}\nu^2 = 0.82$ eV. The soliton of the second type determines the transition from $\psi = \pi - \psi_0$ to $\psi = \psi_0$ via $\psi = \pi/2$ in terms of ψ and recovery from $\sigma = \nu$ to $\sigma = -\nu$ via $\sigma = 1$ in terms of σ (which is a consequence of the multi-valuedness of the function $\arcsin\sigma$): $\sigma_2(x) = \nu\coth[(\pm x + x_0)/\xi]$ (the signs $\pm$ in front of x correspond to $x > 0$ and $x < 0$). The energy of this soliton is substantially lower, $E_2 = \frac{2^{2/3}\lambda^{1/2}}{3} V_{1,1}(2\nu^3 - 3\nu^2 + 1) = 0.055$ eV (MacKenzie et al., 1991). Note, however, that, due to the sharp change of the profile of the kink at $x = 0$, the criterion of applicability of the continuum approximation is not fulfilled; therefore, the solution itself needs some refinement.

3.1.6 Approximation of Immobile Neighbour Chains

The approximation of an isolated chain does not take into account weak interchain interactions and the topological restrictions imposed on the displacements in the chain. The simplest way to consider these factors is to use the approximation of immobile neighbor chains, according to which all the chains except the chosen one are treated as "frozen". This model makes it possible to study quasi-one-dimensional processes that are mostly localized on the selected single chain. We noted that, with allowance made for weak interchain interaction, substantial difference from the approximation of an isolated chain can be expected in the cases only when the displacements corresponding to a localized excitation have a shape of a kink and the chain can only be incorporated into the crystal at displacements that are multiples of the chain period (i.e., the soliton has a topological charge). Therefore, topological restrictions are most important.

First of all, it is natural to consider the transformation (due to interchain interactions) of supersonic soliton-like elementary excitations in an isolated chain, which have a kink-like shape (for displacements) and correspond to extension or compression deformations localized in a fragment of the chain.

As was shown in (Zubova et al., 2005), quasi-one-dimensional supersonic extension solitons do not survive in the chain surrounded by immobile neighbors, because, for the actual topological restrictions and inertia and rigidity characteristics, the region of spatial localization of these solitons is so narrow that the continuum approximation becomes invalid. However, as was found in (Balabaev et al., 2001), even in the case of weak interchain interaction, localized elementary excitations can form "soliton molecules" with an effective width large enough to render them highly mobile. Such bound solitons were found by solving the nonlinear differential equation in partial derivatives for the longitudinal displacement of the chain which left-hand side L(u) coincides with the left part of Eq. (3.6), while the right-hand side,

$$L(u) = -\frac{1}{m}\frac{\partial U}{\partial u}$$

takes into account the effect of interchain interaction. In this case, the validity of the continuum approximation is justified by a sufficiently large spatial localization region of bound solitons. The interchain interaction potential is a periodic function (due to the symmetry of a crystal), which can be obtained by summing up the Lennard-Jones potentials describing weak van-der-Waals interchain interactions. Since the supersonic propagation of a soliton is a consequence of intramolecular anharmonicity, no detailed description of interchain anharmonicity is required in this case.

The equation presented above is considerably more complicated than the equations of nonlinear dynamics for an isolated chain, because, in addition to gradient terms (in the KdV equation), it contains a nongradient component (similar to the sin-Gordon equation). Nevertheless, numerically solving three nonlinear algebraic equations (obtained after analytical transformations) at actual values of the parameters, we found two solitons in the supersonic region (for crystalline PE), which have the velocities v= 1.385 v_s (20 km/s) and 1.012 v_s (14.89 km/s).

In contrast to the isolated-chain case, for which a continuum spectrum of extension soliton velocities exists in the supersonic region, we obtained solitons with discrete propagation velocities and other parameters (Balabaev et al., 2001). The first solution is close to the extension soliton in an isolated chain. In this case, however, the soliton width is small, the continuum approximation is invalid, and this regime is not realized in the discrete model (the conclusion supported by a numerical analysis). On the contrary, the parameters of the second solution corresponding to a soliton molecule are consistent with the conditions of validity of the continuum approximation. The results of a computer simulation (Balabaev et al., 2001) confirm this conclusion. The simulation was carried out using a simulation box in the form of a rectangular parallelepiped. Periodic boundary conditions were imposed in all three directions. The box contained 23 $(\text{–}CH_2\text{–})_{300}$ chains and a chain shorter by two repeating units. Because of the imposed periodic boundary conditions, the molecules were formally infinite. Thus, any localized excitation in the designated molecule could travel any distance along the chain.

The molecules had a *trans*-zigzag shape (Fig. 3.7), and CH_2 groups were modeled by compound atoms with a mass of 14 amu. The valence bond length was $l = 1.53$ Å. The interatomic interaction was described by the potential

$$U(r) = \sum U_3(\theta_i) + \sum U_4(\varphi_i) + \sum U_{nb}(|r_i - r_j|) \tag{3.31}$$

Here, $\mathbf{r} = (\mathbf{r}_1, \ldots, \mathbf{r}_N)$ are the position vectors of all the atoms in the simulation box. The first term in Eq. (3.31) is the sum of all valence angles; the second term, of all torsional angles; and the third term, of all pairs of particles not linked by valence bonds and angles. The interactions between the particles in the simulation box and with their images in the neighbor boxes were taken into account. The valence and conformational angle potentials are given by

$$U_3(\theta) = \frac{1}{2}K_\theta(\theta - \theta_0)^2$$

$$U_4(\varphi) = \alpha \cos\varphi + \beta \cos^3\varphi + \gamma$$

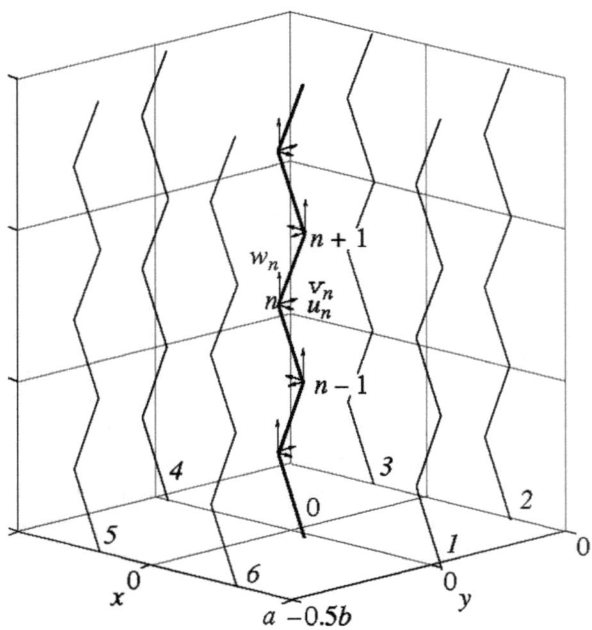

Fig. 3.7 Schematic representation of crystalline PE. The central *trans*-zigzag (0) and the six nearest chains (1,2,. . .,6) are shown. The local coordinate systems are shown for the central chain

where $\theta_0 = 113°$, $K_\theta = 331.4$ kJ/(mol rad^2), $\alpha = 18.41$ kJ/mol, $\beta = 26.79$ kJ/mol, and $\gamma = 8.37$ kJ/mol. Here, $\varphi = 0$ corresponds to the eclipsed conformation, and $\varphi = 180°$, to the unfolded conformation. All atoms separated by more than two neighbors along the chain or belonging to different molecules interacted according to the law

$$U_{nb}(r) = \begin{Bmatrix} U_{LJ}(r) - U_{LJ}(R), & r \leq R \\ 0, & r \geq R \end{Bmatrix} \tag{3.32}$$

where $U_{LJ}(r) = 4\varepsilon\left[\left(\frac{\sigma}{r}\right)^{12} - \left(\frac{\sigma}{r}\right)^{6}\right]$ is the Lennard-Jones potential and R is the radius of interaction. The values $\varepsilon = 0.50$ kJ/mol, $\sigma = 3.2$ Å, and $R = 2.25\sigma$ were used in the calculations.

In Cartesian coordinates, the equations of motion of the considered macromolecular system with the account for geometric constraints imposed on the valence bond lengths take the form of the Lagrangian equations of first kind

$$m_i \frac{d^2 r_i}{dt^2} = -\frac{\partial U}{\partial r_i} + \sum \lambda_\upsilon \frac{\partial f_\upsilon}{\partial r_i},\ i = 1, \ldots, N.$$

Along with the algebraic equations for the valence bonds

$$f_\upsilon(r_i) = 0,\ i = 1, \ldots, N$$

the Lagrangian equations fully determine the dynamic behavior of the system, provided that the initial coordinates and velocities of all particles are known. The equations of motion were integrated numerically using the approach developed in (Balabaev et al., 2001). The simulation box had the sizes $3a \times 4b \times 150c$, where a, b, and c are the crystallographic cell sizes corresponding to the minimum potential energy of the system. At the initial moment of time, an extension defect corresponding to an extension soliton was set on the shorter chain. For better perception, this situation was illustrated by a two-dimensional picture (Fig. 3.8).

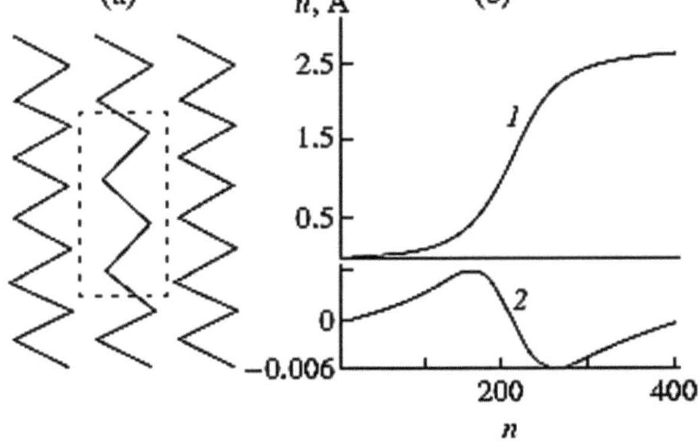

Fig. 3.8 An extension defect – a monomeric unit vacancy (without chain rupture) – in a zigzaglike chain of a polymer crystal: (**a**) a general view of the defect (*grotesque*) and (**b**) displacements of the chain atoms from their equilibrium positions u (*1*) for a chain with defect and (*2*) for the neighbor chains; n is the number of the atom

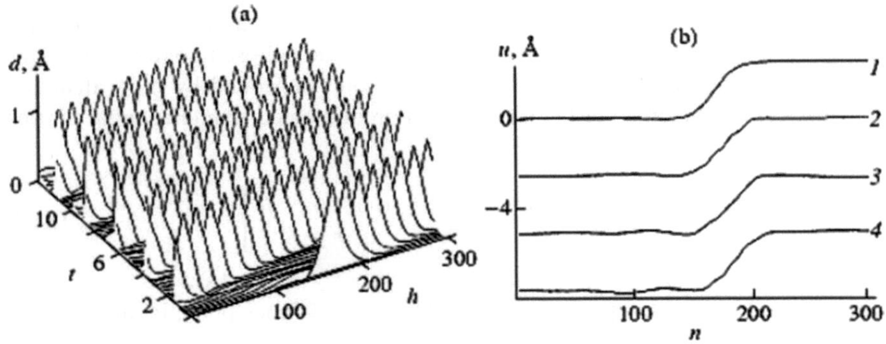

Fig. 3.9 Soliton motion along a chain surrounded by immobile neighbor chains: (**a**) evolution of a solitary wave of deformation and (**b**) distribution of displacements u in the wave over the chain length at various moments of time $t = 0$ (*1*), 2.65 (*2*), 5.1 (*3*), and 7.65 ps (*4*)

In the approximation of immobile neighbors, all the molecules except that carrying the soliton were frozen in their equilibrium positions. In the soliton-carrying molecule, which was shorter by two repeating units than the others, the initial displacements of CH_2 groups along the z axis were chosen in accordance with the analytical solution. The displacements of the atoms in the perpendicular direction (along the y axis) were set simultaneously to retain the values of the bond lengths. The initial velocities of all the atoms were also chosen based on the analytical solution. The subsequent evolution of the system was fully determined by the equations of motion. The motion of the extension defect is demonstrated in Fig. 3.9, where the chain deformation $d = ls - |u - ls|$ moving along the chain as a solitary wave (soliton) is plotted along the vertical axis. When the interchain interactions are artificially "switched off" at some moment of time, the soliton-like excitation breaks down into a set of four solitons with parameters typical of solitons in an isolated chain. Therefore, the supersonic localized excitations predicted theoretically and observed in a computer experiment are the bound states of a set of solitons. In an isolated chain, these bound states are impossible.

As for subsonic solitons of tension–torsion or compression-torsion, they are studied, e.g., in (Savin and Manevitch, 1998a; Savin et al., 2005; Manevitch, 2001).

3.2 Dynamics of Essentially Nonlinear and Vibro-Impact Chains

In Chap. 2 we have demonstrated the simplifications one can achieve by using the vibro-impact models and approximations. Here, we demonstrate the use of the same ideas for the study of a mechanical system – chain with rigid barriers. Then we demonstrate how one can obtain exact solutions for discrete breathers in the framework of the vibro – impact models.

3.2.1 Oscillatory Chain with Rigid Barriers

Oscillatory systems with essential nonlinearities occupy a very special niche in the theory of vibrations. On one side, it is very important to understand their behavior, both from an academic viewpoint and in view of numerous possible applications. On the other hand, they can be very difficult for analysis. Besides an extremely narrow class of integrable systems (Arnold, 1989; Arnold et al., 1997), these systems cannot be described exactly. If one of these integrable cases is in a certain sense close to the system under consideration, then often some kind of perturbation procedure can be realized (Arnold et al., 1997; Kevorkian and Cole, 1996), yielding at least a qualitative understanding of the global dynamics.

In many interesting cases, however, such options are not available. For these cases, one is forced to restrict himself to a search for particular solutions of interest. In absence of the general picture, these particular solutions convey valuable information about the dynamics of the system and are also useful for testing the numeric approaches. An important example of these particular solutions available in many essentially nonlinear systems are nonlinear normal modes (NNMs), discussed in Chapter 2 in more details.

Vibro-impact systems, which are described in detail in many studies (Babitsky, 1978; Manevitch and Gendelman, 2008), demonstrate the most severe nonlinearity – in fact, the strongest possible one. These nonlinear interactions are concentrated at the points of impact; therefore one can substitute computation of the complete trajectory by matching of the parts via the impact conditions (Babitsky, 1978; Zhuravlev and Klimov, 1988; Azeez et al., 1999). Thus, the extreme nonlinearity is "pushed" to the boundary or matching conditions. This is an essential simplification compared to the solution of the complete dynamic problem, especially if one is interested in particular (e.g. periodic) solutions. Consequently, the extremity of the nonlinear interaction may itself simplify the system.

Simplification arising from the extreme impact nonlinearity has inspired numerous attempts to extend it for systems with less extreme behavior (Andrianov and Awrejcewicz, 2003; Gendelman, 2006). These works try to construct special asymptotic approaches for different systems which are in some sense "close" to the vibroimpact ones. In other cases, the situation turns out to be rather opposite – it is convenient to substitute the impact potential by some smooth approximation and then to solve the appropriate smooth problem (Gendelman, 2006).

To clarify the sense of simplification provided by the use of the vibro-impact model, let us consider the chain of nonlinear oscillators with rigid barriers. Corresponding equations of motion can be written as follows:

$$m\frac{d^2w_j}{dt^2}+c_1w_j+c_3w_j^3+c(2w_j-w_{j=1}-w_{j-1})+P\left(w_j,\frac{dw_j}{dt}\right)=0,\quad |\mathrm{w_j}|\leq\Delta \quad (3.33)$$

If $|\mathrm{w_j}| \leq \Delta$, the system (3.33) has only a smooth potential of interaction. When $|\mathrm{w_j}|=\Delta$, the impact interaction appears which can be replaced by periodic external impulses.

The specificity of the vibro- impact model is that if the conditions

$$|w_j| < \Delta \tag{3.34}$$

are satisfied, the system is described by nonlinear smooth equations because

$$P\left(w_j, \frac{dw_j}{dt}\right) = 0 \tag{3.35}$$

One can describe several important particular cases.

The first possibility is a "small" periodic or clamped chain in which differences of the natural linear frequencies are comparable with the frequencies themselves. In this case only, the system is an essentially discrete one. The particular case $c = 0, c_3 = 0$ (a weightless string with discrete masses between rigid limiters) for the string with two and three uniformly situated particles (in the former case $j = 1,2$ and $w^0 = w^3 = 0$; in the latter one $j = 1,2,3$ and $w^0 = w^4 = 0$) has been considered in a paper (Vedenova et al., 1985).

The spatial transformation introduced by V. F. Zhuravlev (1976) was used in this paper. This transformation reflecting the form of an exact solution for uncoupled particles interacting with rigid barriers mentioned above does not remove, but changes a type of discontinuities (the equations of motion after transformation do not contain functions with discontinuities of the second kind). In the case of intensive impact interaction the coupling is demonstrated to be relatively small.

It means that the transformed system contains a small parameter that allows to introduce a slow time and to use a procedure of averaging with respect to fast time. On the basis of such an approach the existence of $(3^n - 1)/2$ nonlinear normal modes that essentially exceed the number of degrees of freedom is revealed. This part of them, including the symmetric one (all particles have similar displacements) and normal modes localized on one or two neighboring particles, are stable with respect to small perturbations. The rest of the nonlinear normal modes are unstable and this is confirmed by computer simulation.

It is also shown how one can analyze a smooth but strongly nonlinear system on the basis of a vibro-impact solution as a starting approximation. These results are illustrated in Figs. 3.10 and 3.11

The system is schematically described in Fig. 3.10. The shapes of possible modes are illustrated in Fig. 3.11. The in-phase mode is stable (denoted by "+" sign,

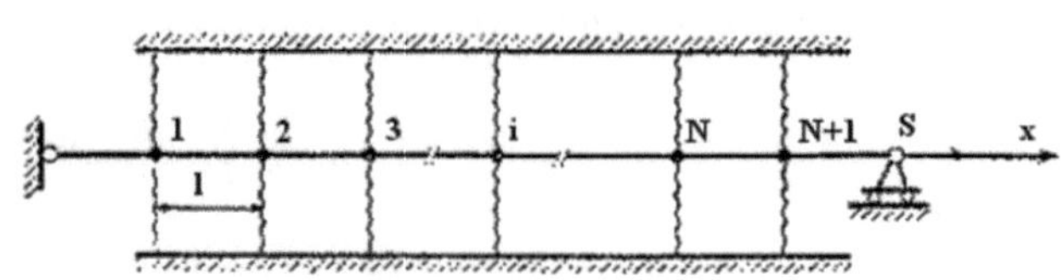

Fig. 3.10 General sketch of the system

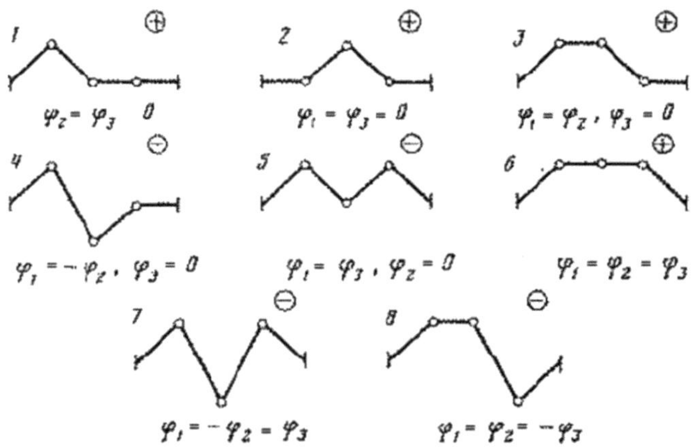

Fig. 3.11 Modal shapes for n=3

solution 6 in Fig. 3.11). The antiphase mode is unstable and denoted by "–" sign (solution 7). Other modes, in which the displacements of two adjacent masses have different signs (solutions 4, 8) and also the mode in which the fixed mass separates two deformed springs (solution 2) that are unstable.

The question arises: what will happen when the coupling parameter grows? To get an answer, it is reasonable to deal with the other tractable vibro-impact model mentioned above and a sequence of impulses. This approach allows revealing much more complicated types of periodic motions than those available for study in the framework of the previous approach. Let us consider the simplest case of a 2DOF system. If the energies are sufficiently small, the stable modes will be the in-phase and antiphase normal modes. In the case of weak coupling with increased energy "saddle-node" bifurcation of antiphase mode leads to formation of rather attractive picture of the nonlinear normal modes presented above. This bifurcation (Mikhlin et al., 1998) will give rise to two pairs of localized modes with two-sided collisions of one of the masses. It turns out that in the case of stronger coupling, increasing the energy parameter gives rise, first of all, to the formation of stable (1*a*) and unstable (2*a*) modes with one-sided impacts of the masses on "unlike" stopping devices (Fig. 3.12). Only a further increase in the energy parameter gives rise to two stable and two unstable modes of type *b* (only one pair is shown in Figs. 3.13, 3.14 and 3.15: one stable and one unstable mode). As to modes of type *c* with one-sided impacts on "like" stopping devices, these are unstable.

This figure presents the branches of the "total energy $E-(w = e$ amplitude $w_2(0)$" plane for a value $\varepsilon = 1$ solution of the coupling parameter. Branches 1 and 2 represent stable periodic states (the solid curves) and unstable periodic states (the dashed curves) with one side impacts of one of the masses with its left barrier, while

Fig. 3.12 Branches of solution on E – u_2 plane (ε=1)

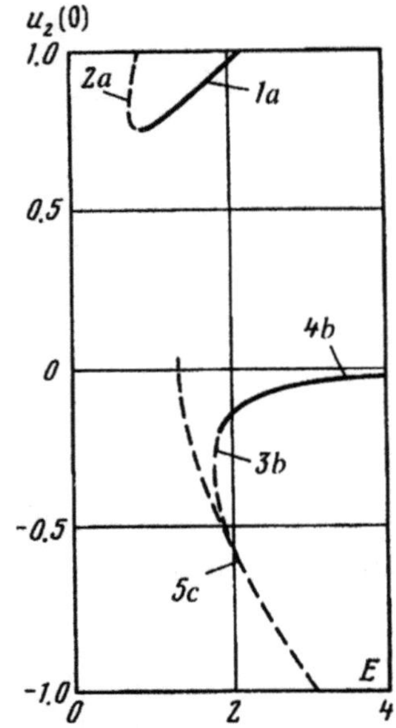

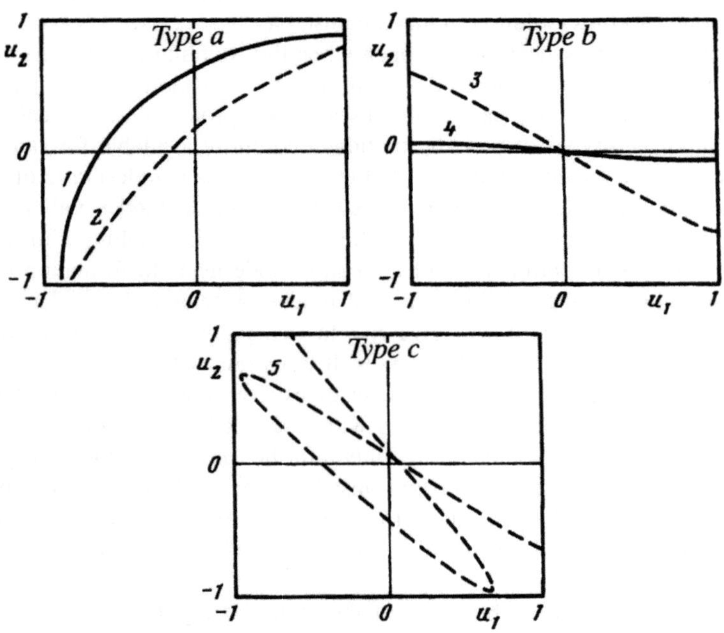

Fig. 3.13 Modes of oscillation on configuration plane u_1–u_2 (ε=1)

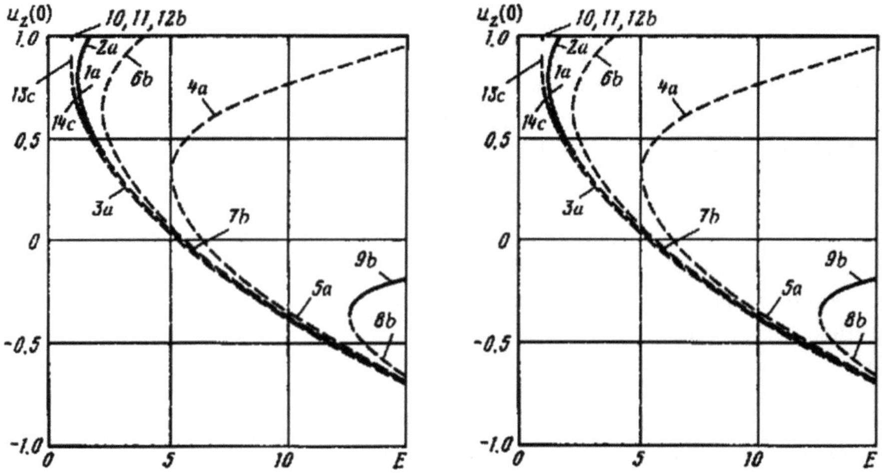

Fig. 3.14 Branches of solution on E – u_2 plane (ε=10)

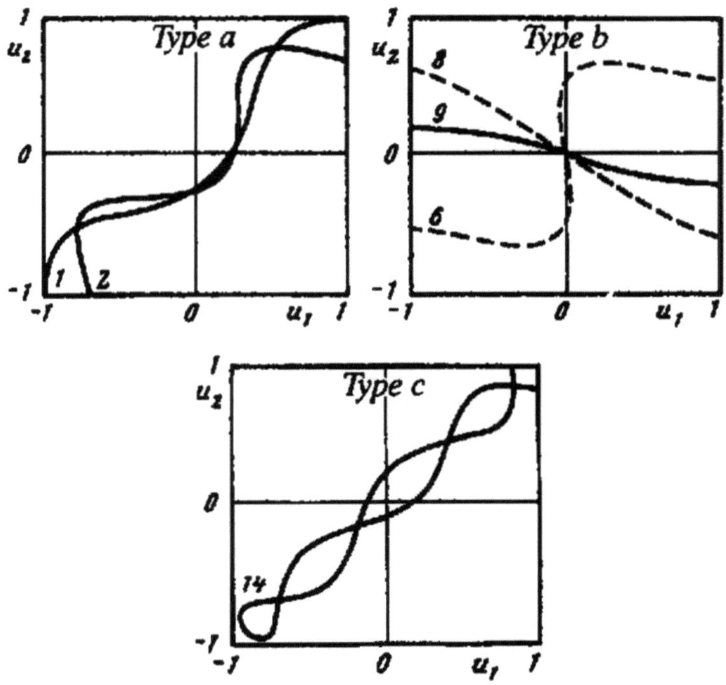

Fig. 3.15 Modes of oscillation on configuration plane u_1–u_2 (ε=10)

the other mass collides with the right barrier (type*a*), that is, the collisions take place with “unlike” stopping devices. The unstable branch 3 and stable branch 4 represent oscillations with two-sided collisions of the first mass (type*b*), the second mass experiencing no collisions. Finally, a solution 5 of type*c* (one-sided collisions of each of the masses with “similar” stopping devices) is unstable. Some modes of oscillation are shown in the configuration plane ($w_{1,}, w_2$) for $\varepsilon = 1$ in Fig. 3.13; stable modes are denoted by the solid curves and unstable modes by the dashed curves. Comparison of the curves in Figs. 3.12 and 3.13 enables us to draw important conclusions as to the change in the order in which the different types of periodic states are “born” as the energy parameter of the system is increased.

In order to demonstrate the increasingly complex behavior of the system at large values of the coupled parameter, we present also the data for periodic vibro-impact states at $\varepsilon = 10$ (Fig. 3.14 – in the “total energy E-amplitude $w_2(0)$” plane; Fig. 3.15 – in the configuration plane (w_1,w_2)). Here, the branches 1*a* and 2*a* (stable, shown in Fig. 3.11 by the solid curves), 3*a*–5*a* (unstable, shown by the dashed curves) represent oscillations of type *a* (collisions with “unlike” stopping devices). Figure 3.12 shows some of the stable and unstable modes of types *a,b,c* that have been found. The numbers on the curves in Fig. 3.15 correspond to the numbering of the modes in Fig. 3.14. Analysis of the results presented in Figs. 3.14 and 3.15 indicates that the order in which the modes of different types are “born” is changed. In particular, stable localized modes, which correspond to two-sided collisions of one of the masses and are most important in the case of small ε values, may be realized here at very large energies. However, modes with one-sided collisions of the masses with “like” stopping devices arise even at small energies.

3.2.2 Discrete Breathers in a Vibro-Impact Chain

Discrete breathers (DB), or spatially localized time – periodic solutions in Hamiltonian classic nonlinear lattices were discovered in late 1960s and have attracted a lot of attention (Ovchinnikov, 1969; Aubry, 1997, 2006; Flach and Gorbach, 2008). These solutions appear both in Klein-Gordon (KG) lattices with nonlinear on-site potential (Campbell and Peyrard, 1990) and in Fermi-Pasta-Ulam (FPU) lattices with nonlinear interaction between the particles (Sievers and Takeno, 1988; Takeno and Sievers, 1988). For both types of models mentioned above, the DB have a well-developed theory (Flach and Gorbach, 2008) and a wide range of applications, including Josephson contacts, nanomechanical systems, Bose-Einstein condensates, carbon nanotubes etc.

Despite all these developments, to the best knowledge of the authors, there exist only two non-trivial models which allow an exact computation of the DB. The first one is the well-known integrable Ablowitz-Ladik model (Ablowitz and Ladik, 1976), one of the discrete counterparts of the nonlinear Schrödinger equation. The other model was suggested by Ovchinnikov and Flach (1999). This model explores the DB in the lattices with homogeneous potentials. It should be mentioned that neither of these models belongs to the most common KG or FPU type.

We are here going to construct the exact solutions for discrete breathers in one-dimensional chains with a nonlinearity of impact type (Gendelman and Manevitch, 2008).

3.2.2.1 System of Klein-Gordon (KG) Type

Let us consider a one-dimensional linear chain with every particle placed between on-site impact barriers. The equations of motion are:

$$\ddot{u}_n + c(2u_n - u_{n-1} - u_{n+1}) = 0, \ \ |u_n| < \Delta, \ n = 0, \ \pm 1, \ \pm 2,... \tag{3.36}$$

Scalar u_n denotes the displacement of the n-th particle, the mass of each particle is adopted to be a unit, c is the rigidity of the linear coupling. The distance between the barriers on each site is equal to 2Δ. An interaction of every particle with the barrier as the displacement achieves $\pm\Delta$ is described as a purely elastic impact. It means that if the impact occurs at time t_0, then the following condition holds for all n:

$$\lim_{t\to t_0-0} \dot{u}_n = - \lim_{t\to t_0+0} \dot{u}_n \Big|_{u_n=\pm\Delta} \tag{3.37}$$

System (3.36) can be considered as a particular case of discrete Klein-Gordon lattices. It should be stressed that System (3.36) is homogeneous, i.e. the impact barriers exist at every site. System (3.36) is obviously non-integrable; still, we are going to demonstrate that, due to its simplicity, it is possible to obtain exact solutions for the DB.

Let us look for the solution of (3.36 and 3.37) with only one particle subject to periodic impacts with the barriers. Without loss of generality, we suggest that this particle corresponds to n = 0. Such impacts are equivalent to the action of periodic external δ- pulses on this particle. In other terms, particular solutions of System (3.36 and 3.37) we are looking for are equivalent to the solutions of the following equations:

$$\begin{aligned} &\ddot{u}_n + c(2u_n - u_{n-1} - u_{n+1}) = \\ &= 2p\delta_{n0} \sum_{k=-\infty}^{\infty} \left[\delta\left(t - \frac{T}{4} + kT\right) - \delta\left(t + \frac{T}{4} + kT\right) \right] \end{aligned} \tag{3.38}$$

where T is the period of the impacts, 2p is the unknown change of the particle moment in the course of the impact, δ_{ij} is the Kronecker symbol.

At this point, the crucial advantage of the vibro-impact model reveals itself – Eq. (3.38) is linear and may be solved exactly. Once the solution is obtained, one should check whether it satisfies the following conditions of self-consistence:

- Maximal displacement of the particle n = 0 is equal to Δ;
- Maximal displacements of all particles with n ≠ 0 are less then Δ (no other impacts occur).

If both these conditions will be satisfied, then the solution of forced linear equation (3.38) is a genuine solution of the initial System (3.36).

It is convenient to rewrite the right-hand side of Eq. (3.38) as a Fourier series (in the sense of distributions):

$$\ddot{u}_n + c(2u_n - u_{n-1} - u_{n+1}) = \delta_{n0}\frac{4p\omega}{\pi}\sum_{j=1}^{\infty}(-1)^j \sin((2j-1)\omega t) \tag{3.39}$$

Here $\omega = 2\pi/T$. Thus, the conditions of the impact are equivalent to a local forcing of the chain with multiple frequencies. The dispersion relation for traveling waves in the linear chain is well-known:

$$\Omega^2 = 2c(1 - \cos q) \tag{3.40}$$

where Ω is the wave frequency, q is the wavenumber. Consequently, the frequency spectrum of any periodic localized solution must be situated in the attenuation zone – above the maximum frequency

$$\Omega_{\max} = 2\sqrt{c} \tag{3.41}$$

The forcing terms in Eq. (3.39) have frequencies ω, 3ω, 5ω etc. Consequently, the forced solution of (3.39) will be localized if

$$\omega > \Omega_{\max} \tag{3.42}$$

A stationary solution of Eq. (3.39) may be easily found with the help of Z transform. It can be written down in the following form:

$$\begin{aligned} u_n(t) &= \frac{(-1)^n p\omega}{\pi c}\sum_{j=1}^{\infty}(-1)^j \frac{g(\mu)^{|n|}}{\sqrt{\mu^2(2j-1)^4 - \mu(2j-1)^2}}\sin((2j-1)\omega t) \\ \mu &= \frac{\omega^2}{4c},\ \mathrm{g}(\mu) = 2\mu(2j-1)^2 - 1 - 2\sqrt{\mu^2(2j-1)^4 - \mu(2j-1)^2} \end{aligned} \tag{3.43}$$

Maximum displacement of the particle n = 0 should be equal to the impact threshold Δ. It is achieved when t = T/4 + kT/2. In other terms:

$$|u_0(T/4)| = \frac{p\omega}{\pi c}\sum_{j=1}^{\infty}\frac{1}{\sqrt{\mu^2(2j-1)^4 - \mu(2j-1)^2}} = \Delta \tag{3.44}$$

From Eq. (3.44) one obtains the value of the unknown coefficient p. With account of (3.44), Eq. (3.43) is reduced to the following form:

$$u_n(t) = (-1)^n \Delta \frac{\sum_{j=1}^{\infty} (-1)^j \dfrac{g(\mu)^{|n|}}{\sqrt{\mu^2(2j-1)^4 - \mu(2j-1)^2}} \sin((2j-1)\omega t)}{\sum_{j=1}^{\infty} \dfrac{1}{\sqrt{\mu^2(2j-1)^4 - \mu(2j-1)^2}}} \tag{3.45}$$

Expression (3.45) is the exact solution for the DB in System (3.36 and 3.37). First of all, it should be mentioned that the series converge both in the numerator and in the denominator. In the numerator, the coefficients for the Fourier series decay like $(2j-1)^{-(n+2)}$ for large j; in the same limit the series in the denominator behaves like $\Sigma(2j-1)^{-2}$.

Maximum displacement of the n-th particle is expressed as:

$$|u_n(T/4)| = \Delta \frac{\sum_{j=1}^{\infty} \dfrac{\left(2\mu(2j-1)^2 - 1 - 2\sqrt{\mu^2(2j-1)^4 - \mu(2j-1)^2}\right)^{|n|}}{\sqrt{\mu^2(2j-1)^4 - \mu(2j-1)^2}}}{\sum_{j=1}^{\infty} \dfrac{1}{\sqrt{\mu^2(2j-1)^4 - \mu(2j-1)^2}}} \tag{3.46}$$

It is easy to demonstrate that function

$$F(x) = 2x - 1 - 2\sqrt{x^2 - x} \tag{3.47}$$

for $x > 1$ obeys $1 > F(x) > 0$ and decreases monotonously when x grows. Thus, the following inequalities hold:

$$|u_n(T/4)| = \Delta \frac{\sum_{j=1}^{\infty} \dfrac{\left(2\mu(2j-1)^2 - 1 - 2\sqrt{\mu^2(2j-1)^4 - \mu(2j-1)^2}\right)^{|n|}}{\sqrt{\mu^2(2j-1)^4 - \mu(2j-1)^2}}}{\sum_{j=1}^{\infty} \dfrac{1}{\sqrt{\mu^2(2j-1)^4 - \mu(2j-1)^2}}} <$$

$$< \left(2\mu - 1 - 2\sqrt{\mu^2 - \mu}\right)^{|n|} \Delta = \left(2\mu - 1 - 2\sqrt{\mu^2 - \mu}\right)^{|n|} |u_0(T/4)| \tag{3.48}$$

This inequality implies that if $\mu > 1$ (i.e. the basic frequency of the impacts ω is in the attenuation zone), then the solution (3.45) is exponentially localized, i.e. the maximum amplitude of the particles decreases exponentially as one moves from the central particle. Such behavior is exactly as one should expect for the DB. Besides, for any n the maximum displacement of the particles is less than Δ, i.e. they are not engaged in the impacts. This observation concludes the proof of consistency for solution (3.45).

It seems not possible to compute closed expressions for a series in expression (3.45). Still, the series converge fast enough, so no special computation difficulties are encountered.

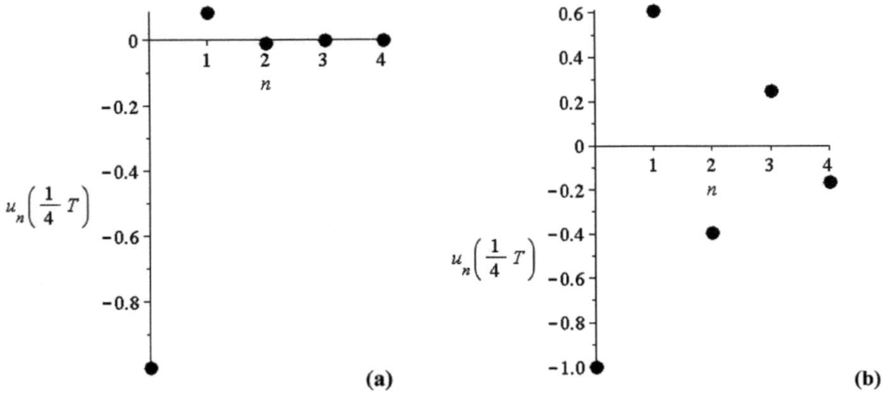

Fig. 3.16 Profile of the discrete breather (model of KG type). Maximum amplitudes of the particles are plotted, **(a)** μ=3, **(b)** μ=1.05

In order to illustrate the solution (3.45), we plot the breather profile – maximum displacement for each particle – for μ=3 (basic frequency far from the boundary of the attenuation zone, Fig. 3.16) and μ=1.05 (basic frequency close to the boundary of the attenuation zone, Fig. 3.16b). Δ is adopted to be unity. From obvious symmetry considerations, it is enough to plot the particles with $n \geq 0$ only.

One can see that, as expected, the breather with a basic frequency far from the boundary of the propagation zone is strongly localized, whereas the DB relatively close to this boundary is much wider. In order to assess the type of motion exhibited by different particles, it is instructive to plot the time dependence of displacement for $n = 0$ and $n = 1$ for the same values of Δ and μ as in Fig. 3.16a, b.

One can see that even for a moderately high basic frequency of the DB (the case $\mu = 3$ the frequency is only 1.73 of the gap value) the displacement of the particle $n = 0$ resembles the triangular wave and its shape is very different from n=1. Quite obviously, the continuum approximation would be completely unsuitable for this case. Alternatively, close to the gap boundary, for $\mu = 1.05$ the "impact" part reveals itself only near the maximum (Fig. 3.17).

In this connection, let us investigate the limit cases of solution (3.45). For the limit of high frequencies $\mu \to \infty$ one obtains:

$$u_n(t) = \begin{cases} 0, n \neq 0 \\ \dfrac{8\Delta}{\pi^2} \sum\limits_{j=1}^{\infty} (-1)^j \dfrac{\sin((2j-1)\omega t)}{(2j-1)^2}, n = 0 \end{cases} \tag{3.49}$$

For $n = 0$ sum (3.49) indeed describes a triangle wave with frequency ω. This situation corresponds exactly to the "antiintegrability" limit well known in the theory of the DB (MacKay and Aubry, 1994), where the oscillations are concentrated on a single particle.

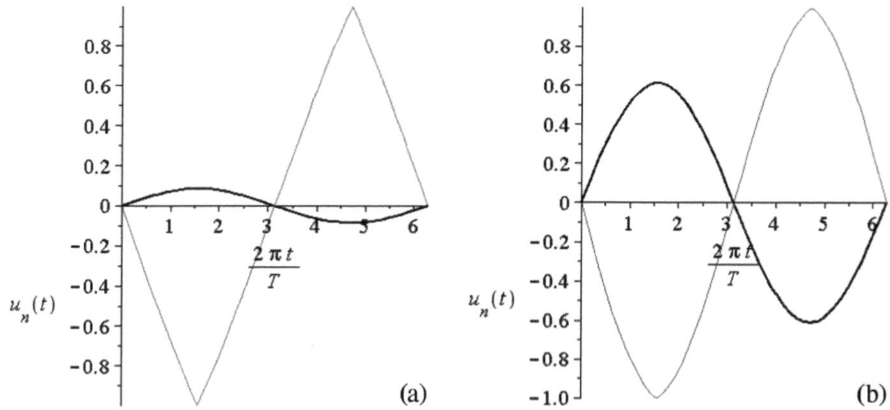

Fig. 3.17 Time history of the particles in KG-type model, *thin line* – $u_0(t)$, *thick line* – $u_1(t)$, (**a**) $\mu=3$, (**b**) $\mu=1.05$

The other limit, $\mu \to 1$, physically corresponds to close vicinity of the boundary of the attenuation zone. Let us consider the case:

$$\mu = 1 + \varepsilon, 0 < \varepsilon << 1 \tag{3.50}$$

For this case

$$\frac{1}{\sqrt{\mu^2(2j-1)^4 - \mu(2j-1)^2}} = \begin{cases} \dfrac{1}{\sqrt{\varepsilon}} + O(\sqrt{\varepsilon}), j = 1 \\ \dfrac{1}{2(2j-1)\sqrt{j^2 - j}} + O(\varepsilon), j > 1 \end{cases} \tag{3.51}$$

From estimation (3.51) it is clear that in the lowest order of approximation only the term with $j=1$ should be kept in all sums of Eq. (3.10). Consequently, the approximate solution will read

$$u_n(t) = -(-1)^n \Delta (1 - 2\sqrt{\varepsilon})^{|n|} \sin \omega t + O(\sqrt{\varepsilon}), \varepsilon \to 0 \tag{3.52}$$

These solutions correspond to rather wide DBs and are usually obtained when considering continuum approximations in terms of modulated variables.

3.2.2.2 System of Fermi-Pasta-Ulam (FPU) Type

The method for obtaining the exact DB solutions described above can be extended for a model with gradient nonlinearity without the on-site potential – chain of the FPU type. If the potential of interaction between the neighboring particles is $V(u_n - u_{n-1})$ then the equations of motion will be

$$\ddot{u}_n + \frac{\partial V(u_n - u_{n-1})}{\partial u_n} + \frac{\partial V(u_{n+1} - u_n)}{\partial u_n} = 0 \tag{3.53}$$

In the symmetric vibro-impact model, the impacts occur when the displacement between the neighboring particles achieves a certain limit value (from above and from below). Thus, the potential V(x) is defined as:

$$V(x) = \frac{1}{2}cx^2, |x| < D \tag{3.54}$$

When the relative displacement achieves its limit value D at time t_0, the impact (actually, a pair of impacts) occurs and the relative velocity changes its sign. Similarly to (3.37), one can formulate this condition as:

$$\lim_{t\to t_0-0}(\dot{u}_n - \dot{u}_{n-1}) = -\lim_{t\to t_0+0}(\dot{u}_n - \dot{u}_{n-1})\Big|_{u_n-u_{n-1}=\pm D}, n = 0, \pm 1, \pm 2... \tag{3.55}$$

The simplest situation which corresponds to the DB will occur if only one inter-particle bond has elongations large enough to cause impacts. Without the loss of generality, let us suppose that this bond is one between the particles n = 0 and n = 1. The action of impacts may be substituted by the action of two series of δ-pulses, acting in opposite directions at the particles 0 and 1. Consequently, this particular solution will satisfy the following equations of the motion:

$$\begin{aligned} &\ddot{u}_n + c(2u_n - u_{n-1} - u_{n+1}) = \\ &= 2p(\delta_{n0} - \delta_{n1}) \sum_{k=-\infty}^{\infty} \left[\delta(t - T/4 + kT) - \delta(t + T/4 + kT)\right] \end{aligned} \tag{3.56}$$

The left-hand side of Eq. (3.56) is linear and so there is no need to solve it once more – the solution can be obtained by appropriate superposition. Based on (3.43), the solution will be

$$\begin{aligned} &u_n(t) = Q(n,t) - Q(n-1,t) \\ &Q(n,t) = \frac{(-1)^n p\omega}{\pi c} \sum_{j=1}^{\infty} (-1)^j \frac{g(\mu)^{|n|}}{\sqrt{\mu^2(2j-1)^4 - \mu(2j-1)^2}} \sin((2j-1)\omega t) \end{aligned} \tag{3.57}$$

In order to determine the unknown coefficient p, one should normalize solution (3.57) according to the impact condition (3.55). By denoting the relative displacement

$$w_n = u_{n+1} - u_n \tag{3.58}$$

we get from (3.57):

$$w_0(t) = u_1(t) - u_0(t) = Q(-1,t) + Q(1,t) - 2Q(0,t) =$$
$$= -\frac{4p\omega}{\pi c}\sum_{j=1}^{\infty}\left[\begin{array}{l}\dfrac{(-1)^j\left(\mu(2j-1)^2 - \sqrt{\mu^2(2j-1)^4 - \mu(2j-1)^2}\right)}{\sqrt{\mu^2(2j-1)^4 - \mu(2j-1)^2}}\times \\ \times\sin((2j-1)\omega t)\end{array}\right] \tag{3.59}$$

The normalization condition will then read

$$|w_0(T/4)| = \frac{4p\omega}{\pi c}\sum_{j=1}^{\infty}\frac{\left(\mu(2j-1)^2 - \sqrt{\mu^2(2j-1)^4 - \mu(2j-1)^2}\right)}{\sqrt{\mu^2(2j-1)^4 - \mu(2j-1)^2}} = D \tag{3.60}$$

Finally, the solution for the DB in the FPU-type vibro-impact chain will be

$$u_n(t) = Z(n,t) - Z(n-1,t)$$
$$Z(n,t) = \frac{(-1)^n D\sum_{j=1}^{\infty}(-1)^j\dfrac{g(\mu)^{|n|}}{\sqrt{\mu^2(2j-1)^4 - \mu(2j-1)^2}}\sin((2j-1)\omega t)}{4\sum_{j=1}^{\infty}\dfrac{\left(\mu(2j-1)^2 - \sqrt{\mu^2(2j-1)^4 - \mu(2j-1)^2}\right)}{\sqrt{\mu^2(2j-1)^4 - \mu(2j-1)^2}}} \tag{3.61}$$

Convergence of all series is easily established by considerations literally similar to those presented above for solution (3.45). The only additional element for proof of consistency is the fact that no other bond besides w_0 is engaged in the impacts. According to (3.61), the deformation of the n-th bond is expressed as:

$$w_n(t) = Z(n+1,t) + Z(n-1,t) - 2Z(n,t) \tag{3.62}$$

Functions Z(n,t) have opposite signs for the neighboring particles; consequently, one obtains:

$$\begin{aligned}|w_n(t)| &= |Z(n+1,t) + Z(n-1,t) - 2Z(n,t)| = \\ &= |Z(n+1,t)| + |Z(n-1,t)| + 2\,|Z(n,t)| \le \\ &\le |Z(n+1,T/4)| + |Z(n-1,T/4)| + 2\,|Z(n,T/4)|\end{aligned} \tag{3.63}$$

However,

$$\begin{aligned}D &= \max|w_0(t)| = |w_0(T/4)| = \\ &= |Z(-1,T/4)| + |Z(1,T/4)| + 2\,|Z(0,T/4)|\end{aligned} \tag{3.64}$$

By virtue of (3.48) and (3.61), quite obviously, for any $n \neq 0$ and $\mu > 1$ the sum of terms in right-hand side of (3.64) is strictly larger than the last sum in (3.63). Therefore

$$|w_n| < D, n \neq 0, \mu > 1 \tag{3.65}$$

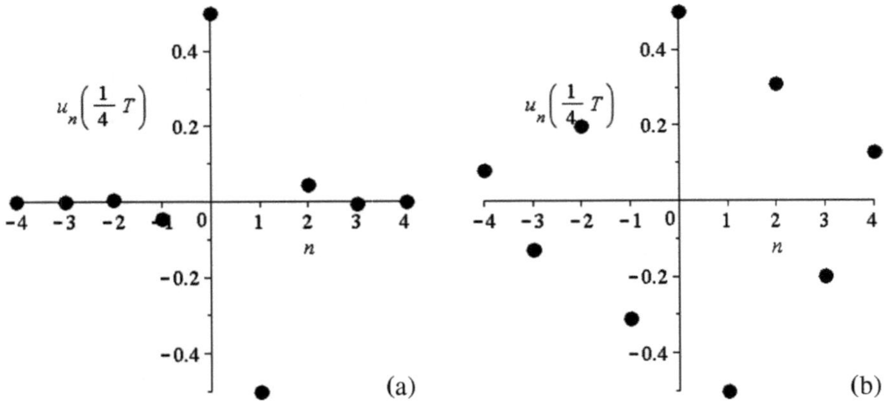

Fig. 3.18 Profile of the discrete breather (model of FPU type). Maximum amplitudes of the particles are plotted, (**a**) $\mu=3$, (**b**) $\mu=1.05$

Inequality (3.65) proves the consistency of solution (3.61) for the DB in FPU-type model. Interestingly, the consistency of this solution follows from the consistency of solution (3.45) for the DB in a KG-type chain. Plots for maximum displacements $u_n(T/4)$ for two different values of μ according to solution (3.61) are presented in Fig. 3.18a, b.

The solutions presented above can be significant as the benchmarks suitable for testing the approximations in the theory of the DB. Besides, the impact interaction is high-energy limit for common models of the nonlinear lattices, such as the Toda lattice (Toda, 1989) or systems with Lennard-Jones or Morse potentials.

The stability of solution (3.45) was verified by means of direct numeric simulations with parameters used for the generation of Fig. 3.16. No detectable instability was revealed within more than 10,000 periods of oscillations in both cases. Yet, such simulation does not prove the stability rigorously. In order to analyze the stability in a rigorous manner, one should check the spectral properties of the linear dynamics around the DB. Such a problem seems to be rather complicated, due to both the singular nature of the problem and the infinite number of harmonics involved in the exact solution.

One can find exact analytic solutions for the discrete breathers both in the non-integrable chains of Klein-Gordon and Fermi-Pasta-Ulam types with vibro-impact potentials. These solutions are possible since the vibro-impact interaction can be rigorously reduced to an action of periodic external force on the linear lattice. Thus, these solutions can also be easily generalized for higher dimensions of the lattices, provided that the linear lattice is combined with the appropriate impact interaction. Moreover, the method described above can allow the construction of more complicated solutions, like coupled DB or DB with internal oscillating modes.

3.3 The Problem of Heat Conduction in Dielectrics

We would like to show that the vibroimpact model gives an unique possibility to obtain analytical results in one of the "hot" problems of modern Solid State Physics – substantiation of the phenomenological theory of heat conductivity.

Heat conductivity in one-dimensional (1D) lattices is a well known classical problem related to the microscopic foundation of Fourier's law. The problem started from the famous work of Fermi, Pasta, and Ulam (FPU) (Fermi et al., 1955), where an abnormal process of heat transfer was detected for the first time. Non-integrability of a system is a necessary condition for normal heat conductivity. As it was demonstrated recently (Lepri et al., 2003) for the FPU lattice, the disordered harmonic chain, diatomic 1D gas of colliding particles, and the diatomic Toda lattice, non-integrability is not sufficient in order to get normal heat conductivity. It leads to a linear distribution of temperature along the chain for a small gradient, but the value of heat flux is proportional to $1/N^{\alpha}$, where N is the number of particles in the chain and the number exponent $0 < \alpha < 1$. Thus, the coefficient of heat conductivity diverges in the thermodynamic limit $N \to \infty$. Analytical estimations have demonstrated that any chain possessing an acoustic phonon branch should have infinite heat conductivity in the limit of low temperatures.

Probably the most interesting question related to heat conductivity of 1D models (which actually inspired the first investigation of Fermi, Pasta and Ulam) is whether small perturbation of an integrable model will lead to a convergent heat conduction coefficient. One supposes that for the one-dimensional chains with conserved momentum the answer is negative (Lepri et al., 1997). Still, normal heat conduction has been observed in some special systems with conserved momentum (Gendelman and Savin, 2000; Giardina et al., 2000), but it may be clearly demonstrated only well apart from the integrable limit. It means that mere non-integrability is insufficient to ensure normal heat conduction if an additional integral is present.

It seems that computational difficulties of the investigation of heat conduction in vicinity of integrable limit are not just an issue of weak computers or ineffective procedures. In systems with conserved momentum, divergent heat conduction is fixed by power-like decrease of the heat flux autocorrelation function with power less than unity. Still, for the systems with on-site potential exponential, decrease is more typical (Savin and Gendelman, 2003). For any fixed value of the exponent the heat conduction converges; if the exponent tends to zero with the value of the perturbation of the integrable case, then for any finite value of the perturbation the characteristic correlation time and length will be finite but may become very large. Consequently, they will exceed any available computation time or size of the system and still, no conclusion on the convergence of heat conduction will be possible.

The way to overcome this difficulty is to construct a model (Gendelman and Savin, 2004) which will be, at least to some extent, analytically tractable and will allow one to predict some characteristic features of the heat transfer process and the behavior of the heat conduction coefficient. Afterwards, the numerical simulation may be used to verify the assumptions made in the analytic treatment. To the best of our knowledge, no models besides pure harmonic chains were treated in such a way to date.

We are going to demonstrate that there exist models which have an integrable system as their natural limit case, small perturbation of the integrability immediately leading to convergent heat conduction. The mechanism of energy scattering in this kind of system is universal for any temperature and set for the model parameters. The simplest example of such a model is a one-dimensional set of equal rigid particles with nonzero diameter ($d > 0$) subjected to periodic on-site potential. This system is completely integrable only if $d = 0$. It will be demonstrated that any $d > 0$ leads to effective mixing due to unequal exchange of energy between the particles in each collision. This mixing leads to a diffusive mechanism of the heat transport and, subsequently, to convergent heat conduction.

Let us consider the one-dimensional system of hard particles with equal masses subject to periodic on-site potential. The Hamiltonian of this system will read

$$H = \sum_n \left\{ \frac{1}{2} M \dot{x}_n^2 + V(x_{n+1} - x_n) + U(x_n) \right\}, \tag{3.66}$$

where M – mass of the particle, x_n – coordinate of the center of the n-th particle, $\dot{x}_n$ – velocity of this particle, $U(x)$ – periodic on-site potential with period a $[U(x) \equiv U(x + a)]$. Interaction of absolutely hard particles is described by the following potential

$$V(r) = \infty \text{ if } r \le d \quad \text{and} \quad V(r) = 0 \text{ if } r > d, \tag{3.67}$$

where d is the diameter of the particle. This potential corresponds to a pure elastic impact with unit recovery coefficient. A sketch of the model considered is presented in Fig. 3.19.

It is well-known that the elastic collision of two equal particles with collinear velocity vectors leads to the exchange of their velocities. An external potential present does not change this fact, since the collision takes zero time and thus the effect of the external force on the energy and momentum conservation is absent.

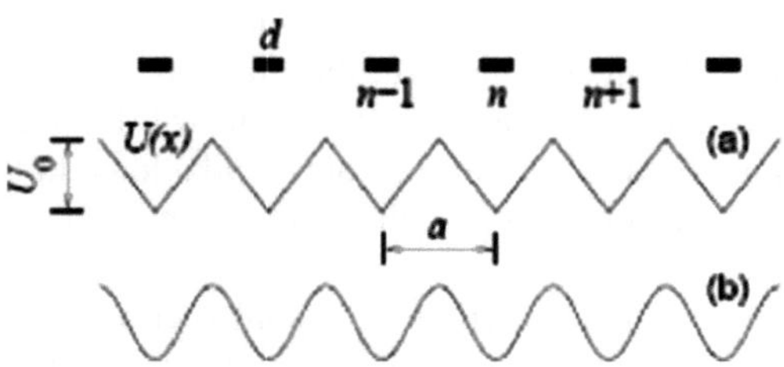

Fig. 3.19 Sketch of the hard-disk chain exposed to a periodic on-site potential $U(x)$ (A is the period of the potential, U_0 – its height, d is the diameter of the disks). Piecewise-linear potential (3.75) (**a**) and sinusoidal potential (3.85) (**b**) are plotted

The one-dimensional chain of equivalent hard particles without external potential is a paradigm of the integrable nonlinear chain, since all interactions are reduced to an exchange of velocities. In other words, the individual values of velocities are preserved and simply transferred from particle to particle. It is natural therefore to introduce quasiparticles associated with these individual values of velocities. They will be characterized by a pair of parameters $(E_k, \mathbf{n}_k)$, where $E_k = v_k^2/2$ is an energy of the quasiparticle, $\mathbf{n}_k$ is a unit vector in the direction of its motion. Every particle at every moment "carries" one quasiparticle. The elastic collision between the particles leads to a simple exchange of parameters of the associated quasiparticles, therefore the quasiparticles themselves should be considered as non-interacting.

The situation changes if the external on-site potential is present. It is easy to introduce similar quasiparticles (E_k will be a sum of kinetic and potential energy). The unit vector $\mathbf{n}$ of each quasiparticle between subsequent interactions may be either constant (motion in one direction) or periodically changing (vibration of the particle in a potential well). In every collision the particles exchange their velocity vectors, but do not change their positions. Consequently two quasiparticles interact in a way described by the following relationships:

$$
\begin{aligned}
E_1' &= E_1 + U(x_c + d/2) - U(x_c - d/2) \\
E_2' &= E_2 - U(x_c + d/2) + U(x_c - d/2) \\
\mathbf{n}_1' &= \mathbf{n}_1, \ \mathbf{n}_2' = \mathbf{n}_2
\end{aligned}
\tag{3.68}
$$

The values denoted by the apostrophe correspond to the state after the collision, x_c is a point of contact between the particles. It should be mentioned that in the case of a nonzero diameter, the quasiparticles are associated with the centers of the carrying particles.

If the diameter of the particles is zero, then the additives to the energies in the first two equations of System (3.68) compensate each other and the energies of the quasiparticles are preserved in the collision. Therefore, the interaction between the quasiparticles effectively disappears and the chain of equal particles with zero size subject to any on-site potential turns out to be a completely integrable system. Thus, contrary to many previous statements, it is possible to construct an example of a strongly nonlinear one-dimensional chain without momentum conservation, which will have clearly divergent heat conductivity (even a linear temperature profile will not be formed).

The situation differs if the size of the particles is not zero, as the energies of the particles are not preserved in the collisions. In order to consider the effect of such an interaction we propose a simplified semi-phenomenological analytical model.

After l collisions the energy of the quasiparticle will be

$$
E(l) = E_0 + \sum_{j=1}^{l} \Delta E_j, \quad \Delta E_j = U(x_j + d/2) - U(x_j - d/2), \tag{3.69}
$$

where j-th collision takes place in point x_j, E_0 is the initial energy of the quasiparticle. Now we suppose that the coordinates of subsequent contact points $\{..., x_{j-1}, x_j, x_{j+1}, ...\}$, taken by modulus of the period of the on-site potential, are not correlated. Such proposition is equivalent to fast phase mixing in a system close to an integrable one and is well-known in various kinetic problems (Gendelman and Savin, 2004). The consequences of this proposition will be verified below by direct numerical simulation.

Average energy of the quasiparticle is equal to $< E_0 >$ over the ensemble of the quasiparticles, as obviously $< \Delta E_j >= 0$. Still, the second momentum will be nonzero:

$$< (E(l) - E_0)^2 >= l < (U(x + d/2) - U(x - d/2))^2 >_x \tag{3.70}$$

The right-hand side of this expression will depend only on the exact shape of the potential function

$$\begin{aligned} &< (E(l) - E_0)^2 >= lF(d), \\ &F(d) = \tfrac{1}{a} \textstyle\int_0^a [U(x + d/2) - U(x - d/2)]^2 dx, \end{aligned} \tag{3.71}$$

The last expression is only correct at the limit of high temperatures; it neglects the fact that the quasiparticle spends more time near the top of the potential barrier due to a lower velocity.

Let us consider the quasiparticle with initial energy $E_0 > U_0$, where U_0 is the height of the potential barrier. Therefore, vector $\mathbf{n}$ is constant. Equations (3.103) and (3.104) describe random walks of the energy of the quasiparticle along the energy scale axis. Therefore, after a certain number of steps (collisions), the energy of the quasiparticle will enter the zone below the potential barrier $E(l) < U_0$. In this case, the behaviour of the quasiparticle will change, as constant vector $\mathbf{n}$ will become oscillating, as described above. After some additional collisions the energy will again exceed U_0, but the direction of motion of the quasiparticle will be arbitrary. It means that the only mechanism of energy transfer in the system under consideration is associated with the diffusion of the quasiparticles, which are trapped by the on-site potential and afterwards released in an arbitrary direction. Such traps-and-releases resemble Umklapp processes of phonon-phonon interaction, but occur in a purely classic system.

The diffusion of the quasiparticles in the chain is characterized by the mean free path, which may be evaluated as

$$\lambda \sim \frac{2a < (U_0 - E_0)^2 >}{n_c F(d)} \sim \frac{2a[2(k_B T)^2 - 2U_0 k_B T + U_0^2]}{n_c F(d)}, \tag{3.72}$$

where n_c is a number of particles over one period of the on-site potential (concentration). Coefficient 2 appears due to the equivalent probability of a positive and negative energy shift in any collision, T – temperature of the system, k_B – Boltzmann constant.

The average absolute velocity of the quasiparticle may be estimated as

$$< |v| > \sim \frac{a}{a - n_c d} \sqrt{\frac{\pi k_B T}{2}} \tag{3.73}$$

Here, the first multiplier takes the nonzero value of d and the absolute rigidity of the particles into account. The second one is due to the standard Maxwell distribution function for a 1D case.

The heat capacity of the system over one particle is unity, as the number of degrees of freedom (i.e. the number of quasiparticles) coincides with the number of the particles and does not depend on the temperature and other parameters of the system. Therefore, the coefficient of heat conductivity may be estimated as

$$\kappa \sim \lambda < |v| > \sim \frac{2a^2}{n_c(a - n_c d)} \frac{2(k_B T)^2 - 2U_0 k_B T + U_0^2}{F(d)} \sqrt{\pi k_B T/2} \tag{3.74}$$

It is already possible to conclude that, according to (3.74) and regardless of the concrete shape of potential $U(x)$ in the limit $d \to 0$, we have $F(d) \to 0$ and therefore $\kappa \to \infty$, although for every nonzero value d the heat conductivity will be finite. Therefore, unlike known models with conserved momentum the small perturbation of the integrable case $d = 0$ immediately brings about convergent heat conductivity.

It is convenient for the following numerical simulation to introduce dimensionless variables. Let us to set the mass of each particle $M = 1$, on-site potential period $a = 2$, its height $U_0 = 1$, and Boltzmann constant $k_B = 1$ in all above relationships. We suppose that the chain contains one particle per each period of the potential, i.e. that $n_c = 1$, and the particle diameter $0 < d < 2$.

Let us consider the periodic piecewise linear on-site potential

$$\begin{aligned} &U(x) = x, x \in [0, 1] \\ &U(x) = 2 - x, x \in [1, 2] \\ &U(x + 2\,l) = U(x), x \in [0, 2]\,, l = 0, \pm 1, \pm 2, ... \end{aligned} \tag{3.75}$$

(the shape of the potential is presented in Fig. 3.19). Then it follows from (3.72) that the non-dimensional heat conduction coefficient is expressed as

$$\kappa = \frac{8(2T^2 - 2T + 1)}{(2 - d)F(d)} \sqrt{\pi T/2}, \tag{3.76}$$

where function

$$\begin{aligned} &F(d) = d^2 - 2/3d^3, \quad \text{for } \; 0 < d \le 1 \\ &F(d) = -4/3 + 4d - 3d^2 + 2/3d^3, \quad \text{for } \; 1 \le d < 2. \end{aligned} \tag{3.77}$$

The dynamics of the system of particles with potential of the nearest-neighbor interaction (3.67) and piecewise linear on-site potential (3.75) may be described exactly. Between the collisions each particle moves under constant force with a sign dependent on the position of the particle. Therefore, the coordinate of each particle depends on time t as a piecewise parabolic function which may be easily computed analytically. If the particle centers are situated at a distance equal to d, elastic collision occurs. The particles exchange their momenta as described above and afterwards, the particle motion is again described by piecewise parabolic functions until the next collision.

Let us consider a finite chain of N particles with periodic boundary conditions. Let one particle be at each potential minimum at the moment $t = 0$ and let us choose Boltzmann's distribution of the initial velocity. Solving the equations of motion, we find a time t_1 of the first collision between some pair of adjacent particles, next a time t_2 of the second collision, in general between another pair of the adjacent particles, and so on. As a result, we obtain a sequence $\{t_i, n_i\}_{i=1}^{\infty}$, where t_i is the time of the ith collision in the system, and n_i and $n_i + 1$ are the particles participating in this collision. First, we incorporate the energy change of the n_ith particle during the Ith collision as

$$\Delta E_{n_i} = \frac{1}{2}({v'_{n_i}}^2 - {v_{n_i}}^2) = \frac{1}{2}({v_{n_i+1}}^2 - {v_{n_i}}^2).$$

Next, we introduce a time step Δt, which is significantly less than the simulation time, but satisfies the inequality $\Delta t \gg t_0$, where $t_0 = lim_{i\to\infty}(t_i/i)$ is the mean time between successive collisions. Then, for each $k = 0, 1, ...,$ we define the local energy flow as a piecewise constant (in time) function

$$j_n(t) = \frac{a}{\Delta t} \sum_{i \in I_{kn}} \Delta E_{n_i}, \quad k\Delta t \le t < (k+1)\Delta t, \tag{3.78}$$

where the integer sets I_{kn}'s are defined by

$$I_{kn} = \{i |\ k\Delta t \le t_i < (k+1)\Delta_t,\ n_i = n\}.$$

The set I_{kn} takes those collisions into account that occur between particles n and $n+1$ during the time interval $k\Delta t \le t < (k+1)\Delta_t$. Equilibration times were typically occurring in the system of the order 10^6. After these times have passed, we define the time-averaged local thermal flow

$$J_n = < j_n(t) >_t \equiv \lim_{t\to\infty} \frac{1}{t} \int_0^t j_n(\tau) d\tau \tag{3.79}$$

and the temperature distribution $T_n = < v_n^2(t) >_t$, where $v_n(t)$ is the velocity of particle n calculated at a time t. To find these averaged quantities, we have used times up to 10^7.

To find the flow-flow correlation function $C(t)$ numerically, we calculated the time average $< J(\tau)J(\tau - t) >_\tau /NT^2$, with $J(t) = \sum_n j_n(t)$ being the total heat flow through the gas/chain system consisting of $N = 500$ particles and temperature $T = \sum_n T_n/N$ averaged over 10^4 realizations of initial thermalization.

Numerical simulation of the dynamics demonstrates an exponential decrease of the autocorrelation $C(t) \sim \exp(-\alpha t)$ for all values of the diameter $0 < d < 2$ and temperature $T > 0$ where the simulation time is plausible from a technical viewpoint. For low temperatures however, the exponential decrease is accompanied by oscillations with a period corresponding to the frequency of the vibrations near the potential minima (Fig. 3.20). The reason for this is that when the temperatures are low, the concentration of transient particles decreases exponentially and a majority of the particles vibrates near the potential minima. It means that the 1D gas on the on-site potential has a finite heat conductivity. The coefficient of the exponential decrease of the autocorrelation function

$$\alpha = -\lim_{t\to\infty} \frac{lnC(t)}{t} \tag{3.80}$$

and the coefficient of the heat conduction

$$\kappa = \int_0^\infty C(t)dt. \tag{3.81}$$

are computed numerically.

Dependence of α and κ on the particle diameter d is presented in Fig. 3.21. The minimum of α and the minimum of κ is attained at $d = 1.4$. As the temperature grows, α decreases (Fig. 3.22a), and heat conduction κ increases.

Theoretical analysis of the heat conductivity presented above allows only approximate [although rather reliable, see Fig. 3.22] prediction of the numerical value of the heat conduction coefficient κ. Still, the other question of interest is the asymptotic dependence of the heat conduction on the parameters of the model. Formulae (3.76 and 3.77) lead to the following estimations:

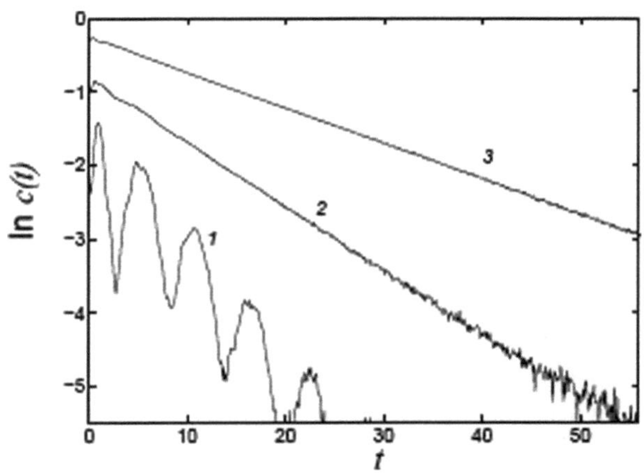

Fig. 3.20 Correlation functions of the system of particles with $d = 0.5$ under temperatures $T = 0.24, 0.45$ and 0.75 (*curves* 1, 2 and 3)

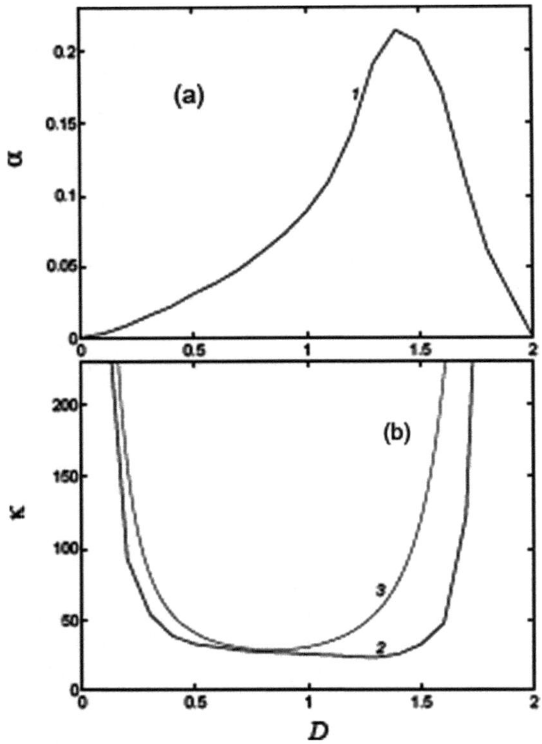

Fig. 3.21 Dependence of the coefficient of the exponential decrease on the autocorrelation function α (**a**) and the coefficient of the heat conduction κ (**b**) on the particle diameter d of 1D gas at T=1. *Curves* 1 and 2 correspond to piecewise linear on-site potential (3.75), *curve* 3 represents theoretical predictions according to formula (3.76)

$$\kappa \sim T^{5/2}, \quad \text{for} \quad T \to \infty, \tag{3.82}$$

$$\kappa \sim d^{-2}, \quad \text{for} \quad d \to +0, \tag{3.83}$$

$$\kappa \sim (2-d)^{-3}, \quad \text{for} \quad d \to 2-0. \tag{3.84}$$

These estimations should be compared to numerical results.

In order to check estimation (3.82), we consider the dependence of the logarithm of the heat conduction lnκ on the logarithm of the temperature ln$T$. From Fig. 3.23 it is clear that in accordance with (3.82) lnκ grows as 2.5ln$T$ as $T \to \infty$. Figure 3.24a demonstrates that as $d \to +0$, the logarithm lnκ grows as $-2\ln d$, in accordance with (3.83). Figure 3.24b demonstrates that as $d \to 2-0$, the logarithm lnκ grows as $-3\ln(2-d)$, in accordance with (3.84). So, it is possible to conclude that analytical estimations (3.82, 3.83, and 3.84) correspond fairly well to the numerical simulations data.

The analytical estimations above imply that the type of dependence of characteristic exponent α and heat conductivity κ on diameter d and temperature T does

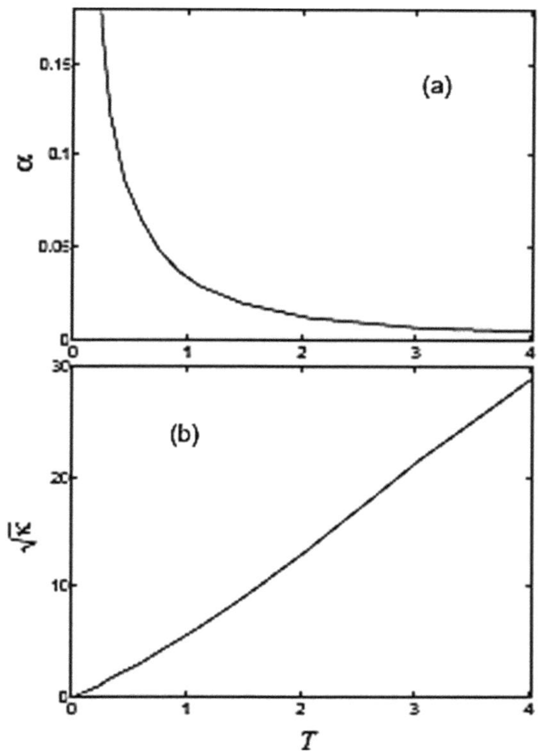

Fig. 3.22 Temperature dependence of exponent coefficient α (**a**) and heat conduction coefficient κ (**b**) for particle diameter $d = 0.5$

not depend on the concrete shape of on-site potential $U(x)$ – actually, only its finiteness and periodicity do matter. Piecewise linear periodic potential (3.75) was chosen since it allowed essential simplification of the numerical procedure. In comparison, we have also considered the smooth sinusoidal periodic potential

$$U(x) = [1 - cos(\pi u)]/2 \tag{3.85}$$

with period 2 and amplitude $U_0 = 1$, similarly to potential (3.75).

Potential (3.85) does not allow exact integration and requires standard numerical procedures. Therefore it is also convenient to replace the rigid wall potential (100) by the smooth Lennard-Jones potential

$$V(\varepsilon; r) = \varepsilon \left(\frac{1}{r-d} - \frac{1}{2-d} \right)^2 . \tag{3.86}$$

Parameter $\varepsilon > 0$ characterizes the rigidity of the potential, the hard-particle potential being the limit case:

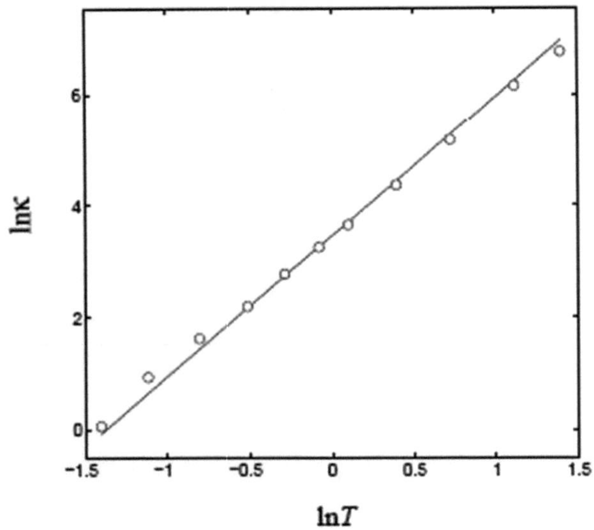

Fig. 3.23 Dependence of heat conduction coefficient on the temperature. The *markers* correspond to numerical results (lnκ versus ln$T$), the straight line is lnκ = 2.5ln$T$ + 3.45, corresponding to (3.82). Particle diameter $d = 0.5$

$$V(r) = \lim_{\varepsilon \to +0} V(\varepsilon; r).$$

Methods of computing the autocorrelation function C(t) and the heat conduction coefficient κ in a 1D chain with analytic potentials of interaction are described in (Gendelman and Savin, 2004). It should be mentioned that in order to get close to the limit of the hard particles small values of ε should be used (at the temperature $T = 1$ value $\varepsilon = 0.01$ was used). It implies a rather small value of the integration step. (We used the standard Runge-Kutta procedure of the fourth order with a constant integration step $\Delta t = 0.0001$). Therefore, for the case of hard (or nearly hard) particles, the simulation with smooth on-site potential (3.85) is far more time-consuming than the simulation with piecewise linear potential (3.75).

In the case of hard particles with smooth on-site potential, the autocorrelation function C(t) decreases exponentially as $t \to \infty$ for all range $0 < d < 2$, $T > 0$, i.e. the heat conduction converges. Figure 3.25 demonstrates that the type of dependence of and κ on parameters d and T is similar for a piecewise linear potential (3.75) and a sinusoidal potential (3.85) (although numerical values α and κ vary slightly). For this potential function $F(d) = \frac{1}{2}\sin^2(\pi d/2)$. It confirms that the type of heat conduction does not depend on a concrete choice of an on-site potential function.

The results above mean that there exists a special class of universality of 1D chain models with respect to their heat conductivity. The limit case of zero-size particles is integrable, but the slightest perturbation of this integrable case by introducing the nonzero size leads to a drastic change of the behavior – it becomes diffusive and the heat conduction coefficient converges. It should be stressed that this class of

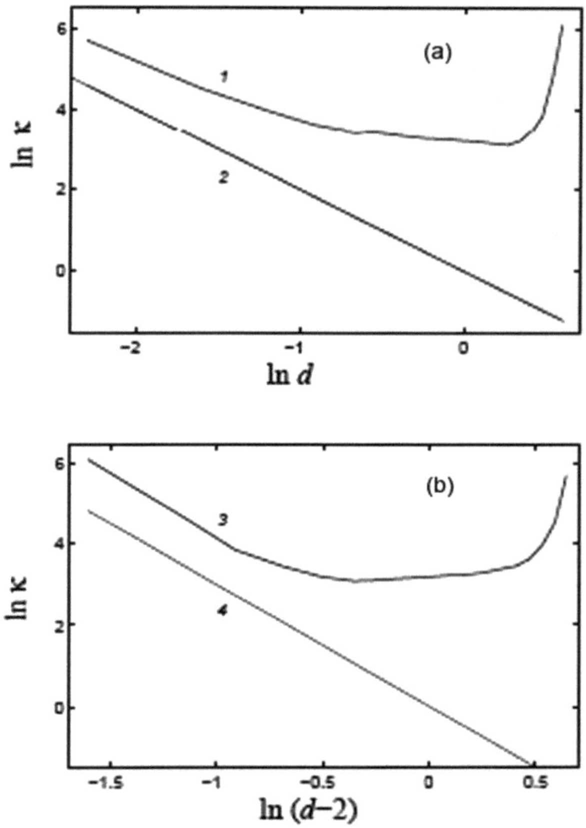

Fig. 3.24 Dependence of the heat conduction coefficient on the particle diameter (logarithmic coordinates, ln κ versus ln d (**a**) and versus ln(2 − d) (**b**), *curves* 1 and 3). *Lines* ln $\kappa = -2 \ln d$ (*curve* 2) and ln $\kappa = -3 \ln (2 - d)$ (*curve* 4) correspond to relationships (3.83) and (3.84). Temperature $T = 1$

universality, unlikely the systems with conserved momentum, cannot be revealed by sole numerical simulation. The reason is that the correlation length (as well as the heat conduction coefficient) diverges as the system approaches the integrable limit. Therefore, any finite capacity of the numerical installation will be exceeded. That is why the analytical approach is also necessary.

3.4 Solitons in Energetically Nondegenerate Quasi-One-Dimensional Models

Beginning with the works of (Fermi et al., 1955) and (Krumhansl and Schriefer, 1975), the study of the soliton mechanisms of energy and state transfer became one of the most relevant and intensely developed directions in many fields of physics, chemical physics, and biophysics. However, in most of the works, it is assumed that the system has either a single stationary state or several such states with the same

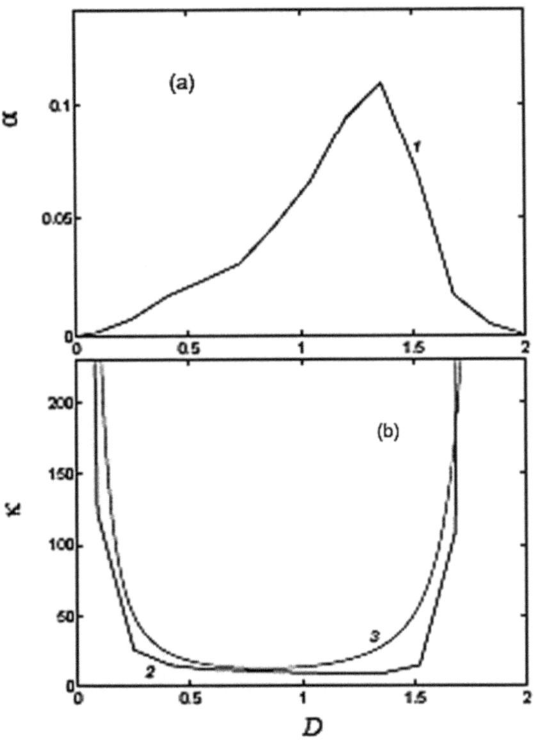

Fig. 3.25 Dependence of the coefficient of the exponential decrease on the autocorrelation function α (**a**) and the coefficient of the heat conduction κ (**b**) on the particle diameter d of 1D gas at T=1. *Curves* 1 and 2 correspond to smooth on-site potential (3.85), *curve* 3 represent theoretical predictions according to formula (3.76)

energy. This assumption is made in almost all works on the dynamics and thermodynamics of structural transitions. This means that all chemical reactions with nonzero thermal effects and all types of exo- and endo-thermal structural transitions are excluded from consideration. This situation in the field is not coincidental, because, until recently, no analytical models were developed for describing stationary processes with the consumption or release of energy on the atomic or molecular level.

To perform a quantitative analysis, we considered a chain of bound bistable (i.e., having two equilibrium positions) oscillators (Manevitch et al., 1994; Manevitch and Smirnov, 1995, 1998; Manevitch and Savin, 1995). Let us assume that the chain is in a metastable state, that is, in the state with higher energy. A small thermal perturbation cannot disrupt the initial state, and, therefore, this system is stable in the linear approximation. Next, we assume that the transition of one of the oscillators into the ground state occurs accidentally. In this case, the energy of reaction Q is released, which is transferred to the energy of oscillations about the new equilibrium position. At high barrier energy E, this will not lead to the transition of the other oscillators to the ground state. Still, the energy Q will be distributed gradually over the entire chain. As can be seen, in this case, no reaction front is formed.

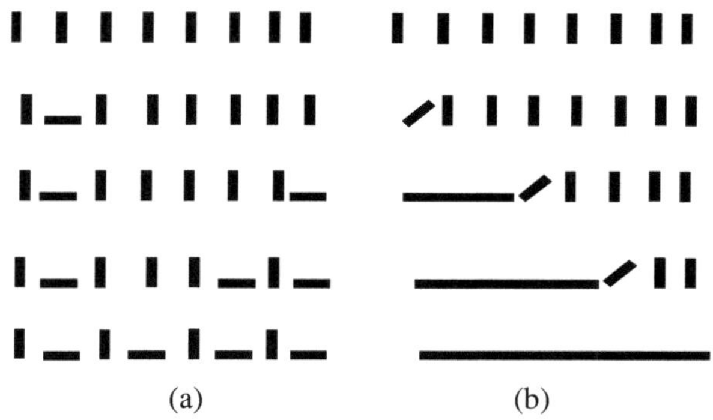

Fig. 3.26 Comparison of the (**a**) local-fluctuation and (**b**) wave mechanisms of propagation of the reaction. The vertical orientation of the *blocks* corresponds to the reagents, whereas the horizontal orientation correponds to the reaction products

If (under proper initial conditions) the transition from the metastable to the ground state occurs in the form of a wave with a front, the situation is substantially different. The stability of the front is provided by the stability of the nonlinear modes comprising the front but not by the feedback mechanism, as in dissipative systems on the macroscopic level. The energy of the wave front is spent only to overcome the energy barrier between the two states and, after that, it is transferred to the next oscillator, whereas the reacted one retains the energy Q and vibrates around the new equilibrium position. Figure 3.26 schematically illustrates the difference between the local-fluctuation and wave mechanisms of the propagation of the reaction. In Fig. 3.26, the vertical orientation of rods corresponds to the metastable state and the horizontal orientation to the ground state. This quantitative model allows an explanation (on the atomic and molecular levels) why the reaction front exists and propagates at a constant velocity. Below, we will show that this scenario can be realized for the motion of the front of an exothermal reaction in a diatomic molecular crystal.

The mathematical apparatus for modeling this process is complicated, because it is unknown how to analyze the governing dynamic equations in order to obtain a solution in the form of a stationary wave describing the transition from the metastable state to a vibrationally excited ground state.

By now, a number of approaches to the analytical description of the propagation of endo- and exothermal processes have been proposed (Manevitch et al., 1994; Manevitch and Smirnov, 1995, 1998; Manevitch and Savin, 1995) that treat elementary events of chemical reactions in solids in terms of nonlinear excitations of arranged molecular systems.

3.4.1 Quasi-One-Dimensional Model of a Molecular Crystal: Soliton Modes of Motion in a Bistable Nonlinear System

Before discussing experimental results, we will consider a simple quasi-one-dimensional model of a diatomic molecular crystal with two uniform equilibrium configurations. This model is schematically displayed in Fig. 3.29 and described in

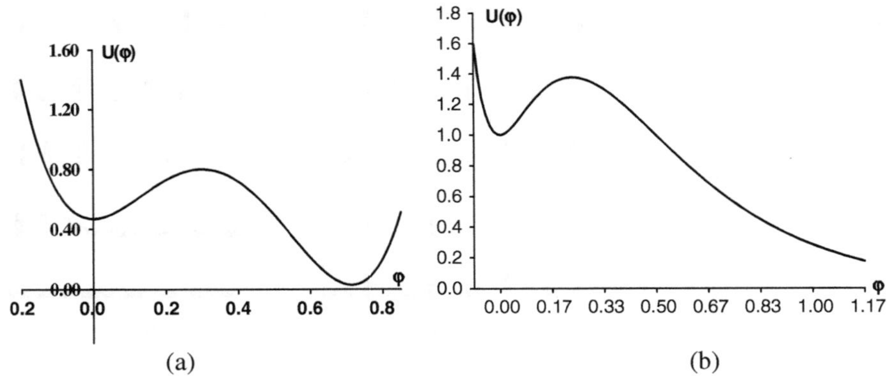

Fig. 3.27 Typical intramolecular potential for a model of diatomic molecular crystal; φ is the reaction coordinate, corresponding to the intramolecular distance. For details, see the text

detail in oscillators (Manevitch et al., 1994; Manevitch and Smirnov, 1995, 1998; Manevitch and Savin, 1995).

The intramolecular interaction is described by the potential U. Figure 3.27 presents some typical examples of this potential.

The potentials shown in Fig. 3.27 correspond to chemical processes of different types. The potential in Fig. 3.27a describes the exothermal (the initial state corresponds to the left minimum of the potential U) or endothermal (the initial state corresponds to the right minimum) processes (a chemical reaction or a structural phase transition) that do not lead to the destruction of the initial system but results in a change in its configuration or in the formation of new intra- and intermolecular bonds. Processes of this type are true topochemical reactions, which occur during solid-phase polymerization processes without the destruction of the sample and formation of side reaction products.

The potential shown in Fig. 3.27b corresponds to reactions of the dissociation type, in the course of which the products of reaction are separated far apart.

Consider the classical description of an elementary act within the framework of the quasi-one-dimensional model presented in Fig. 3.28. The potential energy of a molecular crystal is given by

$$\Pi(\{u,w\}) = \sum_j \left\{ \begin{array}{l} \frac{1}{2}[a(u_{j+1}-w_j)^2 + b(w_{j+1}-u_j)^2 + K(u_{j+1}-u_j)^2 + \\ +k(w_{j+1}-w_j)^2] + U(u_j - w_j) \end{array} \right\}. \quad (3.87)$$

Here, u and w are the displacements of large and small particles from their equilibrium positions; a, b, K, and k are the parameters of the intermolecular interaction; and U is the intramolecular potential.

It is convenient here to introduce new variables for the description of the deformation of the crystal lattice and of the reaction coordinate. In the case studied, these variables are the differences of the displacements of the centers of mass of neighbor

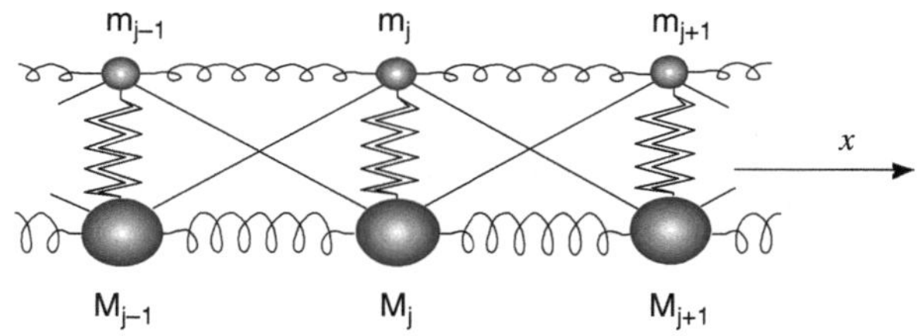

Fig. 3.28 Scheme of a quasi-one-dimensional diatomic molecular crystal. The *filled ellipses* represent the atoms (or atomic groups) united into molecules by intramolecular (reagent) bonds (*double zigzag lines*). *Single lines* (both the *straight* and *curved ones*) correspond to intermolecular bonds. The indices M_j and m_j denote the atoms being the parts of the same molecule, and x is the direction of propagation of the wave of chemical reaction

molecules $\varepsilon_j = \chi_{j+1} - \chi_j = (Mu_{j+1} + mw_{j+1})/M_t - (Mu_j + mw_j)/M_t$ (where M and m and the masses of particles comprising the molecule, $M_t = M + m$) and the intramolecular distance $\varphi_j = w_j - u_j$. The potential energy Π expressed in terms of these variables is divided into the inter- and intra-molecular energies and the interaction energy of the inter- and intra-molecular degrees of freedom. In particular, the density of energy for the uniform state of the crystal ($\varepsilon_j = \varepsilon =$ const, $\varphi_j = \varphi =$ const) reads

$$\frac{\Pi(\varepsilon, \varphi)}{N} = \left[\frac{1}{2}K\varepsilon^2 + U(\varphi) + \varepsilon F(\varphi)\right], \tag{3.88}$$

where $\varepsilon F(\varphi)$ is the interaction energy and U is the effective intramolecular energy. A dynamic system with this potential energy and a two-well potential $U(\varphi)$ has two uniform static states, with one of them corresponding to reagents ($\varphi = 0$, $\epsilon = 0$) and the other, to the reaction products ($\varphi = \varphi_{pr}$, $\varepsilon = \varepsilon_{pr}$). The transition between these states is treated as the reaction.

In the continuum approximation (when the wave length is sufficiently large, as compared to the crystal lattice parameter), the dynamics of this system is described by the system of equations in partial derivatives (Manevitch and Smirnov, 1998):

$$\begin{aligned} &\chi_{tt} - \chi_{xx} - \frac{\partial}{\partial x}F(\varphi, \varphi_x) = 0, \\ &\varphi_{tt} - \gamma\varphi_{xx} + U'(\varphi) + \tfrac{M_t}{\mu}(\alpha\chi + \beta\chi_x) \\ &\varepsilon = \chi_x. \end{aligned} \tag{3.89}$$

Here, α, β, and γ are determined by the intermolecular interaction parameters, and a linear approximation is used for the function F ($F = \alpha\varphi + \beta\varphi_x$). The subscripts denote partial derivatives with respect to the corresponding independent variables, while the prime symbol denotes the differentiation of the potential function U with respect to its argument.

In consideration of the reaction front motion, we search for a solution in the form of a wave with a stationary profile (i.e., a wave dependent only on the variable

$z = x - vt$, where x is a spatial coordinate, v is the propagation velocity of the wave, and t is time) describing the transition from the initial state (reagents) to the final state (reaction products). However, as follows from the analysis of the phase pattern of the system of equations (3.89), direct transitions between the stationary points ($\varepsilon = 0$, $\varphi = 0$) and ($\varepsilon = \varepsilon_{pr}$, $\varphi = \varphi_{pr}$) are impossible because of the difference in the energies of the indicated uniform stationary states. Nevertheless, it is possible to show that the system of equations (Eq. 3.89) has a stationary solution as a solitary wave with the asymptotic values ($\varphi = 0$, $\varepsilon = 0$) at $x - vt \to +\infty$ and ($\varphi = \varphi^*$, $\varepsilon = \varepsilon^*$) at $x - vt \to -\infty$ (the reaction front moves to the right). This solution describes the transition from the initial state ($\varphi = 0$, $\varepsilon = 0$) to an intermediate dynamic state in the region of attraction of the final state ($\varphi = \varphi_{pr}$, $\varepsilon = \varepsilon_{pr}$). In the case where the potential $U(\varphi)$ is closely approximated by a fourth-degree polynomial (Fig. 3.27a), the solution can be expressed in elementary functions:

$$\varphi(x,t) = \frac{\varphi_k}{2}\left[1 - \tanh\left(\frac{x - v_k t}{W}\right)\right], \tag{3.90}$$

$$\varepsilon(x,t) = -\frac{\varphi_k}{2(v_k^2 - s_0^2)}\left[\alpha\left(1 - \tanh\left(\frac{x - v_k t}{W}\right)\right) + \frac{\beta}{W}\sec h^2\left(\frac{x - v_k t}{W}\right)\right]. \tag{3.91}$$

Consider a number of results important for a further discussion. First, the velocity of propagation of the transition wave (soliton) is unambiguously determined by the parameters of the system, and, in the case of an exothermal process, it exceeds the sound velocity s_0. Second, the asymptotic values of the reaction coordinate $\varphi = \varphi^*$ and lattice deformation $\varepsilon^* = \alpha\varphi^*/(s_0^2 - v_k^2)$ correspond to a uniform dynamic state close to, but not coinciding with the final state (products of the reaction). The energy of the intermediate state is higher than that of the final state (products of the reaction). Therefore, it can be treated as a classical analog of an excited state of a growing chain, accompanied by lattice deformation (Manevitch et al., 1994; Manevitch and Smirnov, 1998). The transition from this excited state to the final state is accompanied by a release of the reaction energy and occurs sufficiently far from the wave front. However, a detailed description of this process is beyond the scope of the considered model.

Another important feature of the obtained solution is that there is no principal difference between the potentials of the two-well type (Fig. 3.27a) and dissipative type (Fig. 3.27b). This is due to the fact that dynamical renormalization of the energy is of order φ^2, whereas the potential decreasing at infinitely large φ (Fig. 3.27b) exhibits the exponential ($\sim\exp(-k\varphi)$) or inverse-power ($\sim\varphi^{-k}$, $k>1$) behavior. So, the Lagrangian extremum corresponding to the intermediate state shifts by a finite distance along the reaction coordinate φ^*. Although the solution cannot be expressed in terms of elementary functions, an analysis of the phase pattern of the system shows that a solution of topological-soliton type exists for the front of the transition to an intermediate state. The same is true for endothermic processes (Manevitch and Savin, 1995). However, in this case, the velocity of the front is lower than the speed of sound.

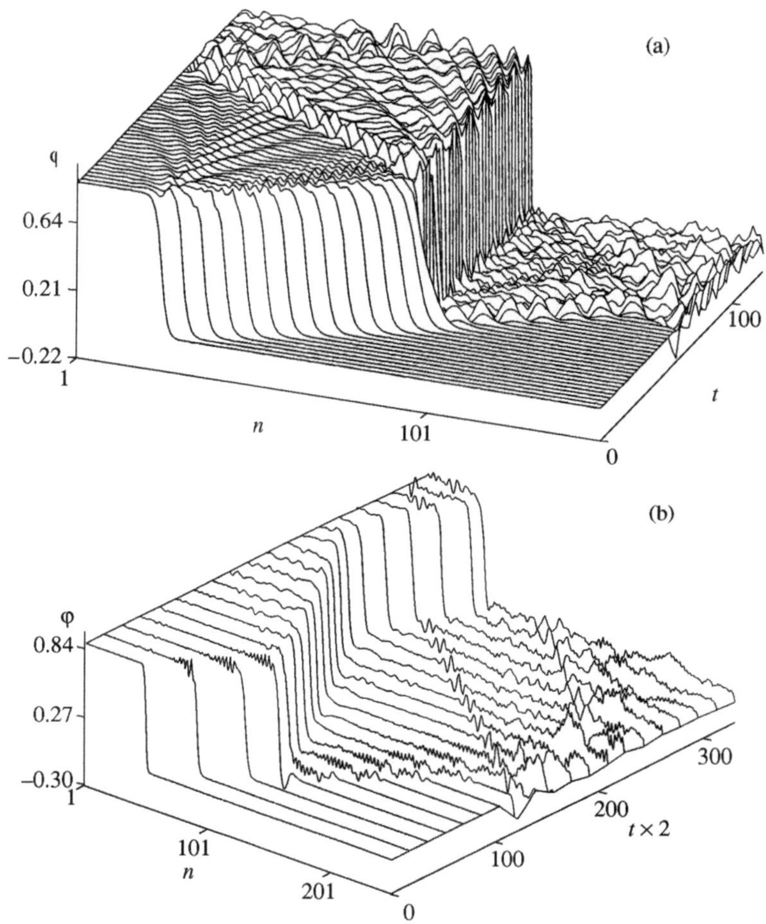

Fig. 3.29 Evolution of a topological soliton in the first-type (**a**) and second-type (**b**) lattices (see text for details); φ is the reaction coordinate

It is important to note the difference between the solution describing the reaction-front dynamics on the atomic and molecular level and the wave propagation of chemical reactions in dissipative media. In the latter case, the wave front is stationary due to the energy exchange between the system and the thermostat, a process characterized by a sufficiently long time of thermalization. In the case under consideration, the front moves at a constant velocity, because the energy of the conservative system is subjected to dynamic renormalization. Equations (3.90) and (3.91) describe the transition between two stationary points of the phase space corresponding to uniform states with equal values of the Lagrangian.

Consider the computer simulation data briefly. Note that the stability and lifetime of soliton solutions (3.90 and 3.91) are strongly dependent on the model lattice

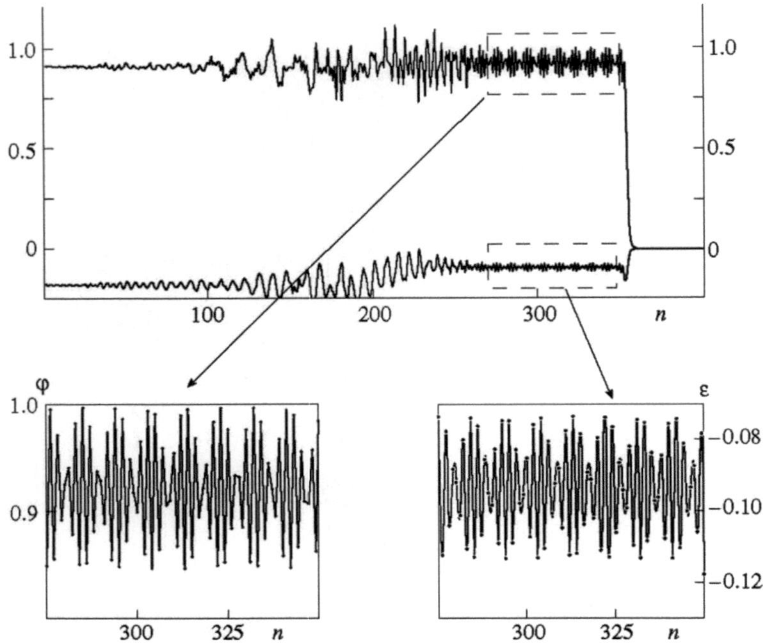

Fig. 3.30 Profile of a topological soliton with radiation. Insertions show the radiation "tail" in the backside reaction front

parameters. We can distinguish two types of lattices principally different from the point of view of the possibility of existence of a static interface. In the first-type lattices (in the approximation of an isolated chain), the interface covers one or two unit cells; it was found to be stable with respect to small perturbations (e.g., thermal noise). Lattices of the second type have no interface (Manevitch and Smirnov, 1995, 1998). In particular, this means that the local-fluctuation mechanism of reaction propagation, according to which an elementary act is a random event occurring at a node of the lattice, cannot be realized. Therefore, the propagation of the reaction over lattices of this type proceeds according to the wave (front) mechanism.

Figure 3.29 exhibits typical examples of motion of the reaction front in lattices of the first and second types. As can be seen, the slowing down and stopping of the soliton in the first case result in the formation of a narrow interface; in the second case, the stopped reaction front is removed outside the sample. The lifetime of the soliton solutions ranges typically from 10 to 200 time units (the unit of time was taken equal to d/s_0, where d is the lattice parameter).

For the reaction front under consideration, the particles behind the front are at rest with respect to the frame bound to the front, because the heat of reaction is converted into the energy of deformation of the lattice. Note, however, that, it is natural to assume that a reaction front should exist behind the heat of reaction equilibrium position. At the present time, no analytical solutions of this type are known.

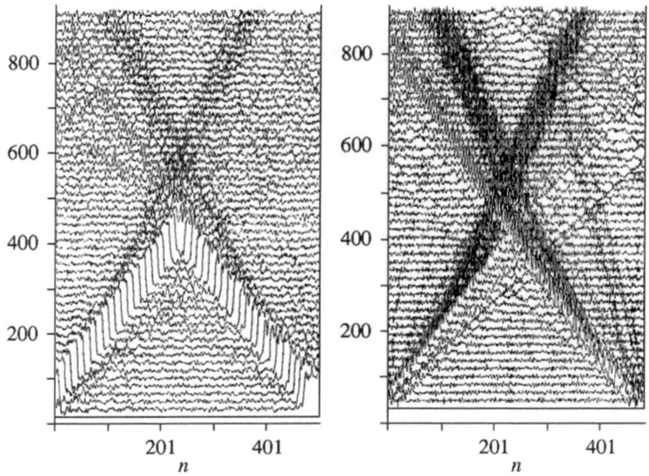

Fig. 3.31 Propagation of the reaction fronts in the second-type lattice at the initial temperature T=0.2. The ends were heated to T=1.0: (**a**) the reaction coordinate φ; (**b**) the lattice deformation ε. Two fronts propagating from the chain ends and the radiation formed behind them (*dark rays* radiating outward) are clearly seen

However, numerical experiments show that, under certain constraints imposed on the parameters of the system, the so-called; "radiating" topological soliton can be initiated. This solution has a narrow front width (on the order of several lattice periods) and predicts that the reaction front emits quasi-monochromatic radiation. The motion of this front is stable and can be closely described by

$$\varphi(x,t) = \frac{\varphi}{2}\left(1 - \tanh\left(\frac{x - vt}{W}\right)\right)(1 + c\sin(kx - \omega t + \delta)),$$

where $W \leq 1$ is the front width and the values φ^*, v, k, ω, and δ depend on the lattice parameters. The condition of stationarity $\omega/k = v$ relates the velocity of the soliton motion with the phase velocity of the radiated wave and expresses constancy of the phase shift of radiation relative to the soliton center. Figure 3.30 presents the structure of the reaction front accompanied by radiation. Figure 3.31 illustrates a birth of two fronts at the ends of a "heated" chain.

3.5 Dynamics of Ensembles of Interacting Nonlinear Chains

The model of coupled oscillatory chains is the next level of complexity. Besides numerous possible applications, the reason for the interest in this model is a richness of its dynamical properties, especially of those related to the beating phenomenon (Khusnutdinova, 1992; Khusnutdinova and Pelinovsky, 2003; Manevitch, 2007; Jensen, 1982; Uzunov et al., 1995; Akhmediev and Ankiewicz, 1993; Kosevich et al., 2008).

The complexity of the problem does not allow solving it exactly. For this reason, we consider this problem in the framework of asymptotic analysis, that allows us to distinguish two limiting regimes: the case of dominating coupling (over nonlinearity) – the weak coupling limit, and the case when the coupling and nonlinearity have the same order – the limit of superweak coupling. In the first case, the local approach turns out to be possible, while in the second one there is a necessity in the consideration of integral quantities that allows a reduction to the system of two coupled nonlinear oscillators.

Let us consider two non-linear oscillatory chains with a weak harmonic interaction. The equations of motion in the short-wave continuum limit are (Kosevich et al., 2008):

$$\begin{aligned} &\frac{\partial^2 u_j}{\partial \tau^2} + \frac{\partial^2 u_j}{\partial x^2} + u_j + 16\beta u_j^3 - \varepsilon\gamma u_{3-j} = 0 \\ &\tau = \omega t,\ \omega^2 = 4 + \varepsilon\gamma \end{aligned} \tag{3.92}$$

where u_j is a modulation of local displacement of the j-th chain ($j = 1,2$), β – parameter of nonlinearity and γ – coupled constant. Small parameter ε characterizes a weakness of inter-chain coupling, τ is the normalized time variable, x – dimensionless space variable, ω – eigenfrequency with inter-chain coupling taken into account. It is useful for further analysis to introduce the complex variables:

$$\Psi_j = \frac{1}{\sqrt{2}}\left(\frac{\partial u_j}{\partial \tau} + iu_j\right),\quad \bar{\Psi}_j = \frac{1}{\sqrt{2}}\left(\frac{\partial u_j}{\partial \tau} - iu_j\right) \tag{3.93}$$

The line over the symbol means a complex conjugation. So, the starting point of our analysis will be Eq. (3.94) for the complex amplitudes Ψ:

$$\begin{aligned} &i\frac{\partial}{\partial \tau}\Psi_j + \Psi_j + \tfrac{1}{2}\frac{\partial^2}{\partial x^2}(\Psi_j - \bar{\Psi}_j) \\ &-4\beta(\Psi_j - \bar{\Psi}_j)^3 - \varepsilon\tfrac{\gamma}{2}(\Psi_{3-j} - \bar{\Psi}_{3-j}) = 0 \end{aligned} \tag{3.94}$$

Let us consider the case of small amplitude oscillation when the complex amplitude $|\Psi|\sim\varepsilon$. It means that the coupling forces ($\sim\varepsilon^2$) in Eq. (3.94) are stronger than nonlinear ones ($\sim\varepsilon^3$). Now we can construct a series of Ψ by parameter $\varepsilon \ll 1$:

$$\Psi_j = \varepsilon(\psi_j + \varepsilon\psi_{j,1} + \varepsilon^2\psi_{j,2} + \ldots) \tag{3.95}$$

and define, alongside with "fast" time, the "slow" time and space variables

$$\begin{aligned} &\tau_0 = \tau,\quad \tau_1 = \varepsilon\tau,\quad \tau_2 = \varepsilon^2\tau \\ &\xi = \varepsilon x \end{aligned} \tag{3.96}$$

After substitution expressions (3.95) and (3.96) into Eq. (3.94), we get equations of main approximation for various orders of ε:

ε^1:

$$
\begin{aligned}
& i\partial_{\tau_0}\psi_j + \psi_j = 0 \\
& \psi_j = \chi_j e^{i\tau_0}
\end{aligned}
\tag{3.97}
$$

ε^2:

$$
\begin{aligned}
& i\partial_{\tau_0}\psi_{j,1} + i\partial_{\tau_1}\psi_j + \psi_j - \frac{\gamma}{2}(\psi_{3-j} - \bar{\psi}_{3-j}) = 0 \\
& \psi_{j,1} = \chi_{j,1}e^{i\tau_0} \\
& i\partial_{\tau_0}\chi_{j,1} + i\partial_{\tau_1}\chi_j - \frac{\gamma}{2}(\chi_{3-j} - \bar{\chi}_{3-j}e^{-2i\tau_0}) = 0
\end{aligned}
\tag{3.98}
$$

The last of Eq. (3.98) lead to the following relations between main amplitude χ_j and first correction amplitude $\chi_{j,1}$:

$$
\chi_{j,1} = \frac{\gamma}{4}\bar{\chi}_{3-j}e^{-2i\tau_0}
\tag{3.99}
$$

The equations defining the dynamics in the time scale τ_1 are following:

$$
i\partial_{\tau_1}\chi_j - \frac{\gamma}{2}\chi_{3-j} = 0
\tag{3.100}
$$

Let us note that Eq. (3.100) are essentially local ones. We can write the solution of Eq. (3.100) in the form:

$$
\begin{aligned}
\chi_1 &= \frac{1}{\sqrt{2}}[X_1\cos(\frac{\gamma}{2}\tau_1) - iX_2\sin(\frac{\gamma}{2}\tau_1)] \\
\chi_2 &= \frac{1}{\sqrt{2}}[X_2\cos(\frac{\gamma}{2}\tau_1) - iX_1\sin(\frac{\gamma}{2}\tau_1)]
\end{aligned}
\tag{3.101}
$$

where function X_1, X_2 depend on “slow” time τ_2 and space variable ξ.

In the next order by small parameter ε, taking into account the relation (3.99), we get the resulting equations for amplitudes χ_j:

$$
i\partial_{\tau_2}\chi_j + \frac{1}{2}\partial_\xi^2\chi_j - \frac{\gamma^2}{8}\chi_j + 12\beta|\chi_j|^2\chi_j = 0
\tag{3.102}
$$

First of all, we can see that Eq. (3.102) are localized on one chain only. Unfortunately, the unknown functions χ_j depend on “fast” time τ_1. To avoid this problem, we have to integrate them with respect to the “fast” time τ_1 over the period $T_1=4\pi/\gamma$. Then we get the resulting equations for the main approximation:

$$
\begin{aligned}
& i\partial_{\tau_2}X_j + \frac{1}{2}\partial_\xi^2 X_j - \frac{\gamma^2}{8}X_j + \\
& \frac{3\beta}{8}(3|X_j|^2X_j + 2|X_{3-j}|^2X_j - X_{3-j}^2\bar{X}_j) = 0
\end{aligned}
\tag{3.103}
$$

where functions X_j are defined by Eq. (3.99). It is a very splendid point, that Eq. (3.103) allow the solutions, concentrated on one chain only. Taking Eq. (3.99) into account, we get a full transition of the initial excitation (e.g., X_1) from the "parent" chain to another one and backwards. One should note that such a result is correct for both localized soliton-like excitations (breathers) as well as for anharmonic plane waves. This conclusion is in good agreement with computer simulation data, some examples being shown in Figs. 3.32 and 3.33.

Let us consider Eq. (3.94) under the condition that coupling and nonlinearity forces have the same order of value. In such a case we have to assume that amplitude $\Psi \sim \varepsilon^{1/2}$. The appropriate expansion of Ψ can be written as

$$\Psi_j = \varepsilon^{1/2}(\psi_j + \varepsilon\psi_{j,1} + \varepsilon^2\psi_{j,2} + \ldots) \tag{3.104}$$

and slow time and space variables have following form:

$$\begin{aligned} &\tau_0 = \tau, \quad \tau_1 = \varepsilon\tau, \quad \tau_2 = \varepsilon^2\tau \\ &\xi = \sqrt{\varepsilon}x \end{aligned} \tag{3.105}$$

Substituting the expansion (3.104) to Eq. (3.94), after some standard manipulations we get the resulting equations for the main approximation amplitude:

$$\begin{aligned} i\partial_{\tau_1}\chi_j + \frac{1}{2}\partial_\xi^2\chi_j - \frac{\gamma}{2}\chi_{3-j} + \\ 12\beta|\chi_j|^2\chi_j = 0 \end{aligned} \tag{3.106}$$

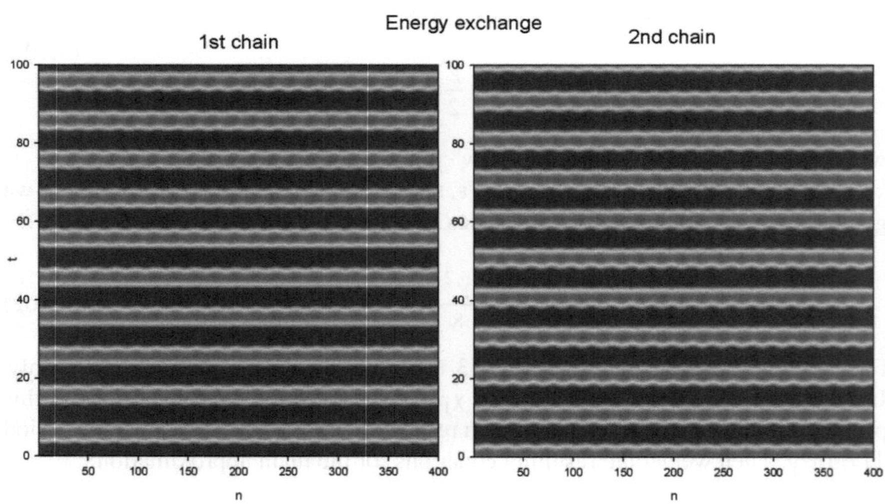

Fig. 3.32 Energy map of small amplitude anharmonic waves in the system of two coupled chains. Initial conditions correspond to a wave, located in the second chain. *Light* (*red online*) stripes correspond to large value energy and *dark* (*blue online*) stripes to small one

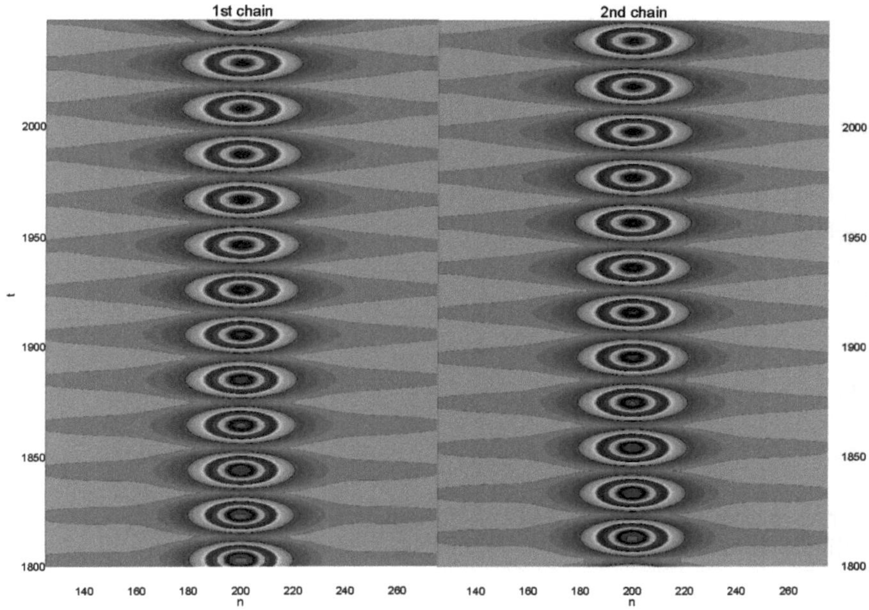

Fig. 3.33 Energy map of standing breather. Fragment of computer simulation. *Bright circles* correspond to the breather location

Contrary to the previous case of weak coupling (see Eq. 3.102), the equations obtained are bound up. This leads to competition between a process of energy exchange and a process of excitation (localized or not) formation. Equation (3.106) have two symmetric solutions:

$$\chi_1 = \chi_2, \ \chi_1 = -\chi_2 \tag{3.107}$$

The first of them corresponds to the in-phase mode, and the second one to the anti-phase mode. It is clear that there is no solution localized on one chain only.

Let us consider the solutions of Eq. (3.106) in the form:

$$\chi_j(\xi, \tau_1) = A_j(\xi - v\tau_1)\exp(i(\omega\tau_1 - q\xi)) \tag{3.108}$$

The assumption $A_j = a_j = \text{const}$ leads to anharmonic "dispersion relations":

$$-(2\omega + q^2)a_j - \gamma a_{3-j} + 24\beta a_j^3 = 0 \tag{3.109}$$

that define the dependence between the frequency of plane wave ω and its amplitudes a_j. The amplitudes of the stationary points of the system when ω and q are fixed are the solutions of Eq. (3.109).

To analyze a temporary evolution of the plane wave we assume that the amplitudes Aj are the functions of time τ_1 only: $A_j = A_j(\tau_1)$. In this case, Eq. (3.106) take the form:

$$i\frac{dA_j}{d\tau_1} - \left(\omega + \frac{q^2}{2}\right)A_j - \frac{\gamma}{2}A_{3-j} + \\ +12\beta|A_j|^2 A_j = 0 \tag{3.110}$$

These equations are fully analogous to equations of two nonlinear oscillators considered in (Manevitch, 2007) in detail.

Eq. (3.110) have two first integrals:

$$H = -\frac{\gamma}{2}(A_1\bar{A}_2 + \bar{A}_1 A_2) - \left(\omega + \frac{q^2}{2}\right)(|A_1|^2 + |A_2|^2) + \\ +6\beta(|A_1|^4 + |A_2|^4) \tag{3.111}$$

and

$$N = |A_1|^2 + |A_2|^2 \tag{3.112}$$

The presence of the integral (3.112) – "occupation number" N – allows to write amplitude A_j in the form:

$$A_1 = \sqrt{N}\cos\theta e^{i\delta_1} \\ A_2 = \sqrt{N}\sin\theta e^{i\delta_2} \tag{3.113}$$

The "angle" variables (θ, δ) are very useful to analyze the phase plane of the system (Kosevich and Kovalyov, 1989). The parameter, controlling the structure of the phase plane of the system is $\kappa = 6\beta N/\gamma$. Up to $\kappa = 0.5$, Eq. (3.110) have only two symmetric stationary points (3.107) that correspond to the in-phase ($A_1 = A_2$ or $\theta = \pi/4$; $\Delta = \delta_1 - \delta_2 = 0$) mode and the anti-phase mode ($A_1 = -A_2$ or $\theta = \pi/4$; $\Delta = \pi$). The attractive area of each of the stationary points is circled by limiting phase trajectories (LPTs). Trhe anti-phase mode becomes instable at $\kappa=0.5$, this leads to the creation two new stationary points, located at $\Delta = \pi$. These new asymmetric stationary points are enclosed by a separatrix passing through the anti-phase stationary point. The distance between the asymmetric points is increased while parameter κ grows. The LPT surrounding the anti-phase stationary point coincides with a separatrix when the parameter κ attains unity. The LPT is broken and the transit-time trajectories appear. Figure 3.34 shows the typical structure of a phase plane in the terms of (θ, Δ)-variables.

Let us discuss these pictures from the point of view of energy exchange. It is clear that no energy exchange process exists in the stationary points of the phase plane. As it follos from representation (3.113), the variable θ defines the amplitude of oscillations of each chain. So any trajectory passing near $\theta = 0$ and $\theta = \pi/2$ describes a process of full energy exchange. Such trajectories can exist near LPTs up to $\kappa=1$. At $\kappa = 1$ the LPT surrounding the anti-phase stationary point is broken and full energy exchange is forbidden. We name this process as confinement of excitation on one chain. But there is an additional case of excitation confinement that is realized

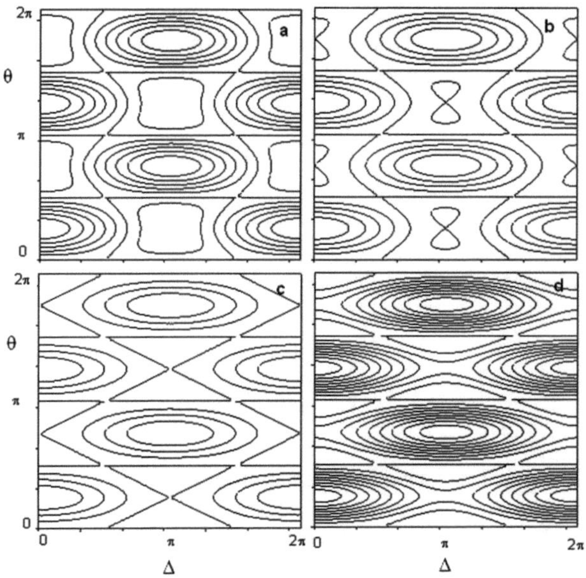

Fig. 3.34 Typical structures of phase plane of Eq. (3.108) at various values of κ: (**a**) κ<0.5; (**b**) 0.5<κ<1; (**c**) κ>1; (**d**) κ=1

before discontinuity of LPT occurs. If the initial conditions of chains excitation are near one of asymmetric stationary points, the corresponding trajectory will be inside the domain closed by the separatrix. In such a case only a small part of energy can be transferred from one chain to another. At a large value of κ only this possibility of energy exchange remains. It should be noted that in the in-phase attractive domain, a large energy exchange is feasible at any values of κ.

What can we say about a localized solution exchange in this system? It is clear, that Eq. (3.106) allow two symmetric soliton-like solutions in the form:

$$\chi_1(\xi,\tau_1) = \frac{1}{4}\sqrt{\frac{2\omega + q^2 \pm \gamma}{3\beta}} \times \\ \times \operatorname{sech}\left(\frac{1}{4}\sqrt{\frac{2\omega + q^2 \pm \gamma}{6\beta}}(\xi + q\tau_1)\right) \exp(i(\omega\tau_1 - q\xi)) \tag{3.114}$$

where sign plus under the square root corresponds to the in-phase state $\chi_1(\xi,\tau_1)=\chi_2(\xi,\tau_1)$ and sign minus the anti-phase state $\chi_1(\xi,\tau_1)=-\chi_2(\xi,\tau_1)$. Because expression (3.114) describes a stationary point, no energy exchange exists. To analyze the beating phenomenon in the presence of a localized solution like (3.114), we assume the solution of Eq. (3.106) as following:

$$\chi_1 = A(\xi)X_1(\tau_1),\ \chi_2 = A(\xi)X_2(\tau_1) \tag{3.115}$$

Such an assumption is good enough near the stationary states of the system. Integrating the Hamiltonian corresponding to Eq. (3.106) with respect to space variables, we get the following:

$$H = -\frac{\gamma}{2}N(X_1\bar{X}_2 + \bar{X}_1X_2) + \frac{1}{2}\mu N(|X_1|^2 + |X_2|^2) + 6\beta\nu N^2(|X_1|^4 + |X_2|^4) \tag{3.116}$$

where new parameters are:

$$N = \int A^2 d\xi, \quad \mu = \int (\partial_\xi A)^2 d\xi / \int A^2 d\xi \\ \nu = \int A^4 d\xi / (\int A^2 d\xi)^2 \tag{3.117}$$

In the framework of such an approach, the only requirement with respect to the space profile A(ξ) is its square integrability. New dependent variables X_1 and X_2 are normalized to unity and the occupation number N has a simple physical meaning. We can see the total analogy between the expressions (3.111) and (3.117). Therefore, all the conclusions made for plane wave solutions are correct for localized soliton-like solutions with a change of the control parameter κ into κ'=νκ.

The computer simulation data is in a good agreement with analytical considerations. Figure 3.35 shows an example of confinement of the breather.

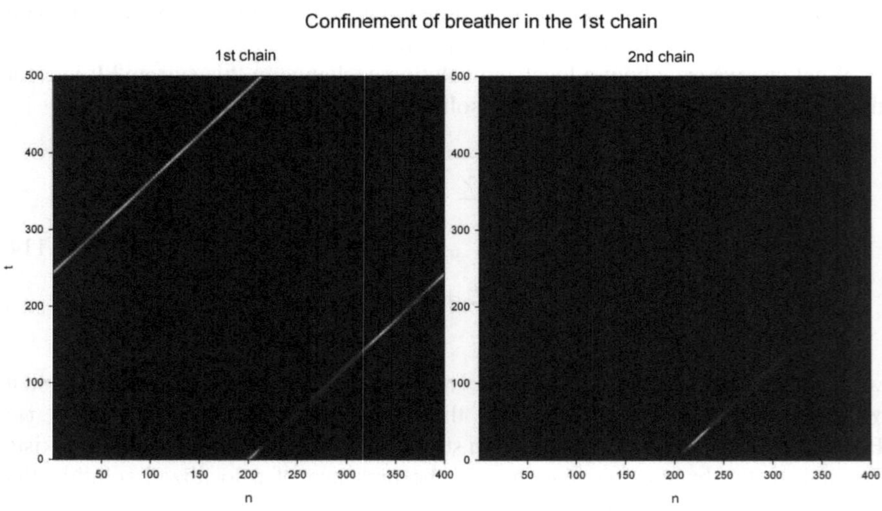

Fig. 3.35 Confinement of localized excitation (breather) on one of coupled chains. *Light areas* show the breather location. n – number of particles, t – time in eigen periods of linear oscillations. Initial condition corresponds to κ'~1.2

3.6 Concluding Remarks

Only a handful of discrete infinite models can be analyzed exactly and completely without any simplifying assumptions. This small family includes linear lattices without defects, as well as a few exactly integrable nonlinear discrete models, such as Toda or Ablowitz-Ladik lattices. For all other cases, simplifications and approximations are required.

The discrete infinite models occupy an intermediate position, between the discrete systems with a low number of degrees of freedom (DOF) and continuous systems. Accordingly, the simplification for obtaining tractable models can proceed in either of these two directions – the discrete infinite system is reduced either to a continuous system or to a low-DOF discrete system.

The "continualization" approach to the tractable modes has been described in Sects. 3.1 and 3.4. The simplification achieved by the transition to a continuum is crucial – it is much easier to analyze partial differential equations (PDEs) than the finite difference equations. In many cases the analytic solution is readily available for the PDE, in other cases it is much easier to be analyzed qualitatively or at least to interpret the numeric results. Of course, this simplification does not come for free: the continuum tractable models represent the initial system only under some conditions initially imposed on the solutions. Such conditions may include closeness to a characteristic frequency of wavelength, smallness of the deformations or the rotations etc. In addition, one discrete model can lead to a number of very different continuous models under different assumptions. Such an ensemble of tractable models is not easy to get, but it is worthwhile since it allows one to classify and characterize possible dynamical regimes for the initial system.

As for the reduction to the low-DOF tractable systems, one can proceed if it is possible to reduce the dynamics to a low number of interacting modes (Sect. 3.5) or quasiparticles (Sects. 3.2 and 3.3). The simplification might be possible due to a relatively low number of the significant interacting modes (thus, we obtain the low-DOF system). Alternatively, the number of modes or quasiparticles can be large, but their interaction will be weak. In this case, one can build the perturbation procedure based on a single-DOF basic approximation. If the former simplifying assumptions are wrong, then the analytic insight into the dynamics is very limited-one can build only particular solutions, such as the discrete breathers. Even such extremely moderate simplification can help to interpret the results of numeric simulations and thus to make system dynamics tractable.

References

Ablowitz, M.J., Ladik, J.F.: Nonlinear differential – difference equations and Fourier analysis. J. Math. Phys. **17**, 1011–1018 (1976)

Akhmediev, N.N., Ankiewicz, A.: Novel soliton states and bifurcation phenomena in nonlinear fiber couplers. Phys. Rev. Lett. **70**, 2395–2398 (1993)

Andrianov, I.V., Awrejcewicz, J.: Asymptotical behavior of a system with damping and high powerform nonlinearity. J. Sound Vib. **267**, 1169–1174 (2003)

Arnold, V.I.: Mathematical Methods of Classical Mechanics. Springer, New York, NY (1989)

Arnold, V.I., Kozlov, V.V., Neishtadt, A.I.: Mathematical Aspects of Classical and Celestial Mechanics. Springer, Berlin (1997)

Aubry, S.: Breathers in nonlinear lattices: existence, linear stability and quantization. Physica D. **103**, 201–250 (1997)

Aubry, S.: Discrete breathers: localization and transfer of energy in discrete Hamiltonian nonlinear systems. Physica D. **216**, 1–30 (2006)

Azeez, M.A.F., Vakakis, A.F., Manevich, L.I.: Exact solutions of the problem of the vibro-impact oscillations of a discrete system with two degrees of freedom. J. Appl. Math. Mech. **63**, 527–530 (1999)

Babitsky, V.I.: Theory of Vibro-Impact Systems: Approximate Methods. Nauka, Moscow (1978) (Revised English translation, Springer, Berlin (1998)

Balabaev, N.K., Gendelman, O.V., Manevitch, L.I.: Supersonic motion of vacancies in a polyethylene crystal. Phys. Rev. E. **64**, 036702 (2001)

Campbell, D.K., Peyrard, M.: CHAOS – Soviet American perspectives on nonlinear science. In: D.K. Campbell (ed.) Nonlinear Science. American Institute of Physics, New York, NY (1990)

Fermi, E., Pasta, J., Ulam, S.: Studies of Non Linear Problems, Los Alamos Rpt LA-1940 (1955)

Flach, S., Gorbach, A.V.: Discrete breathers – advances in theory and applications. Phys. Rep. **467**, 1–116 (2008)

Gendelman, O.V.: Modeling of inelastic impacts with the help of smooth functions. Chaos Solitons Fractals. **28**, 522–526 (2006)

Gendelman, O.V., Manevitch, L.I.: Discrete breathers in vibroimpact chains: analytic solutions. Phys. Rev. E. **78**, 026609 (2008)

Gendelman, O.V., Savin, A.V.: Normal heat conductivity of the one-dimensional lattice with periodic potential of nearest-neighbor interaction. Phys. Rev. Lett. **84**, 2381–2384 (2000)

Gendelman, O.V., Savin, A.V.: Heat conduction in one-dimensional chain of hard discs with substrate potential. Phys. Rev. Lett. **92**, 074301 (2004)

Giardina, C., Livi, R., Politi, A., Vassalli, M.: Finite thermal conductivity in 1D lattices. Phys. Rev. Lett. **84**, 2144–2147 (2000)

Jensen, S.M.: The nonlinear coherent coupler. IEEE J. Quantum Electron. **QE 18**, 1580–1583 (1982)

Kevorkian, J., Cole, J.D.: Multiple Scale and Singular Perturbation Methods. Springer, New York, NY (1996)

Khusnutdinova, K.R.: Non-linear waves in a double row particle system. Vestnik MGU Math. Mech. **2**, 71–76 (1992)

Khusnutdinova, K.R., Pelinovsky, D.E.: On the exchange of energy in coupled Klein-Gordon equations. Wave Motion. **38**, 1–10 (2003)

Kosevich, A.M., Kovalyov, A.S.: Introduction to Nonlinear Physical Mechanics. Naukova Dumka, Kiev (1989) [in Russian]

Kosevich, Yu.A., Manevich, L.I., Savin, A.V.: Wandering breathers and self-trapping in weakly coupled nonlinear chains: classical counterpart of macroscopic tunneling quantum dynamics. Phys. Rev. E. **77**, 046603 (2008)

Krumhansl, J.A., Schriefer, R.: Dynamics and statistical mechanics of one-dimensional model hamiltonian for structural transitions. Phys. Rev. B. **11**, 3535–3545 (1975)

Lepri, S., Livi, R., Politi, A.: Heat conduction in chains of nonlinear oscillators. Phys. Rev. Lett. **78**, 1896–1899 (1997)

Lepri, S., Livi, R., Politi, A.: Thermal conduction in classical low-dimensional lattices. Phys. Rep. **377**, 1–80 (2003)

MacKay, R.S., Aubry, S.: Proof of existence of breathers for time – reversible or Hamiltonian networks of weakly coupled oscillators. Nonlinearity. **7**, 1623–1643 (1994)

MacKenzie, R., Ginder, J.M., Epstein, A.J.: Ring-torsional solitons in polyaniline. Phys. Rev. B. **44**, 2362–2365 (1991)

Manevitch, L.I.: Solitons in polymer physics. Polym. Sci. C. **4**(2), 117–181 (2001)
Manevitch, L.I.: New approach to beating phenomenon in coupled nonlinear oscillatory chains. Arch. Appl. Mech. **77**, 301–312 (2007)
Manevitch, L.I., Gendelman, O.V.: Oscillatory models of vibro-impact type for essentially nonlinear systems. Proc. IMechE, Part C: J. Mech. Eng. Sci. **222**, 2007–2043 (2008)
Manevitch, L.I., Mikhlin, Yu.V., Pilipchuk, V.N.: The Method of Normal Vibrations for Essentially Nonlinear Systems. Nauka, Moscow (1989) [in Russian]
Manevitch, L.I., Savin, A.V.: Soliton mechanism for propagation of endothermal structural transitions in bistable systems. J. Exper. Theor. Phys. **80**, 706–712 (1995)
Manevitch, L.I., Savin, A.V.: Solitons in a polyethylene crystal: isolated chain in transconformation. Phys. Rev. E. **55**, 4713–4722 (1997)
Manevitch, L.I., Savin, A.V. Nonlinear modes and energy transfer in polymer chains. Polym. Sci. A. **47**, 821–849 (2005)
Manevitch, L.I., Savin, A.V., Lamarque, C.-H.: Analytical study and computer simulation of discrete optical in a zigzag chain. Phys. Rev. B. **74**, 014305 (2006)
Manevitch, L.I., Savin, A.V., Lamarque, C.-H.: Low-frequence breathers in a polyethylene crystal. Physica. D. **237**, 600–612 (2008)
Manevitch, L.I., Savin, A.V., Smirnov, V.V., Volkov, S.N.: Solitons in nondegenerate bistable systems. Phys. Uspekhi. **37**, 859–879 (1994)
Manevitch, L.I., Smirnov, V.V.: Propagation of exothermic reactions in condensed matter. Phys. Lett. A. **165**, 427–432 (1992)
Manevitch, L.I., Smirnov, V.V.: New elementary mechanism of structural transition in bistable non-degenerate systems. J. Phys. Cond. Matter. **7**, 255–268 (1995)
Manevitch, L.I., Smirnov, V.V.: Collective interactions in solid-phase chemical processes expressed through non-linear dynamics of crystals. Chem. Rev. **23**, 1–22 (1998)
Manevitch, L.I., Smirnov, V.V.: Localized nonlinear oscillations of a planar zigzag. Doklady. Phys. Chem. **413**, 69–73 (2007)
Manevitch, L.I., Smirnov, V.V.: Solitons in Macromolecular Systems. Nova Science Publishers, New York, NY (2008)
Mikhlin, Y.V., Vakakis, A.F., Salinger, G.: Direct and inverse problems encountered in vibro-impact oscillations of a discrete system. J. Sound Vib. **216**, 227–250 (1998)
Ovchinnikov, A.A.: Excitation spectrum of an antiferromagnetic Heizenberg chain, Zh. Exp. Theor. Phys. **57**, 263 (1969)
Ovchinnikov, A.A., Flach, S.: Discrete breathers in systems with homogeneous potentials: analytic solutions. Phys. Rev. Lett. **83**, 248–251 (1999)
Savin, A.V., Gendelman, O.V.: Heat conduction in one-dimensional lattices with on-site potential. Phys. Rev. E. **67**, 041205 (2003)
Savin, A.V., Manevitch, L.I.: Solitons in a polyethylene crystal: a chain surrounding by immobile neighbors. Phys. Rev. B. **58**(17), 11338–11400 (1998a)
Savin, A.V., Manevitch, L.I.: Structural transformations in crystalline polyethylene: role of topological solitons in premelting. Polym. Sci. A. **40**, 545–554 (1998b)
Savin, A.V., Manevitch, L.I.: Solitons in spiral polymeric macromolecules. Phys. Rev. E. **61**, 7065–7075 (2000)
Savin, A.V., Manevitch, L.I.: Solitons in spiral polymeric macromolecules. A chain surrounded by immovable neighbors. Phys. Rev. B. **63**, 224303 (2001)
Savin, A.V., Manevitch, L.I.: Discrete breathers in a polyethylene chain. Phys. Rev. B. **67**, 144302 (2003)
Savin, A.V., Manevitch, L.I., Cristiansen, P.L., Zolotarjuk, A.V.: Nonlinear dynamics of zigzag molecular chains. Phys. Uspekhi. **42**(3), 1–16 (1999)
Savin, A.V., Zubova, E.A., Manevitch, L.I.: Survival condition for low-frequency breather quasi-one-dimensional strongly anisotropic crystal. Phys. Rev. B. **71**(22), 224303 (2005)
Sievers, A.J., Takeno, S.: Intrinsic localized modes in anharmonic crystals. Phys. Rev. Lett. **61**, 970 (1988)

Takeno, S., Sievers, A.J.: Anharmonic resonant modes in perfect crystals. Solid State Commun. **67**, 1023–1026 (1988)

Toda, M.: Theory of Nonlinear Lattices. Springer, Berlin (1989)

Uzunov, I.M., Muschall, R., Gölles, M., Kivshar, Y.S., Malomed, B.A., Lederer, F.: Pulse switching in nonlinear fiber directional couplers. Phys. Rev. E. **51**, 2527–2537 (1995)

Vedenova, E.G., Manevitch, L.I., Pilipchuk, V.N.: Normal oscillations of a string with concentrated masses on nonlinearly elastic supports. PMM J. Appl. Math. Mech. **49**, 153–159 (1985)

Zhuravlev, V.F.: Investigation of certain vibro-impact systems by the method of non-smooth transformations. Izvestiva AN SSSR Mehanika Tverdogo Tela (Mech. Solids). **12**, 24–28 (1976)

Zhuravlev, V.F., Klimov, D.M.: Applied Methods in Vibration Theory. Nauka, Moscow (1988) [in Russian]

Zubova, E.A., Savin, A.V., Manevitch, L.I.: Dynamics of quasi-1D topological soliton in 2D strongly anisotropic crystal. Physica D. **211**, 294–310 (2005)

Chapter 4
Continuous Systems

Continuous models are, by definition, described by partial differential equations (PDE). Historically, their simplification and reduction to tractable models were accomplished in two different principal ways. The first way to the simplification is to replace the original PDE by simpler PDEs (linear, of lower order, etc.). This method is used in various exact or asymptotic factorization procedures including linearization. The second possible simplification is based on a reduction of the PDE to ordinary differential equations with relatively low number of degrees of freedom. This idea is realized in various versions of modal decomposition and analysis. Sometimes these ideas are used in combination.

Relatively recently, the third approach to the analysis of continuous systems was developed. Some nontrivial PDEs were found to be exactly integrable and thus allowing complete analysis and understanding. Then, new perturbative approaches were developed based on the closeness of particular models to these integrable cases (Arnold, 1980). We have considered some such nonlinear models in Chap. 3; the PDEs were obtained there as asymptotic approximations of discrete systems in various limit cases. In this chapter we deal with the models based predominantly on the linear theory of elasticity.

The continuous systems abound and prevail in science, and no systematic survey of tractable models based on the PDEs seems to be possible within the framework of one book. Instead, we would like to present some examples of reducing the initial complicated PDE model to a tractable one. Such reduction sometimes allows approximate analytic solution of the problem, or at least a reasonable interpretation of numeric results.

4.1 One-Dimensional Models

4.1.1 Bolotin Model

The problems of the linear theory of elasticity for infinite domains can be usually considered as tractable ones due to their high symmetry. Such symmetry is also preserved for finite domains, if boundary conditions of the particular problem allow

L.I. Manevitch, O.V. Gendelman, *Tractable Models of Solid Mechanics*, Foundations of Engineering Mechanics, DOI 10.1007/978-3-642-15372-3_4,

continuation of the solution over the infinite domain. For other boundary conditions an additional analysis is necessary. To illustrate this approach let us consider first a simple dynamical problem for a linear beam with finite length described by the following PDE:

$$EI\frac{\partial^4 w}{\partial x^4} + \mu\frac{\partial^2 w}{\partial t^2} = 0 \tag{4.1}$$

where EI is the bending rigidity, μ is the linear mass density and w is the transversal displacement.

In the case of a simply supported beam the boundary conditions can be written as follows:

$$x = 0,\ L: \quad w = 0; \quad \frac{\partial^2 w}{\partial x^2} = 0 \tag{4.2}$$

and normal oscillations have the form

$$w(x,t) = W_m \sin\frac{m\pi x}{L} e^{i\omega_m t}; \quad m = 1,\ 2\ \ldots \tag{4.3}$$

where

$$\omega_m = \sqrt{\frac{EI}{\mu}}\left(\frac{m\pi}{L}\right)^2 \tag{4.4}$$

It is easy to see that this solution may be continued on an infinite beam.

If boundary conditions are different from (4.2), it may be impossible to perform such simple modal reduction exactly. V.V. Bolotin has shown (Bolotin, 1961; Andrianov et al., 2004) how it is possible to use solution (4.3) if the boundary conditions are different from (4.2). For example, in the case of clamped edges the boundary conditions are

$$x = 0,\ L: \quad w = 0; \quad \frac{\partial w}{\partial x} = 0 \tag{4.5}$$

Then, the normal vibrations can be presented in the form

$$w(x,t) = W(x)\, e^{i\omega t} \tag{4.6}$$

and we get the following ordinary differential equation for the determination of normal modes

$$\frac{d^4 W}{dx^4} - a^2 W = 0,\ a^2 = \frac{\mu}{EI}\omega^2 \tag{4.7}$$

In the case of a simply supported beam considered above

$$W = W_m \sin\frac{m\pi x}{L}; \quad m = 1,\ 2\ \ldots \tag{4.8}$$

Naturally, this solution is not compatible with boundary conditions (4.5). However, the deflections from solution (4.8) are essential only in relatively narrow edge domains. The range where these deflections are essential decreases with growth of m and ω_m. So, the normal modes in this case can be presented as superposition of the sine-like function $W_0 = \sin\frac{\pi(x - x_0)}{\lambda}$ (with still unknown wavelength) and the edge effects. To complete the treatment of the edge effects, the starting equation for the function W(x) may be presented in the form of a combination of two factorized equations

$$\frac{d^2 W_0}{dx^2} + \left(\sqrt{\frac{\mu}{EI}}\omega\right) W_0 = 0, \tag{4.9a}$$

$$\frac{d^2 W_e}{dx^2} - \left(\sqrt{\frac{\mu}{EI}}\omega\right) W_e = 0, \tag{4.9b}$$

The general solution can be written in the form $W = W_0 + W_e$, where W_0, W_e satisfy Eqs. (4.9a, 4.9b) respectively. This is just a consequence of operator factorization of Eq. (4.7):

$$\frac{d^4 W}{dx^4} - a^2 W = \left(\frac{d^2}{dx^2} + a\right)\left(\frac{d^2}{dx^2} - a\right) W = 0 \tag{4.10}$$

In turn,

$$\left(\frac{d^2}{dx^2} - a\right) W = \left(\frac{\partial}{\partial x} + \sqrt{a}\right)\left(\frac{\partial}{\partial x} - \sqrt{a}\right) W = 0 \tag{4.11}$$

and the general solution of (4.11) can be presented as

$$W_e = W_{e,1} + W_{e,2}$$

where $W_{e,1}$ and $W_{e,2}$ satisfy the first order equations, correspondingly

$$\begin{aligned} &\frac{dW_{e,1}}{dx} + \sqrt{a} W_{e,1} = 0 \\ &\frac{dW_{e,2}}{dx} - \sqrt{a} W_{e,2} = 0 \end{aligned} \tag{4.12}$$

The second operator in (4.10) can be used for a more accurate analysis of the solution. If the frequencies are high enough $\left(a = \sqrt{\frac{\mu}{EI}}\omega >> 1\right)$, the derivatives of W_1 and W_2 asymptotically exceed the functions themselves:

$$\frac{dW_e}{dx} \sim \sqrt{a} W_e \tag{4.13}$$

This circumstance allows separating the edge domains in which the functions $W_{e,1}$ or $W_{e,2}$ will dominate. One can present them as

$$W_{e,1} \approx c_1 e^{-\frac{\pi x}{\lambda}}, \quad W_{e,2} \approx c_2 e^{\frac{\pi(x-L)}{\lambda}} \tag{4.14}$$

where $W_{e,1}$ and $W_{e,2}$ should satisfy Eq. (4.12).

The boundary conditions allow finding yet unknown quantities x_0, λ and the arbitrary constants c_1 and c_2:

$$\begin{aligned} &- \text{for } x = 0 \quad W_0 + W_{e,1} = 0, \quad \frac{d}{dx}\left(W_0 + W_{e,1}\right) = 0, \\ &- \text{for } x = L \quad W_0 + W_{e,2} = 0, \quad \frac{d}{dx}\left(W_0 + W_{e,2}\right) = 0. \end{aligned}$$

Using the obtained expressions for W_0, $W_{e,1}$, $W_{e,2}$ one can write

$$\begin{aligned} c_1 - \sin\frac{\pi x_0}{\lambda} = 0, \quad c_1 - \cos\frac{\pi x_0}{\lambda} = 0, \\ c_2 + \sin\pi\frac{L - x_0}{\lambda} = 0, \quad c_2 + \cos\pi\frac{L - x_0}{\lambda} = 0 \end{aligned} \tag{4.15}$$

with the following solution

$$\lambda = \frac{L}{0.5 + m}; \quad x_0 = \lambda(k + 0.25); \quad m = 1,\ 2\ ...; \quad k = 1,\ 2\ ... \tag{4.16}$$

and normal frequencies

$$\omega_m = \pi^2\left(\frac{0.5 + m}{al^2}\right) \tag{4.17}$$

In this simple case, one can compare the result with an exact solution and the error is of order 1% even for the first normal mode.

In the considered case the tractable model consists of the second order equation $\frac{d^2 W_0}{dx^2} + \sqrt{\frac{\mu}{EI}} W_0 = 0$ (the second operator in (4.10)!) with boundary conditions corresponding to a simply supported beam of beforehand unknown length ("basic state" solution of Eq. (4.9a)) and two independent first order equations for edge effects with boundary conditions dependent on a "basic state" solution of the following equations:

$$\frac{dW_{e,1,2}}{dx} \pm \sqrt{\frac{\mu}{EI}} W_{e,1,2} = 0 \tag{4.18}$$

where signs "+" and "–" correspond to functions W_1 and W_2 respectively. The basic state is connected with edge effects only via boundary conditions.

This idea is especially productive in nonlinear case when it is impossible to obtain an exact solution for boundary conditions distinct from those for a simply supported beam (Andrianov et al., 2004).

4.1.1.1 Exercises

With the help of the Bolotin method, calculate the eigenfrequencies and eigenfunctions for the beam which is

1. simply supported at one end and clamped at the other end;
2. a cantilever beam;
3. stretched and clamped at both ends.

4.1.2 Simplification of the Timoshenko Beam

The equations of motion obtained by S.P. Timoshenko (1956) are commonly considered as necessary for an accurate analysis of the transversal beam's vibrations studied earlier exclusively in the framework of Euler–Bernoulli classical theory. But a comparison with exact solutions obtained by use of the two-dimensional theory of elasticity clearly demonstrates that the applicability domain of Timoshenko's theory turns out to be rather wide. It includes the cases when the classical theory does not lead to even qualitatively correct results. For example, the results close to exact ones may be obtained even for very large relative thicknesses of beam ($h/l >> 1$). Explanation of this fact requires further studies in two directions. First, it is necessary to analyze the accuracy from the viewpoint of a two-dimensional theory of elasticity (see also Sect. 4.2). Then, it is desirable to perform an asymptotic analysis of these equations themselves in order to classify the natural vibrations of the beam for all possible geometric relations. This is the subject of this section.

The equations of motion of Timoshenko's beam and physical relations connecting the internal forces with displacements are expressed as:

$$\frac{\partial M}{\partial x} - Q = \rho I \frac{\partial^2 \tilde{\psi}}{\partial t^2} \tag{4.19}$$

$$\frac{\partial Q}{\partial x} = \rho F \frac{\partial^2 \tilde{W}}{\partial t^2} \tag{4.20}$$

$$Q = k' G F \tilde{\gamma}; \quad M = EI \frac{\partial \tilde{\psi}}{\partial x}. \tag{4.21}$$

where Q is transversal force, M is the bending moment, $k'GF$ and EJ are the shear and bending stiffness coefficients respectively, I and F – moment of inertia and cross-section area respectively, *k'* – shear coefficient, depending on the profile of the cross-section, $\tilde{W}$ is the transversal displacement, $\tilde{\psi}$ is the bending angle and $\tilde{\gamma}$ is the shear angle. The slope of the neutral line is a sum of the bending angle and the shear angle:

$$\frac{\partial \widetilde{W}}{\partial x} = \widetilde{\psi} + \widetilde{\gamma} \tag{4.22}$$

Equations (4.19), (4.20), (4.21), and (4.22) can be reduced to the following governing equation

$$EI\frac{\partial^4 \widetilde{W}}{\partial x^4} + \rho F\frac{\partial^2 \widetilde{W}}{\partial t^2} - \left(\rho I + \frac{\rho EI}{k'G}\right)\frac{\partial^4 \widetilde{W}}{\partial x^2 \partial t^2} + \frac{\rho^2 I}{k'G}\frac{\partial^4 \widetilde{W}}{\partial t^4} = 0 \tag{4.23}$$

This equation allows computing the natural vibrations, in which each normal mode has only one node point in the "thickness" direction. The solution for the normal vibrations can be presented as follows:

$$\widetilde{W} = W^*(x)e^{i\omega t}, \quad \widetilde{\psi} = \psi(x)e^{i\omega t}, \quad \widetilde{\gamma} = \gamma(x)e^{i\omega t}$$

Then, (4.19), (4.21), (4.21), and (4.22) lead to ordinary differential equations (in the case of rectangular transversal section):

$$\frac{1}{12}\frac{E}{G}\varepsilon^2\frac{\partial^2 \psi}{\partial \xi^2} + k'\gamma + \frac{1}{12}\varepsilon^2\omega^{*2}\psi = 0 \tag{4.24}$$

$$\frac{d\gamma}{d\xi} + \frac{1}{k'}\omega^{*2}\,W = 0 \tag{4.25}$$

$$\frac{dW}{d\xi} = \psi + \gamma \tag{4.26}$$

Here $\xi = x/l$, *h*, *l* are dimensionless longitudinal coordinate, thickness and characteristic size (length, wavelength when analyzing the high-frequency modes) respectively. We introduce also the parameters: $\varepsilon = h/l$, $\omega^* = \sqrt{\rho\omega^2 l^2/G}$. Parameter ε characterizes relative beam thickness: ε <<1 for "thin" beams and ε >>1 for relatively large h. Certainly, the relations between the functions ψ and γ (i.e. relative role of the bending and shear deformation) as well as the magnitude of the frequency depend on the parameter ε. Let us introduce parameters α, β which quantify these relationships:

$$\gamma \sim \varepsilon^{\alpha}\psi, \quad \omega^* = \varepsilon^{\beta}\Omega, \quad \Omega \sim 1$$

We first consider the thin beams. An asymptotic analysis of Eqs. (4.24), (4.25), and (4.26) with respect to small parameter ε reveals the following possibilities:

(1) $\alpha = 2, \beta = 1$

The limiting equation:

$$\frac{1}{12}\frac{E}{G}\frac{d^4 W}{d\xi^4} + \Omega^2 W = 0 \tag{4.27}$$

is obtained from governing (4.19), if one neglects the terms related to the effect of shear and rotational inertia. The equation describes the natural modes of a "thin" beam with low frequencies and can be obtained within the framework of the common hypothesis of classical beam theory: $\gamma \to 0$, $G \to \infty$. In this case, the physical relations for transversal force can not be used. One should compute them from Eqs. (4.19), (4.20), (4.21), and (4.21) as a reactive force.

(2) $\alpha = 0, \beta = -1$

$$\left(k' - \frac{1}{12}\Omega^2\right)\psi = 0 \tag{4.28}$$

$$-\frac{d\psi}{d\xi} + \frac{1}{k'}\Omega^2 W = 0 \tag{4.29}$$

$$\psi + \gamma = 0 \tag{4.30}$$

The total rotation angle of the transversal section $\frac{\partial \widetilde{W}}{\partial x}$ turns out to be a "small difference of large quantities" $\psi \approx -\gamma$. The frequency of the natural vibrations is determined from Eq. (4.28). We can also find the relationship between the functions ψ and W from expression (4.29). But these functions themselves are not determined in a zero approximation and may be found in the next order of approximation. Corresponding equations of motion are:

$$Q = \rho I \frac{\partial^2 \psi}{\partial t^2}$$

$$\frac{\partial Q}{\partial x} = \rho F \frac{\partial^2 W}{\partial t^2}$$

$$\psi + \gamma = 0$$

where $Q = k'GF\gamma = -k'GF\psi$, $M = -EI\psi$.

The final equation of motion can be written as follows

$$\rho I \frac{\partial^2 \gamma}{\partial t^2} + k'GF\gamma = 0 \tag{4.31}$$

and describes high-frequency vibrations of "thin" beam in which shear deformations compensate the bending ones. We would like to add that existence of such mode was first noted by (Downs, 1976).

Let us consider the "high" beams. In this case, the asymptotic analysis with small parameter ε^{-1} yields two limiting systems.

(3) $\alpha = -2, \beta = 0$

$$\frac{1}{12}\frac{E}{G}\frac{d^2\psi}{d\xi^2} + k'\varepsilon^{-2}\frac{dW}{d\xi} + \frac{1}{12}\omega^2\psi = 0 \tag{4.32}$$

$$k'\frac{d^2 W}{d\xi^2} + \Omega^2 W = 0 \tag{4.33}$$

$$\frac{dW}{d\xi} - \gamma = 0 \tag{4.34}$$

The bending is again a "small difference of large quantities" here and it may be calculated from relation (4.31) after determination of the frequency and displacement W from Eq. (4.33). The equations of motion in this case have the following form:

$$\frac{\partial M}{\partial \xi} + Q = \rho I \frac{\partial^2 \psi}{\partial t^2}$$
$$\frac{\partial Q}{\partial \xi} = \rho F \frac{\partial^2 W}{\partial t^2}$$
$$\frac{\partial W}{\partial \xi} - \gamma = 0$$
$$Q = \kappa' G F \gamma$$

and, as it will be demonstrated below, they describe shear vibrations of a "high" beam with low frequencies.

(4) $\alpha = 0, \beta = 0$

$$\frac{E}{G}\frac{d^2\psi}{d\xi^2} + \Omega^2\psi = 0 \tag{4.35}$$

$$k'\left(\frac{d^2 W}{d\xi^2} - \frac{d\psi}{d\xi}\right) + \Omega^2 W = 0 \tag{4.36}$$

Equation (4.35) provides a possibility to determine the frequency of vibrations and the bending angle. The Eq. (4.36) allows to find the total displacement W. The system Eqs. (4.35 and 4.36) is obtained from the equations of motion

$$\frac{\partial M}{\partial x} = \rho I \frac{\partial^2 \psi}{\partial t^2}$$
$$\frac{\partial Q}{\partial x} = \rho F \frac{\partial^2 W}{\partial t^2}$$
$$\frac{\partial W}{\partial x} = \psi + \gamma$$

and describes the vibrations of a "high" beam with high frequencies, as we will see below. If it is necessary to calculate the corrections for the frequencies obtained from the limiting cases, the perturbation procedure can be used. The perturbation problem is regular in the cases 1, 3, 4 and singular in the case 2.

Below, we have presented the expansions for the frequencies which were found with taking the corrections of first and second order by parameter $\varepsilon^2 = \varepsilon_1,\ \varepsilon^{-2} = \bar{\varepsilon}_1$ into account.

The case (1):

$$\Omega^2 = \frac{k' a^2}{12} k^4 \varepsilon_1 \left[1 - \frac{1+a^2}{12}(1+F_1)\, k^2 \varepsilon_1 + \frac{1}{144}(1+M_1)\, k^4 \varepsilon_1^2 + \dots\right] \tag{4.37}$$

The case (2):

$$\Omega^2 = \frac{12\,k'}{\varepsilon_1} \left[1 - \frac{1+a^2}{12} m^2 \pi^2 \varepsilon_1 - \frac{a^2}{144}(1+M_2)\, m^4 \pi^4 \varepsilon_1^2 + \dots\right] \tag{4.38}$$

The case (3):

$$\Omega^2 = k' m^2 \pi^2 \begin{bmatrix} 1 + \dfrac{12}{1-a^2}(1+F_3)\dfrac{\bar{\varepsilon}_1}{m^2\pi^2} + \\ + \dfrac{144 a^2}{\left(1-a^2\right)^3}(1+M_3)\dfrac{\bar{\varepsilon}_1^2}{m^4\pi^4} + \dots \end{bmatrix} \tag{4.39}$$

The case (4):

$$\Omega^2 = \frac{k' m^2 \pi^2}{b^2} \begin{bmatrix} 1 + \dfrac{12\, b^2}{b^2-1}(1+F_4)\dfrac{\bar{\varepsilon}_1}{m^2\pi^2} + \\ + \dfrac{144 b^2}{\left(b^2-1\right)^3}(1+M_4)\dfrac{\bar{\varepsilon}_1^2}{m^4\pi^4} + \dots \end{bmatrix} \tag{4.40}$$

where $a^2 = \frac{E}{k' G}$, $b = a^{-1}$, $m = 1, 2, \dots$. The values of coefficients k, F_i, M_i are presented in Table 4.1.

Let us note that the characteristic equations for different boundary conditions were first derived on the basis of starting Eq. (4.23) in a paper (Huang, 1961).

The comparison of the frequencies calculated from the limiting systems (with taking the presented corrections into account) with exact solutions of Timoshenko's equations is presented in Figs. 4.1 and 4.2 (it is supposed that $k' = 5/6$ and $\nu = 0.3$, where ν – the Poisson coefficient). The dependence of the low frequencies on the beam thickness is presented in Fig. 4.1, and this of the high frequencies – in Fig. 4.2.

The thick lines correspond to the exact solution, and the dashed lines to approximate solutions which were obtained from the tractable models. As this takes place, the curves obtained from limiting systems themselves, with taking one and two orders of correction into account, are denoted by the numbers 0, 1, 2 respectively.

Table 4.1 The values of coefficients k, F_i, M_i

Equations	Boundary conditions for the beam: Simple support	Clamping support
(4.37)	$k = m\pi$ $F_1 = 0$ $M_1 = 3a^2 = a^4$	$1 - \cosh k = 0$ $F_1 = \frac{2\sin k}{\sinh k - \sin k} - \frac{2}{k}\frac{3a^2-1}{a^2+1}\frac{\cosh k - \cos k}{\sinh k - \sin k}0$ M_1 is unwieldy
(4.38)	$M_2 = 0$	$M_2 = \frac{2}{m^2\pi^2}b^2\sqrt{\frac{1}{3}\left(1+b^2\right)}\frac{\cos\frac{kl}{h} - \cos m\pi}{\sin\frac{kl}{h}}$, where $k^2 = 12\left(1+b^2\right)$
(4.38)	$F_3 = 0$ $M_3 = 0$	$F_3 = \frac{4a}{m\pi\left(1-a^2\right)}\frac{\cos m\pi - \cos a m\pi}{\sin a m\pi}$ $M_3 = 1 + a^2 + \frac{2+4a^2+5a^4+a^2F_3}{4a^2}F_3 + \frac{\left(1+a^2+a^2F_3\right)\left(1-a^2\right)^2}{16a^2}m^2\pi^2F_3^2$
(4.39)	$F_4 = 0$ $M_4 = 0$	$F_4 = \frac{4}{b\,m\pi\left(1-b^2\right)}\frac{\cos m\pi - \cos b\,m\pi}{\sin b\,m\pi}$ $M_4 = 1 + \frac{1}{b^2} + \frac{2-4b^2-6b^4+3b^2F_4+3b^4F_4}{4b^2}F_4 - \frac{\left(1+b^2+b^2F_4\right)\left(b^2-1\right)^{2\,2}}{16a^2}m^2\pi^2F_4^2$

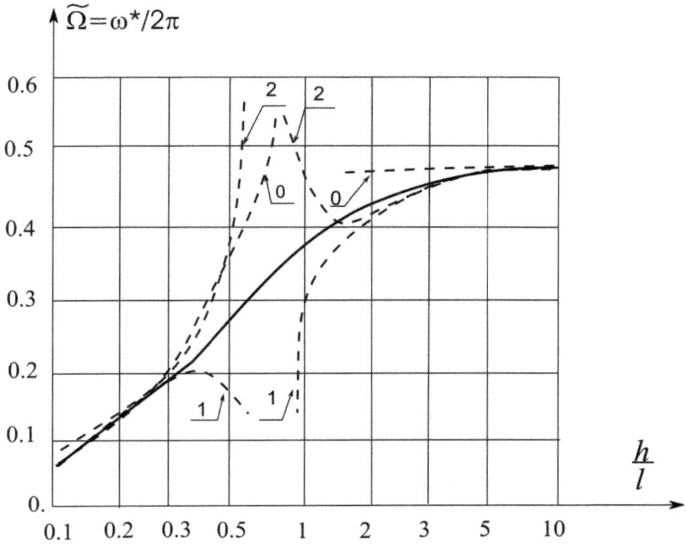

Fig. 4.1 Thickness dependence of the low frequencies – Timoshenko beam

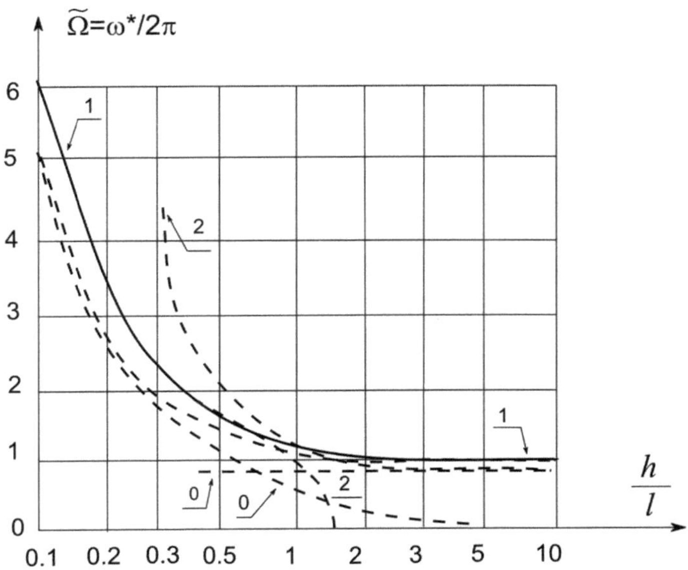

Fig. 4.2 Thickness dependence of the high frequencies – Timoshenko beam

The domains in which the accuracy of the approximation grows with increase of the number of corrections taken into account are distinctly seen.

It is interesting to note that the first approximations for high frequencies, constructed for small *h*/*l* (case 2) and large *h*/*l* (case 4) reasonably approximates the exact solution in all ranges of relative thicknesses. If one takes more terms into account, this property is violated. Such behavior is typical for asymptotic expansions.

In the case of low frequency domains, there is an intermediate range of *h*/*l* in which both approximations are unsatisfactory and the corrections cannot improve their quality due to the reason mentioned above. In this range, the initial Timoshenko's model is useful (in more complex situations, when a general model is unknown, the natural mathematical tool for study of intermediate ranges is the construction of the Pade-approximant that is the rational function having the same power expansions at small and large magnitudes of parameter ε_1 as asymptotic solutions valid for corresponding ε_1).

4.2 The Planar Dynamical Problem and Tractable One-Dimensional Models of an Elastic Solid

The planar dynamical problem of elasticity for a rectangular domain is very important for the substantiation of different approximate models. The latter statement is especially significant for natural vibrations in the case when two opposite longitudinal sides are free from stresses. This case can be considered as a test for the examination of applied theories of bars and beams (Timoshenko, 1922; Guntze, 1969; Prescott, 1942).

We would like to show that even a relatively simple system can demonstrate rich variety of possible qualitatively different dynamical behaviors. As this takes place, every type of behavior may be described by a special tractable model.

The dynamical planar problem in elasticity theory (in the case of a planar stress state) is described by the following equations (Nowacki, 1975):

$$\begin{aligned} &\Delta u - \frac{1+\nu}{2}\frac{\partial^2 u}{\partial z^2} + \frac{1+\nu}{2}\frac{\partial^2 w}{\partial x \partial z} = \frac{\rho\left(1-\nu^2\right)}{E}\frac{\partial^2 u}{\partial t^2} \\ &\Delta w - \frac{1+\nu}{2}\frac{\partial^2 w}{\partial x^2} + \frac{1+\nu}{2}\frac{\partial^2 u}{\partial x \partial z} = \frac{\rho\left(1-\nu^2\right)}{E}\frac{\partial^2 w}{\partial t^2}, \end{aligned} \tag{4.41}$$

where x, z – orthogonal Cartesian coordinates, t – time, $\Delta = \partial^2/\partial x^2 + \partial^2/\partial z^2$, ρ-density, E – elastic modulus, ν – Poisson ratio, u, w – components of the displacements vector in the directions X and Z.

Let us consider natural vibrations of the rectangular domain $0 \le x \le l$; $-\frac{h}{2} \le z \le \frac{h}{2}$ with the boundary conditions:

$$w = 0, \sigma_x = 0, x = 0, l \tag{4.42}$$

Similarly, one can consider other boundary conditions:

$$u = 0, \tau_{xz} = 0, x = 0, l \tag{4.43}$$

admitting exact solutions.

After transition to the dimensionless variables

$$\xi = \frac{x}{l}; \quad \zeta = \frac{2z}{h} \tag{4.44}$$

we look for a solution for natural vibrations as follows:

$$u = \bar{u}\ (\zeta)\cos(m\pi\,\xi)e^{i\omega\,t}, \quad w = \bar{w}\ (\zeta)\sin(m\pi\,\xi)e^{i\omega\,t} \tag{4.45}$$

Substituting expressions (4.45) into Eq. (4.41), one obtains the ordinary differential equations with respect to functions $\widetilde{u}\,(\zeta)$, $\widetilde{w}\,(\zeta)$

$$\begin{aligned} &\left[\frac{1}{\varepsilon^2}\frac{d^2}{d\zeta^2} - \left(\frac{2}{1-\nu} - \delta\right)\right]\widetilde{u} + \frac{1+\nu}{1-\nu}\frac{1}{\varepsilon}\frac{d}{d\zeta}\widetilde{w} = 0; \\ &\left[\frac{2}{1-\nu}\frac{1}{\varepsilon^2}\frac{d^2}{d\zeta^2} - (1-\delta)\right]\widetilde{w} - \frac{1+\nu}{1-\nu}\frac{1}{\varepsilon}\frac{d}{d\zeta}\widetilde{u} = 0, \end{aligned} \tag{4.46}$$

where $\varepsilon = h/2l_x;\ \ \delta = \rho\,\omega^2 l_x^2/G;\ l_x = l/m\pi;\ \ G = E/2\,(1+\nu)$.

It is easy to obtain a general solution of the system (4.46). However, let us consider identical boundary conditions for $\zeta = \pm 1$. In this case, the general solution splits into two uncoupled components and we will restrict ourselves by that corresponding to symmetric $\widetilde{w}$ and antisymmetric $\widetilde{u}$ (with respect to ζ) functions.

As it is known (Nowacki, 1975), the bulk deformation of two-dimensional media is characterized by dilatation

$$\theta = \frac{\partial u}{\partial x} + \frac{\partial w}{\partial z} \tag{4.47}$$

and the small rotation – by rotation vector with the essential component

$$\Omega = \frac{\partial u}{\partial z} - \frac{\partial w}{\partial x} \tag{4.48}$$

In the case of a infinite domain two waves exist: dilatational ($\theta \neq 0;\ \Omega = 0$) and shear ($\theta = 0;\ \Omega \neq 0$). By analogy, let us refer to the natural vibrations as quasi-dilatational, if $|\,\theta\,| >> |\,\Omega\,|$, and quasi-shear, if $|\,\theta\,| << |\,\Omega\,|$. Let us also determine the dimensionless magnitudes $\bar{\sigma}_x,\ \bar{\sigma}_z,\ \bar{\tau}_{xz},\ \bar{\theta},\ \bar{\Omega}$ by the relations

$$\left[(1+\nu)/E\right]\, l_x\sigma_s = \bar{\sigma}_s\,(\zeta)\sin(m\pi\,\xi)e^{i\omega t};$$
$$s = x, z, xz;\quad \theta = \bar{\theta}\, l_x;\ \Omega = \bar{\Omega}\, l_x \tag{4.49}$$

where σ_x, σ_z, τ_{xz}, are the stresses of the plane elastic problem. The general solution of (4.46) with inclusion of the above assumptions can be written as follows:

$$\begin{aligned}
\bar{u} &= \left(c_1/\gamma\right)\,\sinh(\varepsilon\,\gamma\,\zeta) + c_3\sqrt{1-\delta}\,\sinh(\,\varepsilon\sqrt{1-\delta}\,\zeta);\\
\bar{w} &= c_1\,\cosh(\varepsilon\,\gamma\,\zeta) + c_3\,\cosh(\varepsilon\sqrt{1-\delta}\,\zeta);\\
\bar{\sigma}_x &= -c_1\left[(2+\nu)/2\right]\left(\delta/\gamma\right)\cosh(\varepsilon\,\gamma\,\zeta) - c_3\sqrt{1-\delta}\ \sinh(\varepsilon\sqrt{1-\delta}\,\zeta);\\
\bar{\sigma}_z &= c_1\left[(2-\delta)/\gamma\right]\sinh(\varepsilon\,\gamma\,\zeta) + c_3\sqrt{1-\delta}\ \sinh(\varepsilon\sqrt{1-\delta}\,\zeta);\\
\bar{\tau}_{xz} &= -c_1\,\cosh(\varepsilon\,\gamma\,\zeta) + c_3\left[(2-\delta)/2\right]\sinh(\varepsilon\sqrt{1-\delta}\,\zeta);\\
\bar{\theta} &= \left[-c_1(1-\nu)\,\delta/2\gamma\right]\cosh(\varepsilon\,\gamma\,\zeta);\quad \bar{\Omega} = -c_3\,\delta\,\sinh(\varepsilon\sqrt{1-\delta}\,\zeta);\\
\gamma &= \sqrt{1-(1-\nu)\,\delta/2},
\end{aligned} \tag{4.50}$$

where c_1 and c_3 are arbitrary constants.

Keeping in mind some reference problems, we consider first the boundary conditions

$$u = 0,\quad \sigma_z = 0\,\text{for}\,\zeta = \pm 1 \tag{4.51}$$

The corresponding characteristic equation has the form

$$\left(\delta/2\right)\,\gamma\,\sqrt{1-\delta}\ \sinh(\varepsilon\,\gamma)\,\sinh(\varepsilon\sqrt{1-\delta}) = 0 \tag{4.52}$$

This form of the general solution should be reconsidered in the degenerate cases $\delta = 1$, $\delta = 2/(1-\nu)$, $\delta = 0$ – in this case the arguments of the hyperbolic functions become purely imaginary and the latter are transformed to a trigonometric form. In the non-degenerate cases the roots of the characteristic equation split into two groups:

$$\begin{aligned}
&1)\ \delta_{(1)} = 1 + \left(k\pi/\varepsilon\right)^2,\quad k = 0,\,1,\,2,\ldots;\ \bar{\theta}_{(1)} = 0,\ \bar{\Omega}_{(1)} \neq 0\\
&2)\ \delta_{(2)} = \left[2/(1-\nu)\right]\left[1 + \left(k\pi/\varepsilon\right)^2\right],\\
&k = 1,\,2,\,3\ldots;\ \bar{\theta}_{(2)} = 0,\ \bar{\Omega}_{(2)} \neq 0
\end{aligned} \tag{4.53}$$

A similar situation is observed for other types of boundary conditions, except for the case of stressless longitudinal sides. For example under clamping ($u = w = 0$ for $\zeta = \pm 1$), one can obtain the following transcendental equation

$$\sinh(\varepsilon\,\gamma)\cosh(\varepsilon\,\sqrt{1-\delta}\,) - \gamma\sqrt{1-\delta}\ \sinh(\varepsilon\sqrt{1-\delta})\cosh(\varepsilon\,\gamma) = 0 \tag{4.54}$$

Contrary to the reference case above, the analytical representation of the roots is not possible here. However, one can find asymptotic expansions of the roots for the first (corresponding to predominant dilatation) and the second (corresponding to predominant shear) groups and for small and large ε respectively:

$$\begin{aligned}&\varepsilon << 1\\&\quad \delta_{(1)} \approx (k\pi/\varepsilon)^2, \quad k = 1,\ 2,\ 3...; \left|\bar{\theta}_{(1)}\right| << \left|\bar{\Omega}_{(1)}\right|,\\&\quad \delta_{(2)} \approx [2/(1-\nu)]\,[(2k-1)\pi/2\varepsilon]^2, \quad \left|\bar{\theta}_{(2)}\right| >> \left|\bar{\bar{\Omega}}_{(2)}\right|;\\&\varepsilon >> 1\\&\quad \delta_{(1)} \approx 1 + [(2k-1)\pi/2\varepsilon]^2, \quad \left|\bar{\theta}_{(1)}\right| << \left|\bar{\Omega}_{(1)}\right|,\\&\quad \delta_{(2)} \approx [2/(1-\nu)]\left[1 + (k\pi/\varepsilon)^2\right], \quad \left|\bar{\theta}_{(2)}\right| >> \left|\bar{\Omega}_{(2)}\right|.\end{aligned} \tag{4.55}$$

The most interesting case corresponds to free longitudinal sides ($\bar{\sigma}_x = 0,\ \ \bar{\tau}_{xz} = 0$ for $\zeta = \pm 1$), for which a new group of roots of the characteristic equations appears. These roots are strongly connected with an approximate description of beam type solutions. The characteristic equation in this case differs from the previous one by these multipliers: $(2\text{–}\delta)^2$ in front of the first term and 4 in front of the second term.

The asymptotic representation of the root of the first and second groups for $\varepsilon >> 1$ is analogous to the previous one. For $\varepsilon << 1$ it has the form

$$\begin{aligned}&\delta_{(1)} \approx [(2k-1)\pi/2\varepsilon]^2, \quad k = 1,\ 2,\ 3...; \left|\bar{\theta}_{(1)}\right| << \left|\bar{\Omega}_{(1)}\right|,\\&\delta_{(2)} \approx [2/(1-\nu)]\left[1 + (k\pi/\varepsilon)^2\right], \quad \left|\bar{\theta}_{(2)}\right| >> \left|\bar{\Omega}_{(2)}\right|.\end{aligned} \tag{4.56}$$

However, apart from this group of roots, there exists an additional root in interval [0,1]. It is easy to find its asymptotic expansion for small and large values of ε.

$$\begin{aligned}&\varepsilon << 1:\ \delta_{(0)} \approx \frac{2(1+\nu)}{3}\varepsilon^2 - \frac{2(1+\nu)(17+10\nu)}{45}\varepsilon^4 + ...,\\&\varepsilon >> 1:\ \delta_{(0)} \approx \delta_0 + \frac{2(1-0.5\,\delta_0)^5\, e^{-2\sqrt{1-\delta_0}\varepsilon + ...}}{\left[(1+\nu)/4 + (5+\nu)\,\delta_0/8 + (1-\nu)\,\delta_0^2/4\right]}\end{aligned} \tag{4.57}$$

where δ_0 is real positive root of equation

$$\delta_0^3 - 8\delta_0^2 + 8(2+\nu)\,\delta_0 - 8(1+\nu) = 0 \tag{4.58}$$

Depending on the magnitude of ν, δ_0 changes in the interval [0.76, 0.87].

The analysis of a dynamic planar problem in rectangular domain gives us a possibility to characterize a complete vibration spectrum possible in this geometry. Namely, there exist two parts of the spectrum which may be described as quasi-dilatation and quasi-shear ones. Generally speaking, for ε~1 $\left|\bar{\theta}\right|\left|\bar{\Omega}\right|$ in both parts. However, when ε changes, it turns out that for both $\varepsilon >> 1$ and $\varepsilon << 1$ $(\delta_1)\left|\bar{\theta}\right| << \left|\bar{\Omega}\right|$, but for the quasi-dilatation $(\delta_2)\left|\bar{\theta}\right| > \left|\bar{\Omega}\right|$. Asymptotic expansion of the exact solution leads to the conclusion that one of the components (u or w) is dominant for both classes of vibrations. The main term of this component expansion by parameter ε describes shear (dilatation) vibrations and it satisfies the corresponding wave equation if one integrates the initial boundary problem asymptotically. The other component is relatively small; in the first approximation it is a superposition of shear and dilatation components which are in fact main and accompanying standing waves.

Only for boundary conditions that correspond exactly to the periodic solution for a infinite plane, the accompanying standing waves are absent (e.g., for conditions $w = 0,\ \sigma_x = 0\ (x = 0)$ and $u = 0,\ \ \tau_{xz} = 0\ \left(z = \pm h/2\right)$).

The main standing wave is periodic with respect to ζ, and the accompanying one is periodic or an exponentially decreasing function in this direction respectively. A small component of displacement is not present in the boundary conditions in the first approximation and therefore does not influence the main term of the natural frequency expansion. In the case of free longitudinal boundary sides ($\zeta = \pm 1$), a qualitatively new component of the spectrum $\left(\delta_{(0)}\right)$ appears. As this takes place, $\delta_{(0)} \to 0$ for $\varepsilon \to 0$, and $\delta_{(0)}$ tends to a finite value for $\varepsilon \to \infty$. Corresponding solutions are close to beam modes for $\varepsilon <<1$, and to standing surface Rayleigh waves for $\varepsilon >> 1$.

Comparison of the obtained results with Timoshenko's theory, giving a spectrum naturally separated into two parts (see the previous section), shows that the first (lower) part of the Timoshenko beam spectrum corresponds to $\delta_{(0)}$ for $\varepsilon <<1$ and $\delta_{(1)}$ for $\varepsilon >> 1$. As for the second (higher) part, this corresponds to $\delta_{(1)}$ for $\varepsilon <<1$ and $\delta_{(2)}$ for. $\varepsilon >> 1$.

The asymptotics of quasi-dilatations and quasi-shear vibrations can be also found for more complicated boundary conditions (clamping on all sides, for example, or a stressless boundary). The exact solutions of the characteristic equation are not accessible here; the expansions of unknown eigenfunctions by parameter ε or ε^{-1} lead to recurrent relations for the coefficients of these expansions. As this takes place, if the main components of displacement are satisfying, in the first approximation the wave equation may also be revealed. Using these equations, one can find the main terms of the expansions of the eigenfrequencies. In the case of a quasi-beam model, the complicated boundary conditions can be taken into account within the framework of classical theory or the Timoshenko theory.

Similarly, one can study the solutions of the starting system with symmetric u and symmetric w functions (with respect to variable ξ). Here, it is also possible to reveal the quasi-dilatational and quasi-shear vibrations. In the case of stressless sides $\zeta = \pm 1$ a new component of the spectrum appears, which corresponds to longitudinal vibrations of the bar (for $\varepsilon <<1$) and to standing surface Rayleigh waves (for $\varepsilon >> 1$).

The discussed regularities of the vibration spectrum of planar elasticity problem turn out to be significant not only for the substantiation of applied engineering theories, but also for the construction of correct applied theories in the cases when commonly used approaches are not valid (e.g. for vibrations of bars, beams and plates with relatively large thickness or a large number of waves, for the case of impact loading, etc.).

4.2.1 Exercises

1. Find quasi-dilatation and quasi-shear vibrations with symmetric u and W functions (with respect to the ξ-variable) for the accepted boundary conditions.
2. Find surface (Rayleigh) waves for $\varepsilon >> 1$ and stressless sides $\xi = \pm 1$.
3. Find quasi-dilatation and quasi-shear vibrations in the case when

   (a) all sides are clamped;
   (b) the boundary is stressless.

4. Solve the planar problem in the case of two identical opposite periodic normal forces applied to opposite sides of a rectangular infinite strip using initial equations and a tractable model. Compare the results.
5. Perform the same procedure for the case of periodic shear force.
6. Solve the planar problem for two identical concentrated normal forces applied at opposite sides of a rectangular infinite strip.
7. Solve the same problem for concentrated shear forces.

4.3 The Two-Dimensional Orthotropic Model and Its Application to a Complex Contact Problem

The next nontrivial complication is a consideration of a non-isotropic elastic medium. The exact governing equations are linear, but still too complex to yield an exact solution of boundary value problems except the simplest ones. We are going to demonstrate that asymptotic factorization based on an essential anisotropy of the medium leads to the construction of tractable models. Then, many complicated boundary problems may be solved with satisfactory accuracy. Thus, the anisotropy may be rather advantageous for the formulation of tractable models. The derivation of such models can be a nontrivial problem but having succeeded here, an efficient analytical solution turns out to be possible (Manevitch et al., 1982).

4.3.1 Basic Asymptotic Decomposition of the Orthotropic Plate Problem

Let us consider the planar stress state of an orthotropic plate for which the main lines of orthotropy are collinear to the coordinate axes. The solution of boundary problems for such a plate can be reduced to the integration of the equations of equilibrium of the planar orthotropic media in displacements

$$\begin{aligned} B_1\frac{\partial^2 u}{\partial x^2} + G\frac{\partial^2 u}{\partial y^2} + eG\frac{\partial^2 v}{\partial x \partial y} &= 0,\\ B_2\frac{\partial^2 v}{\partial y^2} + G\frac{\partial^2 v}{\partial x^2} + eG\frac{\partial^2 v}{\partial x \partial y} &= 0. \end{aligned} \tag{4.59}$$

under given boundary conditions. Here

$$B_j = \frac{E_j h}{1 - \nu_{12}\nu_{21}}, \quad j = 1,\ 2, \quad G = G_{12}h, \quad e = 1 + \nu_{21}\frac{B_1}{G}, \tag{4.60}$$

E_1, E_2 and G are elastic moduli in the directions x, y and shear modulus; ν_{12}, ν_{21} – Poisson coefficients, where $\nu_{12}E_2 = \nu_{21}E_1$; h – thickness of the plate.

The normal and shear reduced forces in the plane of the plate are determined by the expressions

$$\begin{aligned} T_1 &= B_1\frac{\partial u}{\partial x} + G\,(e-1)\,\frac{\partial u}{\partial y}, \\ T_2 &= B_2\frac{\partial v}{\partial y} + G\,(e-1)\,\frac{\partial u}{\partial x}, \\ S &= G\left(\frac{\partial u}{\partial y} + \frac{\partial v}{\partial x}\right). \end{aligned} \tag{4.61}$$

In order to complete the formulation of the boundary problem for Eq. (4.59), the boundary conditions have to be added and in the case of the third main problem for the orthotropic strip ($0 \le x \le H,\ -\infty < y < \infty$) they have to be formulated as follows: the boundary of the plate is separated into the parts with different boundary conditions, e.g.

$$\begin{aligned} T_1 &= F(y), \quad S = \phi(y), \quad \left(y \in L'\right), \\ u &= u^*(y), \quad v = v^*(y), \quad \left(y \in L''\right). \end{aligned} \tag{4.62}$$

Let us suppose that $B_1 \sim B_2 >> G$, then the quantity $\varepsilon = G^2/B_1B_2$ may considered as a small parameter. This condition explicitly defines the essential anisotropy of the medium in the considered problem – for common isotropic medium the Young modulus and the shear modulus should be of the same order, provided that the Poisson ratio will not be too close to its maximum value.

We introduce the affine transformations of the coordinates and the unknown functions

$$x = \varepsilon^{-1/4}q^{-1/4}x_1, \quad y = y_1, \quad u = U^{(1)}, \quad v = \varepsilon^{-1/4}q^{-1/4}V^{(1)}, \tag{4.63}$$

$$x = \varepsilon^{-1/4}q^{-1/4}x_2, \quad y = y_2, \quad u = \varepsilon q^{-1/4}U^{(2)}, \quad v = \varepsilon^{1/4}V^{(2)} \tag{4.64}$$

where $q = B_2/B_1 \sim 1$.

Substitution of the transformations (4.63) and then (4.64) into Eq. (4.59) leads to systems (4.65) and (4.66) respectively

$$\begin{aligned} U_{xx}^{(1)} + U_{yy}^{(1)} + e\varepsilon\, V_{xy}^{(1)} &= 0, \\ V_{yy}^{(1)} + eU_{xy}^{(1)} + \varepsilon\, V_{xx}^{(1)} &= 0, \end{aligned} \tag{4.65}$$

$$
\begin{aligned}
&U_{xx}^{(2)} + eV_{xy}^{(2)} + \varepsilon\, U_{yy}^{(2)} = 0, \\
&V_{xx}^{(2)} + V_{yy}^{(2)} + e\varepsilon\, U_{xy}^{(2)} = 0.
\end{aligned}
\tag{4.66}
$$

Here and below, the following notations are accepted for brevity:

$$
\frac{\partial \phi^i}{\partial x_i} = \phi_x^i, \quad \frac{\partial \phi^i}{\partial y_i} = \phi_y^i
$$

Evidently, in the case of the planar deformation the corresponding equations are similar.

As it is seen from the transformations (4.63) and (4.64), the solution, obtained by asymptotic integration of the system (4.65), varies along x axis slower than the similar solution of system (4.66) (in the former case $\partial/\partial x = \varepsilon^{1/4} q^{1/4} \partial/\partial x_1$, in the latter $\partial/\partial x = \varepsilon^{-1/4} q^{1/4} \partial/\partial x_2$).

The components of the displacement vector may be presented by superposition of the solutions of both types

$$
u = u_1 + u_2, \qquad v = v_1 + v_2 \tag{4.67}
$$

We will look for the functions $U^{(n)}$, $V^{(n)}$ as series by small parameter $\varepsilon^{1/4}$:

$$
\begin{aligned}
U^{(n)} &= \sum_{m=0}^{\infty} \sum_{j=0}^{3} \varepsilon^{m+j/4} U^{n,4m+j}, \\
V^{(n)} &= \sum_{m=0}^{\infty} \sum_{j=0}^{3} \varepsilon^{m+j/4} V^{n,4m+j} \qquad (n = 1,\ 2)\,.
\end{aligned}
\tag{4.68}
$$

It is convenient to introduce additional transformations of the coordinates

$$
\xi_1 = x_1 \sum_{m=0}^{\infty} \varepsilon^m \alpha_m, \qquad \eta_1 = y_1, \tag{4.69}
$$

$$
\xi_2 = x_2 \sum_{m=0}^{\infty} \varepsilon^m \beta_m, \qquad \eta_2 = y_2 \tag{4.70}
$$

Here the coefficients α_0, β_0 are equal to unity, because the equations of the zeroth approximation have to coincide with the limiting systems which can be obtained from Eq. (4.65) and (4.66) when $\varepsilon \to 0$. The coefficients α_m, β_m ($m = 1, 2, \ldots$) which are calculated in the course of the solution are used for simplification of the equations for higher approximations. Substituting series Eqs. (4.68 and 4.69) for $n = 1$ into the system (4.65) we obtain

$$\sum_{m=0}^{\infty}\sum_{j=0}^{3}\varepsilon^{m+j/4}\left(\sum_{\nu=0}^{m}U_{\xi\xi}^{1,4\nu+j}c_{m-\nu}+U_{\eta\eta}^{1,4m+j}+e\varepsilon\sum_{\nu=0}^{m}V_{\xi\eta}^{1,4\nu+j}\alpha_{m-\nu}\right)=0,$$
$$\sum_{m=0}^{\infty}\sum_{j=0}^{3}\varepsilon^{m+j/4}\left(V_{\eta\eta}^{1,4m+j}+\varepsilon\sum_{\nu=0}^{m}U_{\xi\eta}^{1,4\nu+j}\alpha_{m-\nu}+\varepsilon\sum_{\nu=0}^{m}V_{\xi\xi}^{1,4\nu+j}c_{m-\nu}\right)=0. \quad (4.71)$$

where

$$c_p=\sum_{s=0}^{p}\alpha_s\alpha_{p-s}$$

Similarly, after substitution of the series (4.68) and (4.70) (for n=2) into system (4.66) one can find

$$\sum_{m=0}^{\infty}\sum_{j=0}^{3}\varepsilon^{m+j/4}\left(\sum_{\nu=0}^{m}U_{\xi\xi}^{2,4\nu+j}d_{m-\nu}+e\sum_{\nu=0}^{m}V_{\xi\eta}^{2,4\nu+j}\beta_{m-\nu}+\varepsilon\,U_{\eta\eta}^{2,4m+j}\right)=0,$$
$$\sum_{m=0}^{\infty}\sum_{j=0}^{3}\varepsilon^{m+j/4}\left(\sum_{\nu=0}^{m}V_{\xi\xi}^{2,4\nu+j}d_{m-\nu}+V_{\eta\eta}^{2,4m+j}+e\varepsilon\sum_{\nu=0}^{m}U_{\xi\eta}^{2,4\nu+j}\beta_{m-\nu}\right)=0. \quad (4.72)$$

where

$$d_p=\sum_{s=0}^{p}\beta_s\beta_{p-s}$$

Splitting of Eqs. (4.71) and (4.72) by parameter $\varepsilon^{1/4}$ leads to two infinite systems of equations with respect to the functions $U^{2,4m+j}$, $V^{2,4m+j}$ ($m=0,1,\ldots;j=0,1,2,3$)

$$U_{\xi\xi}^{1,4m+j}c_0+U_{\eta\eta}^{1,4m+j}=-\sum_{\nu=0}^{m-1}\left(U_{\xi\xi}^{1,4\nu+j}c_{m-\nu}+eU_{\xi\eta}^{1,4\nu+j}\alpha_{m-\eta-1}\right), \quad (4.73)$$

$$V_{\eta\eta}^{1,4m+j}=eU_{\xi\eta}^{1,4m+j}\alpha_0-\sum_{\nu=0}^{m-1}\left(eU_{\xi\eta}^{1,4\nu+j}\alpha_{m-\nu}+V_{\xi\xi}^{1,4\nu+j}c_{m-\eta-1}\right) \quad (4.74)$$

for the stress state of the first type and

$$U_{\xi\xi}^{2,4m+j}=-eV_{\xi\eta}^{2,4m+j}\beta_0-U_{\eta\eta}^{2,4(m-1)+j}-$$
$$-\sum_{\nu=0}^{m-1}\left(U_{\xi\xi}^{2,4\nu+j}d_{m-\nu}+eV_{\xi\eta}^{2,4\nu+j}\beta_{m-\eta}\right), \quad (4.75)$$

$$V^{2,4m+j}_{\xi\xi} d_0 + V^{2,4m+j}_{\eta\eta} = -\sum_{\nu=0}^{m-1}\left(V^{2,4\nu+j}_{\xi\xi} d_{m-\nu} + eU^{2,4\nu+j}_{\xi\eta}\beta_{m-\eta}\right) \tag{4.76}$$

for the stress state of the second type.

It should be mentioned that if the upper summation limit for a certain sum is lesser than the lower one, this sum is considered to be zero. Similarly, every one of the functions $U^{n,4m+j}$, $V^{n,4m+j}$ $(n = 1, 2)$ with the negative second index is supposed to be zero.

So, one can write from Eqs. (4.73), (4.74), (4.75), and (4.76) for m = 0

$$\begin{aligned} U^{1,j}_{\xi\xi} + U^{1,j}_{\eta\eta} &= 0, \\ V^{1,j}_{\eta\eta} &= -eU^{1,j}_{\xi\eta} \end{aligned} \tag{4.77}$$

for the state of the first type and

$$\begin{aligned} U^{2,j}_{\xi\xi} &= -eV^{2,j}_{\xi\eta}, \\ V^{2,j}_{\xi\xi} + V^{1,j}_{\eta\eta} &= 0 \end{aligned} \tag{4.78}$$

for the state of the second type.

As it is shown in our book (Manevitch et al., 1982) (a corresponding technique is demonstrated below, in Sect. 4.4.3), the coefficients α_m, β_m $(m = 1, 2, \ldots)$ can be determined in a way that Eqs. (4.73), (4.74), (4.75), and (4.76) will be written as follows

$$U^{1,4m+j}_{\xi\xi} + U^{1,4m+j}_{\eta\eta} = 0 \tag{4.79}$$

$$V^{1,4m+j}_{\eta\eta} = eU^{1,4m+j}_{\xi\eta} - \sum_{\nu=0}^{m-1}\left(eU^{1,4\nu+j}_{\xi\eta}\alpha_{m-\nu} + V^{1,4\nu+j}_{\xi\xi} c_{m-\eta-1}\right) \tag{4.80}$$

for the state of the first type and

$$U^{2,4m+j}_{\xi\xi} = -eV^{2,4m+j}_{\xi\eta} - U^{2,4m+j}_{\eta\eta} - \sum_{\nu=0}^{m-1}\left(U^{2,4\nu+j}_{\xi\xi} d_{m-\eta} + eV^{2,4\nu+j}_{\xi\eta}\beta_{m-\nu}\right), \tag{4.81}$$

$$V^{2,4m+j}_{\xi\xi} + V^{2,4m+j}_{\eta\eta} = 0 \tag{4.82}$$

for the states of the second type.

Therefore the functions $U^{1,4m+j}$, $V^{2,4m+j}$ turn out to be harmonic. Needless to say, it is huge simplification with respect to the initial problem.

Substituting the sums (4.67) and corresponding transformations (4.63, 4.67, 4.68, 4.69, and 4.70) in the expressions for the reduced forces T_1, T_2, S and displacements u, v, we obtain

$$T_1 = B_1\left(u_x + \nu_{21}v_y\right) = B_1\varepsilon^{1/4}q^{1/4}\sum_{m=0}^{\infty}\sum_{j=0}^{3}\varepsilon^{m+j/4}\times$$
$$\times\left[\begin{array}{l}\sum\limits_{\nu=0}^{m}\left(U_{\xi}^{1,4\nu+j}\alpha_{m-\nu} + \varepsilon^{1/2}q^{1/4}U_{\xi}^{2,4\nu+j}\beta_{m-\nu}\right)+\\ +\varepsilon^{1/2}q^{1/4}(e-1)V_{\eta}^{2,4m+j} + \varepsilon(e-1)V_{\eta}^{1,4m+j}\end{array}\right] = 0, \quad (4.83)$$

$$T_2 = B_2\left(u_y + \nu_{12}v_x\right) = B_2\varepsilon^{1/4}\sum_{m=0}^{\infty}\sum_{j=0}^{3}\varepsilon^{m+j/4}\times$$
$$\times\left[\begin{array}{l}V_{\eta}^{2,4m+j} + \varepsilon^{1/2}q^{-1/4}\left(V_{\eta}^{1,4m+j} + (e-1)\sum\limits_{\nu=0}^{m}U_{\xi}^{1,4\nu+j}\alpha_{m-\nu}\right)+\\ +\varepsilon(e-1)U_{\xi}^{2,4\nu+j}\beta_{m-\nu}\end{array}\right] = 0, \quad (4.84)$$

$$S = G\left(u_y + v_x\right) = G\sum_{m=0}^{\infty}\sum_{j=0}^{3}\varepsilon^{m+j/4}\left[U_{\eta}^{1,4m+j} + q^{1/4}\sum_{\nu=0}^{m}V_{\xi}^{2,4\nu+j}\beta_{m-\nu}+\right.$$
$$\left.\varepsilon\left(q^{1/4}U_{\eta}^{2,4\nu+j}\beta_{m-\nu} + \sum_{\nu=0}^{m}V_{\xi}^{1,4\nu+j}\alpha_{m-\nu}\right)\right], \quad (4.85)$$

$$u = U^{(1)} + \varepsilon q^{1/4}U^{(2)} = \sum_{m=0}^{\infty}\sum_{j=0}^{3}\varepsilon^{m+j/4}\left(U^{1,4m+j} + \varepsilon q^{1/4}U^{2,4m+j}\right) = 0, \quad (4.86)$$

$$u = \varepsilon^{3/4}q^{-1/4}V^{(1)} + \varepsilon^{1/4}V^{(2)} =$$
$$= \sum_{m=0}^{\infty}\sum_{j=0}^{3}\varepsilon^{m+\frac{j+1}{4}}\left(V^{2,4m+j} + \varepsilon^{1/2}q^{-1/4}V^{1,4m+j}\right) = 0 \quad (4.87)$$

Substitution of the series (4.84, 4.85, 4.86, and 4.87) for the reduced forces and displacements into boundary conditions (4.62) and following the splitting by parameter $\varepsilon^{1/4}$ allows one to find the boundary conditions corresponding to the boundary problems for the harmonic functions $U^{1,4m+j}$, $V^{2,4m+j}$. The functions $V^{1,4m+j}$ and $U^{2,4m+j}$ may be calculated by simple integration of the expressions (4.80 and 4.81) and also turn out to be harmonic. One can see from (4.79, 4.80, 4.81, 4.82, 4.83, 4.84, 4.85, 4.86, and 4.87) that the stress states of both types are coupled only via the boundary conditions.

The displacement u has the dominant effect on the stress-strain state of the first type (basic state) as well as on a corresponding reduced force T_1 and component of the shear stress S_1 depending on u; the latter can be determined in the basic approximation from the following equations (in initial variables)

$$B_1u_{xx} + Gu_{yy} = 0, \quad T_1 = B_1u_x, \quad S_1 = u_y \quad (4.88)$$

As for the dominant components of the second type stress-strain state, they are (in the initial variables) displacement v, reduced force T_2 and component of the shear stress S_2, which can be found in the basic approximation from the equations

$$B_2v_{yy} + Gv_{xx} = 0, \quad T_2 = B_2v_y, \quad S_2 = Gv_x \quad (4.89)$$

The full shear force is expressed as:

$$S = S_1 + S_2$$

When solving Eq. (4.88), the boundary conditions involving the displacement u or normal force T_1 can be satisfied, but not those involving the displacement v and shear force; the latter are satisfied by the solution of Eq. (4.89). In this case, the boundary conditions with respect to normal displacement u and normal reduced force are not satisfied. But they have a higher order with respect to the small parameter and can be satisfied by the solution of the next approximation for the basic state etc. Indeed, the obtained model is tractable – anyway, much more tractable than the initial exact equations of the orthotropic plane. At every step scalar Laplace equations should be solved exclusively. Therefore, it turns out to be possible to consider very complicated boundary problems analytically.

4.3.2 The Contact Problem for a Planar Orthotropic Strip

As the next example, we would like to present the solution of the contact problem for a planar orthotropic layer ($0 \leq x \leq H,\ -\infty < y < \infty$) which is compressed by two symmetric stamps with corner points; sliding, friction and possible stick are taken into account. The stamps are pulled into the layer by compressive forces P acting along the symmetry of the stamps' axis. It is assumed that there are two sliding regions in every contact area close to the end points of the contact (the conditions $|S| = k\,|T_1|$ are satisfied at these regions), and the stick region is situated between them, at which the displacement of the layer is equal to the displacement of the stamp. As this takes place, the shear stresses in the sliding zones have opposite directions. Due to the symmetry, the boundary points of the sliding zones (which are unknown and have to be found while solving the problem) are placed symmetrically with respect to the x-axis and the sliding zones have the same length in both contact regions.

The problem can be reduced to the integration of equilibrium Eq. (4.59) under the following boundary conditions:

$$\begin{aligned} T_1 = S = 0 \quad & (x = 0,\ |y| > l)\,, \\ T_1 = S = 0 \quad & (x = H,\ |y| > l) \end{aligned} \tag{4.90}$$

at free boundary regions;

$$\begin{aligned} u = f(y) + c_1 \quad & (x = 0,\ |y| > l)\,, \\ u = -f(y) - c_1 \quad & (x = H,\ |y| > l) \end{aligned} \tag{4.91}$$

at the contact zones;

$$\begin{aligned} S = \operatorname{sign}(y) k T_1 \quad & (x = H,\ b < |y| < l)\,, \\ S = -\operatorname{sign}(y) k T_1 \quad & (x = H,\ b < |y| < l) \end{aligned} \tag{4.92}$$

at the sliding regions;

$$v = 0 \quad (x = 0,\ |y| < b)\,, \qquad v = 0 \quad (x = H,\ |y| < b) \tag{4.93}$$

at the stick regions. The boundary $|y| = b$ between the connection and sliding zones will be determined further.

Besides, the equations of equilibrium for the stamp and the condition that the displacement vanishes at infinity should be satisfied.

We use the procedure described above for the solution of the problem. Let us suppose that $k = k_0\varepsilon^{1/4},\ k_0 1$. Then, the problem can be reduced to the subsequent integration of the equations of the first and the second type (4.79, 4.80, 4.81, and 4.82) under corresponding boundary conditions. Let us note that the following relations are valid

$$U^{1,4m+j} = V^{1,4m+j} = U^{2,4m+j} = V^{2,4m+j} = 0 \quad (j = 1,\ 3)$$

The solution of this problem is equivalent to the solution of boundary problems for the analytical functions $\Phi^{1,4m+j}(z_1)$ and $\Phi^{2,4m+j}(z_2)$, which may be defined by the following relations

$$\begin{aligned}\Phi^{1,4m+j}(z_1) &= \frac{\pi}{2H}\varepsilon^{-1/4}q^{-1/4}\alpha^{-1}\left(U_\eta^{1,4m+j} - iU_\xi^{1,4m+j}\right),\\ \Phi^{2,4m+j}(z_2) &= \frac{\pi}{2H}\varepsilon^{1/4}q^{-1/4}\beta^{-1}\left(V_\eta^{2,4m+j} - iV_\xi^{2,4m+j}\right),\end{aligned} \tag{4.94}$$

where

$$\begin{aligned}&z_1 = \eta_1^* + i\xi_1^*, \quad z_2 = \eta_2^* + i\xi_2^*,\\ &\eta_1^* = \frac{\pi}{2H}\varepsilon^{-1/4}q^{-1/4}\eta_1, \quad \xi_1^* = \frac{\pi}{2H}\varepsilon^{-1/4}q^{-1/4}\xi_1,\\ &\eta_2^* = \frac{\pi}{2H}\varepsilon^{-1/4}q^{-1/4}\eta_2, \quad \xi_2^* = \frac{\pi}{2H}\varepsilon^{-1/4}q^{-1/4}\xi_2 \quad \left(0 \le \xi_1^*,\ \xi_2^* \le \frac{\pi}{2}\right).\end{aligned}$$

These functions can be found using the Keldysh-Sedov formula (Sedov, 1980). Because the latter one is presented for the strip with width $\pi/2$, there is a need for additional coordinate transformations.

$$\begin{aligned}\xi_1 &= \frac{2H}{\pi}\varepsilon^{1/4}q^{1/4}\alpha\xi_1^*, \quad \eta_1 = \frac{2H}{\pi}\varepsilon^{1/4}q^{1/4}\alpha\eta_1^*,\\ \xi_2 &= \frac{2H}{\pi}\varepsilon^{-1/4}q^{1/4}\beta\xi_2^*, \quad \eta_2 = \frac{2H}{\pi}\varepsilon^{-1/4}q^{1/4}\beta\eta_2^*,\\ \alpha &= \sum_{m=0}^{\infty}\alpha_m\varepsilon^m, \quad \beta = \sum_{m=0}^{\infty}\beta_m\varepsilon^m.\end{aligned} \tag{4.95}$$

The width of the strip on the new coordinates is equal to $\xi_2^*,\ \eta_2^*\ (n = 1,\ 2)$. The equations of equilibrium (4.79, 4.80, 4.81, and 4.82) for functions $U^{n,4m+j}$, $V^{n,4m+j}$

($n = 1, 2;\ m = 0, 1, 2, \ldots;\ j = 0, 2$) are invariant with respect to transformations (4.95). The boundary conditions in this case will be:

(a) for functions $U^{1,4m+j}$

$$\begin{aligned} U^{1,0} &= \theta\left[c_1 + f_1\left(\eta_1^*\right)\right], \\ U^{1,4m+j} &= -q^{-1/4}U^{2,4(m-1)+j} \quad (4m+j \neq 0) \end{aligned} \tag{4.96}$$

at contact regions $\left(\xi_1^* = 0, \frac{\pi}{2};\ \lfloor\eta_1^*\rfloor < l_1^*\right)$;

$$\begin{aligned} U_{\xi^*}^{1,4m+j} &= q^{1/4}\varepsilon^{1/2}\alpha\beta^{-1}\sum_{\nu=0}^{m}U_{\xi^*}^{2,4\nu+j-2}\beta_{m-\nu} - \sum_{\nu=0}^{m}U_{\xi^*}^{1,4\nu+j}\alpha_{m-\nu} - \\ &- (e-1)\,V_{\eta^*}^{1,4(m-1)+j} - q^{1/4}\varepsilon^{1/2}\alpha\beta^{-1}(e-1)V_{\eta^*}^{2,4m+j-2} \end{aligned} \tag{4.97}$$

out of contact regions $\left(\xi_1^* = 0, \frac{\pi}{2};\ \lfloor\eta_1^*\rfloor < l_1^*\right)$

(b) for functions $V^{2,4m+j}$

$$V^{2,4m+j} = -q^{-1/4}V^{1,4m+j-2} \tag{4.98}$$

out of connection regions $\left(\xi_2^* = 0, \frac{\pi}{2};\ \lfloor\eta_2^*\rfloor < b_2^*\right)$;

$$\begin{aligned} V_{\xi^*}^{2,4m+j} &= \theta\, sign\left(\eta_2^*\right) k_0 \times \\ &\times \left[\begin{array}{l} \sum\limits_{\nu=0}^{m}\left(q^{-1/2}\varepsilon^{-1/2}\beta\alpha^{-1}U_{\xi^*}^{1,4\nu+j}\alpha_{m-\nu} + q^{-1/4}U_{\xi^*}^{2,4\nu+j-2}\beta_{m-\nu}\right) + \\ +q^{-1/2}\varepsilon^{-1/2}(e-1)\beta\alpha^{-1}V_{\eta^*}^{1,4(m-1)+j} + q^{-1/4}(e-1)\,V_{\eta^*}^{2,4(m-1)+j} \end{array}\right] - \\ &- \left[\begin{array}{l} q^{-1/4}\varepsilon^{-1/2}\beta\alpha^{-1}U_{\eta^*}^{1,4m+j} + U_{\eta^*}^{2,4(m-1)+j} + \\ + \sum\limits_{\nu=0}^{m}\left(q^{-1/4}\varepsilon^{-1/2}\beta\alpha^{-1}V_{\xi^*}^{1,4\nu+j}\alpha_{m-\nu} + V_{\xi^*}^{2,4\nu+j-2}\beta_{m-\nu}\right) \end{array}\right] \end{aligned} \tag{4.99}$$

out of connection regions $\left(\xi_2^* = 0, \frac{\pi}{2};\ \lfloor\eta_2^*\rfloor < b_2^*\right)$.

Here

$$b_1^* = \frac{\pi}{2H}\varepsilon^{-1/4}q^{-1/4}\alpha^{-1}l, \quad b_2^* = \frac{\pi}{2H}\varepsilon^{-1/4}q^{-1/4}\beta^{-1}b, \quad \theta = \begin{cases} 1 & \left(\zeta_n^* = 0\right) \\ -1 & \left(\zeta_n^* = \frac{\pi}{2}\right) \end{cases}$$

relations (4.96, 4.97, 4.98, and 4.99) and (4.94) lead to the conclusion that in the contact region the real parts of the functions $\Phi^{1,4m+j}(z_1)$ are given, and at the rest of the boundary the imaginary ones. Similarly, at the stick regions, the real parts of the functions $\Phi^{2,4m+j}(z_2)$ and from them the imaginary ones are given.

The coordinates of the points dividing the regions of sliding and stick are not known a priori and have to be determined while solving the problem. All forces must be continuous in these points and tend to zero for $|y| \to \infty$.

In the case of an orthotropic strip compressed by two rigid stamps with the planar foundations ($f(y) = 0$), one can find (in the principal asymptotic approximation)

$$\Phi^{1,0}(z_1) = A\left(\sinh^2(2z_1) - \sinh^2(2l_1^*)\right)^{-1/2} \tag{4.100}$$

$$\begin{aligned}\Phi^{2,0}(z_2) = {} & \frac{2A}{\pi}\varepsilon^{-1/2}q^{-1/4}\left(\sinh^2(2z_2) - \sinh^2(2b_2^*)\right)^{1/2}\times \\ & \times\left[\int_{-\infty}^{-l_1^*\sqrt{\varepsilon}} \frac{f(t)\,dt}{\sinh(2(t-z_2))} + \int_{l_1^*\sqrt{\varepsilon}}^{\infty} \frac{f(t)\,dt}{\sinh(2(t-z_2))} - \right. \\ & \left. - q^{-1/4}k_0\left(\int_{-l_1^*\sqrt{\varepsilon}}^{-b_2^*} \frac{f(t)\,dt}{\sinh(2(t-z_2))} + \int_{b_2^*}^{l_1^*\sqrt{\varepsilon}} \frac{f(t)\,dt}{\sinh(2(t-z_2))}\right)\right],\end{aligned} \tag{4.101}$$

where

$$f(t) = \left[\left(\sinh^2(2t) - \sinh^2(2b_2^*)\right)\left|\sinh^2(2l_1^*) - \sinh^2(2\varepsilon^{-1/2}t)\right|\right]^{-1/2}$$

The constant A is determined from the condition of the stamp equilibrium

$$A = -\frac{Pq^{-1/4}\varepsilon^{-1/4}\cosh 2l_1^*}{B_1 K \tanh 2l_1^*} \tag{4.102}$$

Here, $K(k)$ is the full elliptic integral of the first kind. The axial reduced force under the stamp will be

$$T_1 = -\frac{Pq^{-1/4}\varepsilon^{-1/4}\pi\cosh 2l_1^*}{2HK\tanh 2l_1^*}\frac{1}{\sqrt{\sinh^2 2l_1^* - \sinh^2 2\eta_1^*}} \tag{4.103}$$

The contact forces T_1 in the main approximation are similar to those in the case of a smooth stamp because the friction coefficient does not affect the function $U^{1,0}$. After transition of $H \to \infty$ in Eq. (4.103), one can find the solution for a smooth planar stamp pressing into the semi-plane:

$$T_1 = -\frac{P}{\pi}\frac{1}{\sqrt{l^2 - y^2}} \tag{4.104}$$

If $H \to 0$, the stress state under the stamp transforms into a homogeneous one

$$T_1 = -\frac{P}{2l} \tag{4.105}$$

In the case of the strip, the condition at infinity leads to the following equation:

$$\int_{l_1^*\sqrt{\varepsilon}}^{\infty} f(t)\cosh(2t)\,dt = q^{-1/4}k_0 \int_{b_2^*}^{l_1^*\sqrt{\varepsilon}} f(t)\cosh(2t)\,dt \tag{4.106}$$

If $H \to \infty$, we obtain the corresponding relation for a semi-plane

$$K\left(b/l\right) = q^{-1/4}k_0 K'\left(b/l\right) \tag{4.107}$$

The considered problem in this limiting case can be also solved by the approximate method proposed by L.A. Galin (1953). The corresponding relation for determination of the connection region was presented as a series by parameter$\varepsilon^{1/4}$. In the principal approximation it coincides with Eq. (4.107).

In the case of a strip, the shear forces at the sliding region can be calculated by the formula

$$S = \mp sign\,(\eta_2^*)kT_1 \tag{4.108}$$

where T_1 is a function of η_1^*. At the connection region:

$$\begin{aligned} S = {} & \frac{GA}{H}\varepsilon^{-1/2}q^{-1/4}\sqrt{\sinh^2 2b_2^* - \sinh^2 2\eta_2^*} \times \\ & \times \left[\int_{-\infty}^{-l_1^*\sqrt{\varepsilon}} \frac{f(t)\,dt}{\sinh(2\left(t-\eta_2^*\right))} + \int_{l_1^*\sqrt{\varepsilon}}^{\infty} \frac{f(t)\,dt}{\sinh(2\left(t-\eta_2^*\right))} - \right. \\ & \left. -q^{-1/4}k_0 \left(\int_{-l_1^*\sqrt{\varepsilon}}^{-b_2^*} \frac{f(t)\,dt}{\sinh(2\left(t-\eta_2^*\right))} + \int_{b_2^*}^{l_1^*\sqrt{\varepsilon}} \frac{f(t)\,dt}{\sinh(2\left(t-\eta_2^*\right))} \right) \right] \end{aligned} \tag{4.109}$$

("–" in (4.51 and 4.52) corresponds to $\xi_2^* = 0$, "+" – to $\xi_2^* = \frac{\pi}{2}$).

The friction dependence of the dimensionless prolongation of the connection region for $\varepsilon^{1/2} = 0.5$ is shown in Fig. 4.3. Curve 2 was obtained in (Sedov, 1980), the curves 1, 3, 4, and 5 correspond to a semi-plane and $2l/H = 1, 2, 5$ respectively. It is seen, that the connection region decreases with a decrease of the friction coefficient k (and disappears when $k = 0$), and increases with decreasing strip width. The distribution of dimensionless shear force $S^* = 2lS/P$ under the stamp is shown in Fig. 4.4 for $\varepsilon^{1/2} = 0.5$, $k = 0.3$, $q = 0.9$.

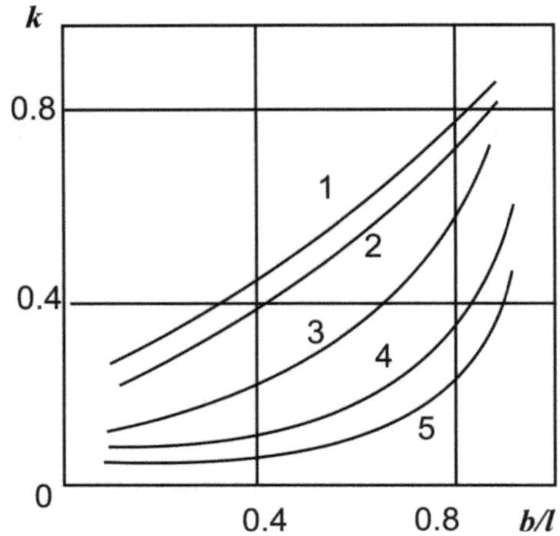

Fig. 4.3 Dependence of the dimensionless length of the connection region on the friction coefficient

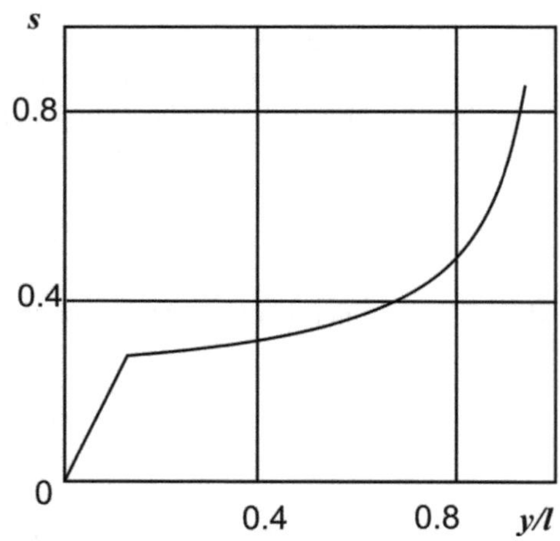

Fig. 4.4 The distribution of dimensionless shear force $S^* = 2lS/P$ under the stamp

4.4 Models of Elastic Foundation

4.4.1 General Equations and Asymptotic Analysis

There are different contemporary models of elastic foundation, which are used for the calculation of elastically supported beams and plates (Vlasov and Leontyev, 1960; Korenev, 1969; Lekhnitsky, 1960; Muravsky, 1967). All these models suggest a certain distribution of stresses (depending on the assumptions made) and can be considered as approximations of a 3D elastic body. But even the most advanced model (Muravsky, 1967) describing elastic foundation directly on the basis of scalar of elasticity (without the very restrictive hypotheses of the Winkler type) does not allow one to describe the stress-strain state near the boundary of elastic semi-space correctly and to satisfy the tangential boundary conditions. We here present the approach based on the asymptotic analysis of the static 3D equations which allows satisfying not only the normal, but also the tangential boundary conditions.

Equilibrium equations for a transversally isotropic elastic body in the case of axisymmetric surface loading have the form:

$$\begin{aligned}&\frac{\partial^2 u}{\partial r^2}+\frac{k\varepsilon}{2}\left(\frac{\partial^2 u}{\partial z^2}+\frac{\partial^2 w}{\partial r\partial z}\right)+\frac{1}{r}\left(\frac{\partial u}{\partial r}-\frac{u}{r}\right)=0,\\&\frac{k}{2}\left(\frac{\partial^2 u}{\partial r\partial z}+\frac{\partial^2 w}{\partial r^2}\right)+\frac{\partial^2 w}{\partial z^2}+\frac{k}{2r}\left(\frac{\partial u}{\partial z}+\frac{\partial w}{\partial r}\right)=0.\end{aligned} \tag{4.110}$$

where r, z – the cylindrical coordinates, u, w – the components of the displacement vectors, $\varepsilon=\frac{E'}{E}$, E, E', G, G' – Young and shear moduli in the plane of isotropy and in the transversal direction; $G=\frac{E}{2}$, $G'=\frac{E'}{2}$; Poisson coefficients are supposed to be equal to zero.

Let us introduce the affine coordinate transformations

$$\text{(a) } r=r_1,\quad z=z_1,\quad u=\varepsilon U_1,\quad w=W_1. \tag{4.111}$$

$$\text{(b) } r=r_2,\quad z=\varepsilon^{1/2}z_2,\quad u=\varepsilon^{1/2}U_2,\quad w=\varepsilon W_2 \tag{4.112}$$

Substitution of (4.111) and then (4.112) into (4.110) leads to two systems of equations

$$\begin{aligned}&\frac{\partial^2 U_1}{\partial r_1^2}+\frac{k}{2}\left(\varepsilon\frac{\partial^2 U_1}{\partial z_1^2}+\frac{\partial^2 W_1}{\partial r_1\partial z_1}\right)+\frac{1}{r_1}\left(\frac{\partial U_1}{\partial r_1}-\frac{U_1}{r_1}\right)=0,\\&\frac{k}{2}\left(\frac{\partial^2 U_1}{\partial r_1\partial z_1}+\frac{\partial^2 W_1}{\partial r_1^2}\right)+\frac{\partial^2 W_1}{\partial z_1^2}+\frac{k}{2r_1}\left(\varepsilon\frac{\partial U_1}{\partial z_1}+\frac{\partial W_1}{\partial r_1}\right)=0,\end{aligned} \tag{4.113}$$

$$\begin{aligned}&\frac{\partial^2 U_2}{\partial r_2^2}+\frac{k}{2}\left(\frac{\partial^2 U_2}{\partial z_2^2}+\varepsilon\frac{\partial^2 W_2}{\partial r_2\partial z_2}\right)+\frac{1}{r_2}\left(\frac{\partial U_2}{\partial r_2}-\frac{U_2}{r_2}\right)=0,\\&\frac{k}{2}\left(\frac{\partial^2 U_2}{\partial r_2\partial z_2}+\frac{\partial^2 W_2}{\partial r_2^2}\right)+\frac{\partial^2 W_2}{\partial z_2^2}+\frac{k}{2r_2}\left(\frac{\partial U_2}{\partial z_2}+\varepsilon\frac{\partial W_2}{\partial r_2}\right)=0\end{aligned} \tag{4.114}$$

We suggest that $E>>E'$, in this case ε can be considered a small parameter. One should mention that, in this case, the definition of essential anisotropy is different from the one used in Sect. 4.3 – the Young moduli are very different there.

In the case of a semi-space, the solution of (4.113) corresponds to a slow varying stress state along the z-axis. This state is realized far enough from the boundary of the semi-space. As for the solutions of system (4.114), these correspond to the stress state, quickly changing along the z-axis, which is localized near the boundary of the semi-space.

The components of the displacement vector can be presented as a sum of these two types of solutions

$$r = u + u_2, \quad w = w_1 + w_2 \tag{4.115}$$

Let us substitute (4.115) into Eqs. (4.110) and into the boundary conditions after transformations (4.111) and (4.112).

Functions U_i, W_i are searched for as an asymptotic series by parameter $\varepsilon^{1/2}$

$$U_i = U_{i0} + \varepsilon^{1/2} U_{i1} + \varepsilon\, U_{i2} + \ldots, \quad W_i = W_{i0} + \varepsilon^{1/2} W_{i1} + \varepsilon\, W_{i2} + \ldots \\ i = 1, 2 \tag{4.116}$$

Here, it is convenient to introduce the preliminary coordinate transformation

$$\zeta_1 = \alpha z_1, \quad \zeta_2 = \beta z_2 \tag{4.117}$$

and to present the values α and β by the following expansions by ε

$$\alpha = \alpha_0 + \alpha_1 \varepsilon + \alpha_2 \varepsilon^2 + \ldots, \quad \beta = \beta_0 + \beta_1 \varepsilon + \beta_2 \varepsilon^2 + \ldots \tag{4.118}$$

Selecting the terms with a similar power of ε, one can find the systems of equations with respect to the functions U_{im}, W_{im} $(m = 0,\ 1,\ 2, \ldots)$ and the corresponding boundary conditions. Coefficients α_0 and β_0 have to be equated to unity because the equations of the main approximations have to coincide with the limiting systems obtained from (4.113) for $\varepsilon \to 0$. As we have shown above, the coefficient $\alpha_k,\ \beta_k$ $(k = 1,\ 2,\ \ldots)$ can also provide a coincidence of all independent equations in the following approximations with the corresponding limiting systems.

As a result, we receive:

(1) for the stress state of the first type

$$\begin{aligned} &\frac{k}{2}\frac{\partial^2 W_{1m}}{\partial r_1^2} + \frac{k}{2r_1}\frac{\partial W_{1m}}{\partial r_1} + \frac{\partial^2 W_{1m}}{\partial \zeta_1^2} = 0, \\ &\frac{\partial^2 U_{1m}}{\partial r_1^2} + \frac{1}{r_1}\left(\frac{\partial U_{1m}}{\partial r_1} + \frac{U_{1m}}{r_1}\right) + \frac{k}{2}\frac{\partial^2 W_{1m}}{\partial r_1 \partial \zeta_1} = f_{1m}, \end{aligned} \tag{4.119}$$

(2) for the stress state of the second type

$$\begin{aligned} &\frac{\partial^2 U_{2m}}{\partial r_2^2} + \frac{1}{r_2}\left(\frac{\partial U_{2m}}{\partial r_2} - \frac{U_{2m}}{r_2}\right) + \frac{k}{2}\frac{\partial^2 U_{2m}}{\partial \zeta_2^2} = 0, \\ &\frac{k}{2}\frac{\partial^2 U_{2m}}{\partial r_2 \partial \zeta_2} + \frac{k}{2}\frac{\partial^2 U_{2m}}{\partial \zeta_2^2} + \frac{\partial^2 W_{2m}}{\partial \zeta_2^2} = f_{2m}. \end{aligned} \tag{4.120}$$

The right hand parts f_{1m} and f_{2m} depend on the functions U_{in}, W_{in} and the quantities α_n, β_n determined in the lower orders of approximation ($n = 0, 1, 2, \ldots, m-1$), i.e. they are the known functions of the coordinates. The boundary conditions for each approximation can be formulated after a selection of the corresponding terms in the expansions of boundary conditions. As this takes place, it is necessary to take the series for stresses together with the expressions (4.116) into account

$$\sigma_z = E'\left[\frac{\partial W_{10}}{\partial \zeta_1} + \varepsilon^{\frac{1}{2}}\left(\frac{\partial W_{11}}{\partial \zeta_1} + \frac{\partial W_{20}}{\partial \zeta_2}\right) + \varepsilon\left(\frac{\partial W_{10}}{\partial \zeta_1}\alpha_1 + \frac{\partial W_{12}}{\partial \zeta_1} + \frac{\partial W_{21}}{\partial \zeta_2}\right) + \ldots\right],$$

$$\tau_{r2} = \frac{E'}{2}\left[\begin{array}{l}\dfrac{\partial U_{20}}{\partial \zeta_2} + \dfrac{\partial W_{10}}{\partial r_1} + \varepsilon^{\frac{1}{2}}\left(\dfrac{\partial U_{21}}{\partial \zeta_2} + \dfrac{\partial W_{11}}{\partial r_1}\right) + \\ +\varepsilon\left(\dfrac{\partial U_{22}}{\partial \zeta_2} + \dfrac{\partial U_{10}}{\partial \zeta_1} + \dfrac{\partial U_{20}}{\partial \zeta_2}\beta_1 + \dfrac{\partial W_{12}}{\partial r_1} + \dfrac{\partial W_{20}}{\partial r_2}\right) + \ldots\end{array}\right].$$

Thus, the solution of the system (4.110) is reduced to the subsequent integration of the equations for the functions W_{im}, U_{im} under corresponding boundary conditions. The integration of the second type equations for every approximation is made after integration of first type equations. Coupling between them is determined by the boundary conditions. Yet, the main constituents of the normal and radial displacements W_{10} and U_{20} are accurate enough for many applications

$$\frac{k}{2}\frac{\partial^2 W_{10}}{\partial r_1^2} + \frac{k}{2r_1}\frac{\partial W_{10}}{\partial r_1} + \frac{\partial^2 W_{10}}{\partial \zeta_1^2} = 0, \qquad (4.121)$$

$$\frac{\partial^2 U_{20}}{\partial r_2^2} + \frac{1}{r_2}\left(\frac{\partial U_{20}}{\partial r_2} + \frac{U_{20}}{r_2}\right) + \frac{k}{2}\frac{\partial^2 U_{20}}{\partial \zeta_2^2} = 0. \qquad (4.122)$$

Suggesting $U_{20} = 0$, one finds the model of elastic foundation proposed in (Muravsky, 1967). The subsequent suggestion

$$W_{10}(r_1, \zeta_1) = \varphi(r_1)\,\psi(\zeta_1)$$

$$\psi(\zeta_1) = \begin{cases} 1 & \text{when } \zeta_1 = 0, \\ 0 & \text{when } \zeta_1 = \mathrm{H} \end{cases}$$

leads to a two-parametric model (Vlasov and Leontyev, 1960) after application of the procedure, corresponding to the Galerkin method applied to Eq. (4.121).

Equations (4.121) and (4.122), which describe the reduced model of elastic foundation, are essentially simpler than the initial system (4.110). Each of them contains only one unknown function. At the same time, they satisfy both normal and tangential boundary conditions on lateral surfaces. It follows from asymptotic analysis that the boundary problem for Eq. (4.121) should be formulated at the beginning. After its solution, the tangential boundary conditions for Eq. (4.122) have to be determined. Naturally, the solution can be corrected by higher-order approximations.

4.4.2 Example – Dynamical Problem

Let us consider, on the basis of the presented approach, a vibration of an infinite plane on the transversally isotropic elastic semi-space subjected to a concentrated impact (Manevitch and Vorob'eva, 1972). This problem was considered in (Vlasov and Leontyev, 1960) using a two-parametric model; the latter does not allow to take the tangential boundary conditions at the contact plane into account.

The equation of motion at cylindrical coordinates looks like

$$D\nabla_r^2\nabla_r^2 v(r,t) = -q(r,t) - m_1 \frac{\partial^2 v(r,t)}{\partial t^2} \tag{4.123}$$

where $\nabla_r^2\nabla_r^2$ – double Laplace operator, D – the cylindrical stiffness of the plane, m_1 – the mass of square surface unity of the plate, $v(r,t)$ – transversal displacement of the plane, $q(r,t)$ – reactive pressure acting on the plane from an elastic foundation (complete adhesion is suggested).

Considering the elastic foundation as a weightless one we describe it by the approximate Eqs. (4.121) and (4.122). The friction between the plate and foundation is not taken into account.

If the plate stiffness is relatively small in comparison to that for the elastic foundation, the initial conditions can be written as follows:

$$v|_{t=0} = 0, \quad \left.\frac{\partial v}{\partial t}\right|_{t=0,\ r=0} = V_0,$$

where V_0 is the velocity of impacting mass at the point of collision.

Then we consider Eq. (4.121) with boundary conditions

$$\left.\frac{\partial W_{10}}{\partial \zeta_1}\right|_{\zeta_1=0} = -\frac{q(r_1,t)}{E'}, \quad W_{10}|_{\zeta_1\to\infty} = 0, \quad W_{10}|_{r_1\to\infty} = 0$$

for computing the function W_{10}. If one presents the solution of (4.121) as

$$W_{10} = \varphi\ (r_1,t)\ \psi\ (\zeta_1) \tag{4.124}$$

and then uses the Galerkin method, the equation for function $\varphi\ (r_1,t)$ obtains the following form:

$$kl_1\left[\frac{\partial^2\varphi\ (r_1,t)}{\partial r_1^2} + \frac{1}{r_1}\frac{\partial\varphi\ (r_1,t)}{\partial r_1}\right] - n_1\varphi\ (r_1,t) = -q\ (r_1,t) \tag{4.125}$$

Here

$$l_1 = \frac{E'}{2}\int_0^\infty \psi^2\ (\zeta_1)\ d\zeta_1, \quad n_1 = E'\int_0^\infty \left(\frac{d\psi\ (\zeta_1)}{d\zeta_1}d\zeta_1\right)$$

and $\psi\ (\zeta_1)$ characterizes the displacement as the function of the dimensionless "thickness" coordinate of the foundation.

Taking the coincidence of the normal displacements of the plate (at point $z = 0$) and the displacements of the elastic foundation into account and excluding the reactive pressure $q(r,t)$ from (4.123) and (4.126), one can find the equation

$$\nabla_{r_1}^2 \nabla_{r_1}^2 v(r_1, t) - 2m^2 \nabla_{r_1}^2 v(r_1, t) + s^4 v(r_1, t) = -m^* \frac{\partial^2 v(r_1, t)}{\partial t^2} \tag{4.126}$$

where

$$m^2 = \frac{kl_1}{2D}, \quad s^4 = \frac{n_1}{D}, \quad m^* = \frac{m_1}{D}.$$

This equation is similar to that of the case of a two-parametric model and has the following solution

$$v(r_1, t) = \frac{V_0}{\omega} J_0(br_1) \sin \omega\, t,$$

with $\omega^2 = \dfrac{s^4 + \lambda^4}{m^*}$, $\quad b = \sqrt{\sqrt{\lambda^4 + \lambda^4} - m^2}$, $\quad \lambda^4$ may be found in accordance with (Vlasov and Leontyev, 1960).

Then, normal displacements in the semi-space (in initial variables) are determined by the formula

$$w(r, z, t) = \frac{V_0}{\omega} J_0(br)\, \psi(z) \sin \omega\, t$$

where $\psi(z)$=1 at z =0 and $\psi(z) \to 0$ when $z \to \infty$, J_n is the Bessel function of the first kind of order n. We do not discuss the higher-order corrections of presentation (4.124) (it may be significant for estimation of the stresses in the semi-space) but will discuss the effect of tangential boundary conditions.

Let us consider Eq. (4.122) in order to compute the shear stresses. Since $\tau_{rz}|_{z=0} = 0$, the corresponding boundary condition has the following form:

$$\left.\frac{\partial U_{20}}{\partial \zeta_2}\right|_{\zeta_2=0} = -\left.\frac{\partial W_{10}}{\partial r_1}\right|_{\zeta_1=0} = -\frac{V_0}{\omega} J_1(br_2) \sin \omega\, t.$$

Besides, $U_{20}|_{\zeta_2 \to \infty} = 0, \quad U_{20}|_{r_2 \to \infty} = 0.$

After application of the Hankel transformation to Eq. (4.122), the solution under the given boundary conditions and inverse transform lead to the following relationship:

$$U_{20} = -\frac{V_0 \sqrt{k}}{\sqrt{2}\omega} J_1(br_2) \exp\left(-b\sqrt{\frac{2}{k}} \zeta_2 \sin \omega\, t\right) \tag{4.127}$$

This solution characterizes the tangential displacements of the semi-space which are not taken into account in the two-parametric model of elastic foundation. The corresponding shear stresses, which decrease quickly apart from the boundary, have the same order of magnitude as the shear stresses dependent on normal z-displacement. Here, the main components of normal stresses σ_r and σ_θ are expressed via function U_{20}.

The next approximations allow to find the corrected values for contact pressure, i.e. estimation of the effect of shear stresses on the main characteristics of the system.

For this, let us consider the second equation of the system (4.120) for $m = 1$. Taking into account (4.127), one can find

$$W_{20} = -\frac{V_0 k^2 b}{4\omega} J_0(br_2) \exp\left(-b\sqrt{\frac{2}{k}}\zeta_2 \sin\omega t\right).$$

Then, one can write the boundary conditions for the first equation of (4.119) for m = 1.

$$\begin{aligned} \left.\frac{\partial W_{11}}{\partial \zeta_1}\right|_{\zeta_1=0} &= -\left.\frac{\partial W_{20}}{\partial \zeta_2}\right|_{\zeta_2=0} = -\frac{V_0 k^{3/2}}{2\sqrt{2}\omega} J_2(br_2) \sin\omega t, \\ W_{11}|_{\zeta_1\to\infty} &= 0, \quad W_{11}|_{r_1\to\infty} = 0. \end{aligned} \tag{4.128}$$

After application of the Hankel transformation to Eq. (4.119) and boundary conditions, we have the following solution:

$$W_{11} = \frac{V_0 k}{2\omega} J_0(br_1) \exp\left(-b\sqrt{\frac{2}{k}}\zeta_1 \sin\omega t\right)$$

The displacement under the plate with account of two approximations is expressed as:

$$w = w_1 + w_2 = W_{i0} + \varepsilon^{1/2} W_{11} = \frac{V_0}{\omega} \left(1 + \varepsilon^{1/2}\frac{k}{2}\right) J_0(br) \sin\omega t.$$

The second term determines the correction to the solution for a corresponding two-parametric model, so that

$$q(r,t) = \frac{E' V_0 b}{\omega} \left(1 - \varepsilon^{1/2}\sqrt{\frac{k}{2}}\right) J_0(br) \sin\omega t.$$

4.4.3 Example – An Axisymmetric Stamp

Let a stamp in the shape of a body of revolution be impressed in a transversely-isotropic half-space (Vorobiova et al., 1979). The contact domain consists of the friction part abutting on the boundary of the contact domain, and the adhesion part. Because of the symmetry, the contact domain and the adhesion part are concentric circles with a common center on the stamp axis. The radius of the circle separating the friction and the adhesion parts are not known beforehand and should be determined during the solution of the problem. It is also required to determine the normal and tangential forces in the contact domain.

The problem is reduced to solving the equilibrium equations for a transversely-isotropic medium

$$\begin{aligned} &\frac{\partial^2 w}{\partial r^2}+\frac{1}{r}\frac{\partial w}{\partial r}+\frac{1}{k}\frac{\partial^2 w}{\partial z^2}+\left(\frac{\partial u}{\partial r}+\frac{u}{r}\right)=0,\\ &\frac{\partial^2 u}{\partial r^2}+\frac{1}{r}\frac{\partial u}{\partial r}-\frac{u}{r^2}+k\varepsilon\frac{\partial^2 u}{\partial z^2}+k\varepsilon\frac{\partial^2 w}{\partial r\partial z}=0,\quad k=\frac{G'}{E'},\quad \varepsilon=\frac{E'}{E}. \end{aligned} \tag{4.129}$$

These equations are more general than (4.110), since they admit an arbitrary Poisson ratio. This system should be solved under the following boundary conditions outside the contact domain *($r > a$),* in the whole contact domain *($r < a$),* on the adhesion part ($0 < r < b$) and on the slip part *($b < r < a$):*

$$\begin{aligned} &\sigma_z=\tau_{rz}=0,\quad r>a,\quad w=-C+f(r),\quad r<a,\\ &u=0,\quad o<r<b,\quad \tau_{rz}=-\rho\sigma_z,\quad b<r<a \end{aligned} \tag{4.130}$$

In addition

$$w\to 0,\quad u\to 0 \text{ when } \sqrt{r^2+z^2}\to\infty.$$

Here, *a.* is the radius of the contact domain, *b* is the radius of the adhesion section, $f(r)$ is a function describing the shape of the stamp, *E* and *E'* are the tension-compression elastic moduli in the plane of isotropy and in the normal direction, *G'* is the shear modulus in a plane normal to the plane of isotropy, *w* and *u* are displacement vector components in the directions of the z and *r* axes, respectively, and σ is the friction coefficient. The Poisson ratio is taken equal to zero.

Let us introduce the transformations of the variables

$$r=r_1,\quad z=k^{-1/2}z_1,\quad w=W^1,\quad u=\varepsilon\,U^1 \tag{4.131}$$

$$r=r_2,\quad z=\varepsilon^{1/2}k^{1/2}z_2,\quad w=\varepsilon\,W^2,\quad u=\varepsilon^{1/2}U^2 \tag{4.132}$$

Substituting (4.131) and (4.132) into (4.129), we obtain, respectively

$$W^1_{rs}+W^1_{zz}+\varepsilon\,k^{1/2}U^1_{sz}=0,\quad U^1_{sr}+\varepsilon\,k^2U^1_{zz}+k^{3/2}W^1_{rz}=0 \tag{4.133}$$

$$\varepsilon\,k^2W^2_{rs}+W^2_{zz}+k^{3/2}U^2_{sz}=0,\quad U^2_{sr}+U^2_{zz}+\varepsilon\,k^{1/2}W^2_{rz}=0 \tag{4.134}$$

Here

$$\varphi_s^l = \frac{\partial \varphi^l}{\partial r_l} + \frac{\varphi^l}{r_l}, \quad \varphi_{sz}^l = \frac{\partial}{\partial z_l}\left(\frac{\partial \varphi^l}{\partial r_l} + \frac{\varphi^l}{r_l}\right), \quad \varphi_{sz}^l = \frac{\partial}{\partial z_l}\left(\varphi_s^l\right),$$

$$\varphi_{rs}^l = \left(\frac{\partial \varphi^l}{\partial r_l}\right)_s = \frac{\partial^2 \varphi^l}{\partial r_l^2} + \frac{1}{r_l}\frac{\partial \varphi^l}{\partial r_l} \qquad (l = 1,\ 2)$$

Let us assume that $E > E' \sim G'$, then ε can be considered as a small parameter in an asymptotic analysis of (4. 133) and (4.134). The solution obtained by asymptotic integration of the system of the first kind (4.133) corresponds to a stress-strain varying relatively slowly along the z axis as compared with the corresponding solution of the system of the second kind (4.134), which is of a boundary layer nature.

Let us represent the displacement vector components in the form of the sum of solutions of both kinds

$$u = u_1 + u_2, \quad w = w_1 + w_2 \tag{4.135}$$

We seek the functions W^l and U^l $(l = 1, 2)$ in the form of an asymptotic series by the parameter $\varepsilon^{1/2}$

$$\begin{aligned} W^l &= \sum_{n=0}^{\infty} \varepsilon^{n/2} W^{l,n} = \sum_{i=0}^{\infty}\sum_{j=0}^{1} \varepsilon^{i+j/2} W^{l,2i+j} \\ U^l &= \sum_{n=0}^{\infty} \varepsilon^{n/2} U^{l,n} = \sum_{i=0}^{\infty}\sum_{j=0}^{1} \varepsilon^{i+j/2} U^{l,2i+j} \end{aligned} \tag{4.136}$$

An additional coordinate transformation is introduced as

$$\zeta_1 = z_1 \sum_{i=0}^{\infty} \varepsilon^i \alpha_i, \quad \zeta_2 = z_2 \sum_{i=0}^{\infty} \varepsilon^i \beta_i \tag{4.137}$$

with undetermined coefficients a_i and β_i $(i = 0, 1, \ldots)$.

Substituting (4.136) and (4.137) into (4.133) and (4.134) and into the boundary conditions, and splitting the expressions obtained in powers of $\varepsilon^{1/2}$, we obtain the following equations and boundary conditions for the functions $W^{l,2i+j}$ and $U^{l,2i+j}$.

The stress-strain state of the first kind is described as:

$$\begin{aligned} W_{rs}^{1,2i+j} + W_{\zeta\zeta}^{1,2i+j} b_0 &= -\sum_{\nu=0}^{i-1}\left(W_{\zeta\zeta}^{1,2\nu+j} b_{i-\nu} + k^{1/2} U_{s\zeta}^{1,2\nu+j} \alpha_{i-1-\nu}\right) \\ U_{sr}^{1,2i+j} &= -k^2 \sum_{\nu=0}^{i-1} U_{\zeta\zeta}^{1,2\nu+j} - k^{1/2} \sum_{\nu=0}^{i} W_{r\zeta}^{1,2\nu+j} \alpha_{i-\nu} \\ b_p &= \sum_{i=0}^{p} \alpha_i \alpha_{p-i} \quad (i = 0,\ 1, \ldots) \end{aligned} \tag{4.138}$$

The boundary conditions are:
for $i = 0, j = 0$

$$W^{1,0} = -C + f(r), \quad r < a, \quad W_{\zeta}^{1,0} = 0, \quad r > a \tag{4.139}$$

for all the remaining i, j

$$\begin{aligned} &W^{1,2i+j} = -W^{2,2(i-1)+j}, \quad r < a \\ &W_{\zeta}^{1,2i+j}\alpha_0 = -\sum_{\nu=0}^{i-1} W_{\zeta}^{1,2\nu+j}\alpha_{i-\nu} - k^{-1}\sum_{\nu=0}^{i} W_{\zeta}^{2,2\nu+j-1}\beta_{i-\nu}, \quad r > a \end{aligned} \tag{4.140}$$

The stress-strain state of the second kind is described as:

$$\begin{aligned} W_{\zeta\zeta}^{2,2i+j}c_0 &= -\sum_{\nu=0}^{i-1} W_{\zeta\zeta}^{2,2\nu+j}c_{i-\nu} - k^{3/2}\sum_{\nu=0}^{i} U_{\zeta}^{2,2\nu+j}\beta_{i-\nu} - k^{2}W_{rs}^{2,2(i-1)+j} \\ U_{sr}^{2,2i+j} + U_{\zeta\zeta}^{2,2i+j}c_0 &= -\sum_{\nu=0}^{i-1} U_{\zeta\zeta}^{2,2i+j}c_{i-\nu} - k^{3/2}W_{r\zeta}^{2,2\nu+j}\beta_{i-1-\nu} \\ c_p &= \sum_{i=0}^{p}\beta_i\beta_{p-i} \end{aligned} \tag{4.141}$$

The boundary conditions are

$$\begin{aligned} U^{2,2i+j} &= -U^{1,2i+j-1}, \quad r < b \\ U_{\zeta}^{2,2i+j}\beta_0 &= -\sum_{\nu=0}^{i-1}\left(U_{\zeta}^{2,2\nu+j}\beta_{i-\nu} + kU_{\zeta}^{1,2\nu+j}\alpha_{i-1-\nu}\right) \\ &\quad - k^{1/2}\left(W_{r}^{2,2i+j} + W_{rs}^{2,2(i-1)+j}\right) \\ &\quad - \rho\sum_{\nu=0}^{i} W_{\zeta}^{1,2\nu+j}\alpha_{i-\nu} + k^{-1}W_{\zeta}^{2,2i+j-1}\beta_{i-\nu}, \quad r > b \end{aligned} \tag{4.142}$$

It should be taken into account that if any upper limit of the summation in (4.138) and (4.140, 4.141, and 142) is negative, then this sum equals zero. Compared, if any function of the second superscript (denoting the number of the approximation) is negative, then this function is zero. For instance, for $i = 0$ we obtain from (4.138) and (4.141):

$$\begin{aligned} &W_{rs}^{1,j} + W_{\zeta\zeta}^{1,j}b_0 = 0, \quad U_{sr}^{1,j} = -k^{3/2}W_{r\zeta}^{1,j} \\ &W_{\zeta\zeta}^{2,j}c_0 = -k^{3/2}U_{s\zeta}^{2,j}\beta_0, \quad U_{sr}^{2,j} + U_{\zeta\zeta}^{2,j}c_0 = 0 \end{aligned}$$

The first equations (for the functions $W^{1,2i+j}$) in system (4.138), as well as the second equations (for the functions $U^{2,2i+j}$) in system (4.141) will be referred to as basic equations.

The following theorem is valid for the basic equations: if the coefficients α_i and β_i are determined by the formulas

$$\begin{aligned}
&\alpha_0 = 1, \quad 2\alpha_{p+1} = k^2\gamma_p + \sum_{m=1}^{p}\left(k^2\gamma_{p-m} - \alpha_{p+1-m}\right)\alpha_m,\\
&\beta_0 = 1, \quad 2\beta_{p+1} = -k^2\delta_p - \sum_{m=1}^{p}\left(k^2\delta_{p-m} + \beta_{p+1-m}\right)\beta_m,\\
&\gamma_0 = \delta_0 = 1, \quad \gamma_n = \alpha_n + k^2\sum_{j=0}^{n-1} b_j\,\gamma_{n-1-j},\\
&\delta_n = \alpha_n + k^2\delta_{n-1} - \sum_{j=0}^{n-1}\delta_j\,c_{n-j}
\end{aligned} \tag{4.143}$$

then the fundamental equations have the form

$$\begin{aligned}
&W_{rs}^{1,2i+j} + W_{\zeta\zeta}^{1,2i+j} = 0, \quad U_{sr}^{2,2i+j} + U_{\zeta\zeta}^{2,2i+j} = 0\\
&i = 0,\ 1,\ \ldots;\quad j = 0,\ 1,\ \ldots
\end{aligned} \tag{4.144}$$

The proof of this statement will be presented below.

We shall henceforth consider the coefficients α_i, β_i to be defined by (4.143). It is seen from the expansions presented that the boundary conditions for the functions $W^{1,2i+j}$ are satisfied in the solution of the first equations of the system (4.138) describing the stress-strain state of the first kind. The functions $U^{1,2i+j}$ are defined as particular solutions of the second equations of this system.

The boundary conditions for the system (4.141), describing the stress-strain state of the second kind, are determined after the appropriate equations of the first kind have been solved. These boundary conditions are satisfied in the solution of the second equations of the system (4.141). The functions $W^{2,2i+j}$ are found as particular solutions of the first equations of this system.

The problem therefore is reduced to a successive integration of (4.144) for the functions $W^{1,2i+j}$ and the functions $U^{2,2i+j}$. Finding the functions $U^{1,2i+j}$ and $W^{2,2i+j}$ is not difficult.

The exact solution of (4.144) can be obtained by using integral transformations (Korenev, 1971). It can be shown that the functions $W^{1,2i+j}$ and $U^{2,2i+j}$ are continuous in the whole domain of definition, and their derivatives are continuous everywhere except at the points $r = b$, $z = 0$ and $r = a$, $z = 0$, where they have integrable singularities. It hence follows that the solution is asymptotic in nature everywhere with the exception of arbitrarily small neighborhoods of the above- mentioned points.

After the functions $W^{l,2i+j}+j$ and $U^{l,2i+j}$ ($l = 1, 2; i = 0, 1, \ldots; j = 0, 1$) have been determined, the stresses σ_z and τ_{rz} are found from the formulas

$$\sigma_z = E' \sum_{i=0}^{\infty} \sum_{j=0}^{1} \varepsilon^{\,i+j/2} \sum_{\nu=0}^{i} \left(k^{\,1/2} W_{\zeta}^{1,2\nu+j} \alpha_{i-\nu} + k^{-1/2} \varepsilon^{\,1/2} W_{\zeta}^{2,2\nu+j} \beta_{i-\nu} \right),$$

$$\tau_{rz} = G' \sum_{i=0}^{\infty} \sum_{j=0}^{1} \varepsilon^{\,i+j/2} \sum_{\nu=0}^{i} \left(k^{\,1/2} \varepsilon\, U_{\zeta}^{1,2\nu+j} \alpha_{i-\nu} + k^{-1/2} U_{\zeta}^{2,2\nu+j} \beta_{i-\nu} \right) + \\ + W_r^{1,2i+j} + \varepsilon W_r^{2,2i+j}$$

The unknown boundary of the adhesion and friction parts are found from the condition of continuity of the tangential stresses on this boundary.

The constant C in (4.139) (the settlement of the stamp) is determined from the stamp equilibrium condition (P is the magnitude of the clamping force)

$$P + 2\pi \int_0^a \sigma_z r dr = 0 \qquad (4.145)$$

As an illustration, let us solve the problem for a stamp with a flat base ($f(r) \equiv 0$). The solution (taking account of one approximation) is reduced by integrating (4.146) with the boundary conditions (4.147).

$$W_{rs}^{1,0} + W_{\zeta\zeta}^{1,0} = 0, \quad U_{sr}^{2,0} + U_{\zeta\zeta}^{2,0} = 0 \qquad (4.146)$$

$$\begin{aligned} &\zeta_1 = 0, \quad W^{1,0} = -C, \quad r < a, \quad W_{\zeta}^{1,0} = 0, \quad r > a, \\ &\zeta_2 = 0, \quad U^{2,0} = 0, \quad r < b, \quad U_{\zeta}^{2,0} = \begin{cases} -\rho W_{\zeta}^{1,0}, & b < r < a \\ -k^{1/2} W_r^{1,0}, & r > a \end{cases} \end{aligned} \qquad (4.147)$$

The function $W^{1,0}$ has the form (the constant C is determined from the condition (4.145))

$$W^{1,0} = -2C\pi^{-1} \arcsin\left[2a \left(\sqrt{\zeta_1^2 + (r+a)^2} + \sqrt{\zeta_1^2 + (a-r)^2} \right)^{-1} \right] \\ C = P\frac{k^{1/2}}{4aE'} \qquad (4.148)$$

We seek the solution of (4.146) in the form ($J_1(x)$ is the first order Bessel function)

$$U^{2,0} = \int_0^{\infty} A(p) \exp(p\zeta_2)\, J_1(pr)\, dp$$

Substituting this expression into the second pair of boundary conditions (4.147) and taking into account (4.148), we obtain a system of dual integral equations to determine the function A(p)

$$\int_0^\infty A(p)J_1(pr)\,dp = 0, \quad r > b$$

$$\int_0^\infty A(p)J_1(pr)p\,dp = \begin{cases} 2C\rho\pi^{-1}\left(a^2 - r^2\right)^{-1/2}, & b < r < a \\ -2Cak^{-1/2}(\pi\, r)^{-1}\left(r^2 - a^2\right)^{-1/2}, & r > a \end{cases}$$

The solution of these equations has the form $K(x)$ which is the complete elliptic integral of the first kind (Korenev, 1971):

$$A(p) = \frac{2C\rho}{\pi^2}\left[-k^{-1/2}\pi\,\sin(px) + B(p)\right]$$

$$B(p) = \frac{2\rho}{a}p\int_b^a K\left(\frac{x}{a}\right)x\sin(px) + pk^{1/2}\int_b^a \ln\left(\frac{a+x}{a-x}\right)\sin(px)\,dx$$

The normal contact stresses and the tangential stresses under the flat stamp are determined by the formulas

$$\sigma_z = -\,(2\pi\,a)^{-1}P\left(a^2 - r^2\right)^{-1/2}, \quad \tau_{rz} = G'k^{-1/2}U_\zeta^{2,0}$$

It is clear from physical considerations that the tangential stresses should be continuous on the interface of the adhesion and slip zones, therefore, the derivative $U_\zeta^{2,0}$ should be continuous on this boundary. We have

$$\zeta_2 = 0, \quad U_\zeta^{2,0} = 2C\rho\,\pi\left(a^2 - r^2\right)^{-1/2}, \quad b \le r \le a$$

$$U_\zeta^{2,0} = \frac{2Ck^{1/2}}{\pi^2}\int_0^\infty \sin(ap)J_1(rp)dp \qquad (4.149)$$

$$+\frac{2C}{\pi}\int_0^\infty B(p)J_1(rp)dp, \quad 0 \le r < b$$

Integrating the inner integral by parts twice, we reduce the last formula to

$$\zeta_2 = 0, \quad U_\zeta^{2,0} = 2C\pi^{-1}\left(G_1 + G_2 + I_1 + I_2\right),$$

$$G_1 = \frac{\rho\pi\,r}{a\sqrt{a^2 - r^2}} + r\left[\frac{k^{1/2}}{a^2 - b^2} - \frac{2\rho}{a}\frac{d}{ab}\left(bK'\frac{b}{a}\right)\right]\left[b + \left(b^2 - r^2\right)^{-1/2}\right]^{-1},$$

$$G_2 = \left[k^{1/2}\ln\frac{a+b}{a-b} - \frac{2\rho}{a}\frac{d}{ab}\left(bK'\frac{b}{a}\right)\right]r\left(b^2 - r^2\right)^{-1/2}\left[b + \left(b^2 - r^2\right)^{-1/2}\right]^{-1},$$

$$I_1 = \int_0^N j_1(pr)\,dp, \quad I_2 = \int_N^\infty f_1(pr)\,dp \quad (N > 0),$$

$$j_1(pr) = 2\left\{\begin{array}{l} \frac{\rho}{a}\int_b^a \frac{d^2}{dx^2}\left[xK\left(\frac{x}{a}\right)\right]\sin(px)\,dx + \\ +a\,k^{1/2}\int_b^a \frac{x\sin(px)}{\left(a^2-x^2\right)^2}\,dx \end{array}\right\}\frac{J_1(pr)}{p}$$

The function G_1 is continuous in r in the interval $0 \le r \le b < a$. The integral I_1 is also continuous in this interval (as a definite integral of the continuous function f_1). The improper integral I_2 converges uniformly (for large p the function j_1 admits the estimate $f_1 < B/p_2$ (Erdelyi, 1956)), and therefore converges to a continuous function. The function G_2 undergoes a discontinuity at $r = b$. Hence, $G_2 \equiv 0$ is necessary for the continuity of the derivative $U_\zeta^{2,0}$. Consequently

$$k^{3/2}\ln\frac{1+b/a}{1-b/a} = 2\rho\frac{b}{a}K\left(\frac{b}{a}\right) \tag{4.150}$$

Relationship (4.150) determines the boundary of the adhesion and friction parts not known earlier.

After simple, but awkward manipulations, we obtain the following formula for the tangential stresses on the adhesion part ($\Pi_1\ (n, x)$ is the complete elliptic integral of the third kind):

$$\tau_{rz} = \frac{1}{2}P\frac{k^{3/2}}{\pi^2a^2t}\left(1-t^2\right)^{-1/2}$$
$$\times\left[\begin{array}{l} \rho\pi\,k^{-1/2}t + \left(\beta_*^2-t^2\right)^{1/2} \\ \times\left[\beta_*^{-1}\ln\frac{(1+/\beta_*)}{1-\beta_*} - 2\frac{\rho k^{-1/2}t^2\Pi_1(t^2-1,\,\beta_*)}{(1-t^2)}\right]\left(1-t^2\right)^{-1/2} \\ -\ln\frac{\left(1-t^2\right)^{-1/2}+\left(\beta_*^2-t^2\right)^{1/2}}{\left(1-t^2\right)^{1/2}-\left(\beta_*^2-t^2\right)^{1/2}} \end{array}\right]$$
$$t = r/a, \quad \beta_* = b/a \quad \left(0 \le t \le \beta_*\right)$$

On the slip part

$$\tau_{rz} = \frac{1}{2}P\frac{\left(1-t^2\right)^{-1/2}}{\pi^2a^2t} \quad \left(\beta_* \le t < 1\right)$$

The dependence of the quantity b/a (the ratio of the adhesion part radius to the stamp radius) on the friction coefficient p for $k = 1/3$ is shown in Fig. 4.5 (curve 1) and the distribution of the dimensionless tangential stresses $T_1 = \tau_{rz}2\pi\,a/P$ in the contact domain is presented in Fig. 4.6 for $k = 1/3$, $\rho = 0.3$, The point $\beta_* = b/a$ separates the adhesion and friction parts.

It should be noted that in the solution obtained (taking just one approximation into account), the singularity in the contact stresses on the boundary or the contact domain has the form $(a - r)^{-1/2}$., while the exact solution of the problem in the presence of Coulomb friction should contain the singularity $(a - r)^{-1/2+\theta(\rho,\varepsilon)}$ exactly as for the plane problem (this follows from the fact that the equations of three-dimensional elasticity theory reduce the plane problem and complex shear in the neighborhood of the singular line (Cherepanov, 1974)). Therefore, the expression for θ is known (Galin, 1953). The series expansion in ε has the form

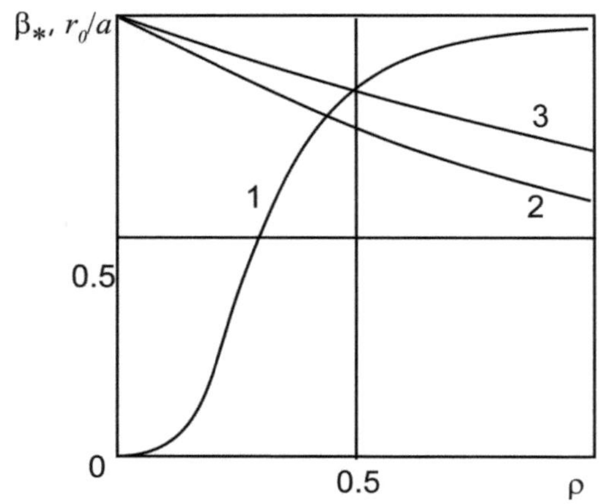

Fig. 4.5 Dependence of the ratio of the adhesion part radius to the stamp radius on the friction coefficient

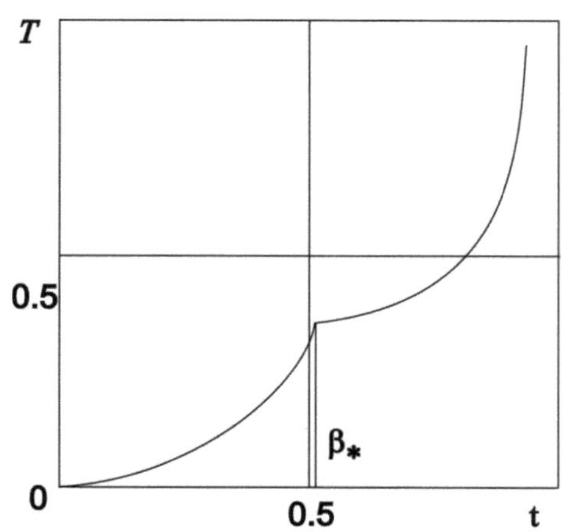

Fig. 4.6 Distribution of the dimensionless tangential stresses in the contact domain

$$(a-r)^{-1/2+\theta} = (a-r)^{-1/2} \times \left\{ \begin{array}{l} 1+\varepsilon^{-1/2}\rho k^{-1/2}\ln(a-r) \\ +\varepsilon\left[\rho k^{3/2}\ln(a-r)+1/2\rho^2 k(a-r)\right]\ldots \end{array} \right\} \tag{4.151}$$

The singularity above is obtained with the first term of the expansion. Subsequent approximations will evidently contain appropriate corrections.

The relative error of any partial sum of series (4.151) becomes arbitrarily large as $r \to a$. Meanwhile, uniform accuracy in the whole contact domain can be achieved by a "matching" of the approximate and "singular" solutions, which has the form

$$\sigma_z^* = A(a-r)^{-1/2+\theta}$$

The constant factor A is determined from the matching conditions which are given as follows: both the approximate and singular solutions and their derivatives with respect to r should agree in a certain neighborhood of $r = r_0$ i. e., for $r = r_0$, $z = 0$

$$\sigma_z = \sigma_z^*, \quad \partial\sigma_z/\partial r = \partial\sigma_z^*/\partial r \tag{4.152}$$

Conditions (4.152) permit the determination of the radius r_0, and the constants of the approximate and singular solutions, in combination with the integral equilibrium condition for the stamp. The dependence of the position of the matching line on the coefficient p is shown in Fig. 4.5 for $k = 1/3$ and $\varepsilon = 1$ (curve 2) and $\varepsilon = 1/3$ (curve 3). To obtain the solution which is uniformly suitable in the whole domain, the singular solution must be used for $r_0 \leq r \leq a$. Incidence of the matching points in the adhesion zone ($r_0 < b$) indicates the need to take account of higher approximations. It follows from Fig. 4.5 that this holds for large ρ ($\rho > 0.4$), and for real friction coefficients ($\rho < 0.3$) the zones in which the use of the singular solution is necessary are small and occupy less than 20% (on the radius) of the contact domain even when taking just one approximation into account.

Let us show that if the coefficients α_i and β_i are determined from (4.143), then the fundamental equations have the form (4.144). We prove this for the system (1.138). The proof is by induction.

For $i = 0$ we have

$$W_{rs}^{1,j} + W_{\zeta\zeta}^{1,j} b_0 = 0, \quad U_{sr}^{1,j} = -k^{3/2} W_{r\zeta}^{1,j} \alpha_0 \tag{4.153}$$

In order for the first equation of the system (4.153) to have the form (4.144), it is sufficient to set $\alpha_0 = 1$ (hence $b_0 = 1$).

For $i = 1$

$$\begin{aligned} W_{rs}^{1,2+j} + W_{\zeta\zeta}^{1,2+j} &= -W_{\zeta\zeta}^{1,j} b_1 - k^{1/2} U_{s\zeta}^{1,j} \alpha_0 \\ U_{sr}^{1,2+j} &= -k^2 U_{\zeta\zeta}^{1,j} - k^{3/2} W_{rs}^{1,j} \alpha_i - k^{3/2} W_{r\zeta}^{1,2j} \alpha_0 \end{aligned} \tag{4.154}$$

Let us integrate the second equation of (4.153) with respect to r and differentiate with respect to ζ. We obtain

$$U_{sr}^{1,j} = -k^{3/2} W_{\zeta\zeta}^{1,j} \alpha_0 \tag{4.155}$$

With account of the condition at infinity, we set the arbitrary function, which appears in the integration, equal to zero. After substituting (4.155) into the first equation of (4.154), we find

$$W_{rs}^{1,2+j} + W_{\zeta\zeta}^{1,2+j} = W_{\zeta\zeta}^{1,j}\left(b_1 - k^2\alpha_0^2\right)$$

Equating coefficients of $W_{\zeta\zeta}^{1,j}$ to zero (and taking into account that $\alpha_0 = 1$ and $b_1 = 2\,\alpha_1$), we obtain $\alpha_1 = k^2/2$.

Therefore, the theorem is valid for $i = 0{,}1$. Let us assume the theorem to be valid for $i \le p$, and let us prove it for $i = p + 1$.

We show that if the functions $W^{1,2i+j}(i = 0, 1, \ldots, p)$ satisfy the first equation in (4.144), then the functions $U^{1,2i+j}$ $(i = 0, 1, \ldots, p)$, determined from the second equations of the system (4.138), will satisfy the following equation

$$U_{rs}^{1,2i+j} + U_{\zeta\zeta}^{1,2i+j} = 0 \tag{4.156}$$

If the functions $U^{1,2i+j}$ are found from the equation

$$U_{rs}^{1,2i+j} = f_2$$

where f_2 satisfies (4.156), then $U^{1,2i+j}$ is the solution of this equation. Indeed,

$$U^{1,2i+j} = r^{-1}\int\left(r\int j_2 dr\right)dr$$

$$U_{rs}^{1,2i+j} + U_{\zeta\zeta}^{1,2i+j} = r^{-1}\int\left(r\int j_{sr} + j_{2\zeta\zeta}dr\right)dr = 0$$

It therefore remains to show that the right sides of the second set of equations of system (4.138) satisfy (4.156) for $i \le p$.

If the function $W^{1,2i+j}$ satisfies (4.144), then its derivative $W_r^{1,2i+j}$ satisfies (4.156). The right side of the second equation of (4.138) now evidently satisfies (4.156) for $i = 0$, and this is proved by induction for $i = 1, 2, \ldots, p$.

In accordance with (4.156), we substitute the quantity $U_{rs}^{1,2i+j}$ into the second set of (4.138) instead of $U_{\zeta\zeta}^{1,2i+j}$, and then we integrate the relationships obtained with respect to r and differentiate with respect to ζ. We obtain

$$U_{s\zeta}^{1,2i+j} = k^2\sum_{\nu=0}^{i-1} U_{s\zeta}^{1,2\nu+j} b_{i-1-\nu} - k^{3/2}\sum_{\nu=0}^{i} W_{\zeta\zeta}^{1,2\nu+j}\alpha_{i-\nu} \quad (i = 0, 1, \ldots, p) \tag{4.157}$$

System (4.157) can be considered as a system of linear algebraic equations in $U_{s\zeta}^{1,2i+j}$. This system has a triangular matrix of coefficients with nonzero elements along the principal diagonal, and is therefore solvable everywhere. The solution of the system (4.157) has the form

$$U_{s\zeta}^{1,2i+j} = -k^{3/2} \sum_{\nu=0}^{i} W_{\zeta\zeta}^{1,2\nu+j} \gamma_{i-\nu} \quad (i = 0,\ 1,\ \ldots, p) \tag{4.158}$$

where the quantities $\gamma_n, n = 0, 1, \ldots$ are determined by the recursion formulas n (4.143).

Substituting (4.158) into the $(p + 1)$-th fundamental equation of the system (4.138), we obtain

$$W_{rs}^{1,2(p+1)+j} + W_{\zeta\zeta}^{1,2(p+1)+j} = -\sum_{\nu=0}^{p} W_{\zeta\zeta}^{1,2\nu+j} \left(b_{p+1-\nu} - k^2 \sum_{m=0}^{p-\nu} \alpha_m \, \gamma_{p-\nu-m} \right) \tag{4.159}$$

Because of the selection of the coefficients α_i mentioned in (4.143), all the coefficients on the right side of (4.158) vanish. Indeed, from the relationship for $\alpha_{p+1-\nu}$ it follows according to (4.143)

$$\sum_{m=0}^{p+1-\nu} \alpha_m \, \alpha_{p+1-\nu-m} - \sum_{m=0}^{p-\nu} k^2 \alpha_m \, \gamma_{p-\nu-m} = 0$$

i.e., the expression in the parentheses in (4.159) vanishes.

Therefore, the basic equations of the system (4.138) have the form of the first relationship of (4.144). The proof for system (4.141) is similar.

4.5 On the Concept of Solids

As it was mentioned in the Chap. 1, the concept of the elastic body has been initially formed in the beginning of nineteenth century on the basis of discrete molecular models and following the transition to the elastic continuum in spite of the vague understanding of atoms and molecules at that time. Afterwards, this approach has been replaced by a purely continuous description well coordinated with our intuitive macroscopic concept of solids. The corresponding field theory is specific to solids namely because it requires a possibility to introduce a rigid reference state and the existence of a static shear modulus (contrary to liquids). The latter property can be considered as a consequence of the former one, and, in turn, may be acceptable as a macroscopic definition of solids. Such a definition includes all types of solids (crystals, glasses, rubbers).

However, the observed macroscopic properties of the mentioned solids are strongly different. From the first view, the behavior of glasses is very similar to that of polycrystals except for the transition to the liquid state (melting for crystals or polycrystals with an abrupt change of density and symmetry seems strongly different from glass-liquid transition without similar phenomena). But, as it is well-known, glasses possess also a series of low-temperature anomalies (Anderson,

1984), such as a linear temperature dependence of the heat capacity for low temperatures, which are absent for crystals or polycrystals. As for rubbers, their most pronounced and evident features are very low Young and shear moduli compared to crystals and glasses. It is clear that transition to the microscopic level is necessary to explain and understand the specific behavior of different types of solids.

It seems that the most natural starting point for microidentification of a glass as a solid is the interpretation of a rigid reference frame as the rigid reference configuration of the constituent particles (Alexander, 1998).

In the case of a crystalline solid such an interpretation turns out to be valid if the constituent particles are atoms. Then, one can write the potential energy of crystal (and corresponding equations of motion), taking interatomic interaction into account. The continuum theory of elasticity is just a long wavelength and low frequency limit of discrete atomic theory. It is not surprising that such a program reflected in a clear and complete form in (Born and Huang, 1954) has been actually realized a short time after the reliable ascertainment of the atomic nature of matter (Born and von Karman, 1912).

However, this approach is not applicable for other types of solids (e.g. rubbers or gels) which are not rigid and are fluid-like at microscopic (molecular) level. Therefore, the approach mentioned above cannot be directly used, and the understanding of rubber elasticity has been achieved in a totally different way, using the methods of statistical physics for macromolecules (Flory, 1953).

As for glasses, it seems that the only principal structural difference from crystal is the randomness of a microscopic reference frame. But from this viewpoint it is impossible to understand the reasons of the low-temperature anomalies mentioned above. From the other side, the commonly accepted understanding of glasses as high-viscous liquids contradicts to an intuitively clear apprehension of both glasses and crystals as solids.

So, a very complicated problem arises: how to coordinate all these contradictive facts and ideas?

In spite of the evident importance of this problem for the understanding of one of the key scientific notions, it seems that the most general and deep analysis which takes into account the main previous achievements was undertaken by Alexander (1998)

First of all, the unification of crystals and glasses as solids is underlined by the realization of the role of permutation as the real active symmetry of liquids, which is destroyed by solidification. Therefore, the solid is described as a unique state which realizes one special permutation of the constituent atoms on the lattice sites – out of N! possible ones. Namely, this uniqueness allows labeling the particles in the solid by their respective equilibrium positions (i.e. to efficiently introduce the Lagrange variables). It also allows to avoid "the complex averaging between different configurations which is essential in the theory of real fluids – and the corresponding complex entropy considerations" (Alexander, 1998). In particular, this difference between solids and liquids "is illustrated by the dramatic difference in the description of the phase transition between the two phases. Melting, the solid–to-liquid transition is predicted quite accurately and almost quantitatively by the simple and very intuitive Lindemann criterion. On the other hand, even the best microscopic descriptions of freezing, the liquid-to-solid transition, are complicated, controversial and inaccurate" (Alexander, 1998).

One can add that a melting crystalline solid can be identified as an instability of the lattice (if one excludes the role of boundaries and spatial inhomogenities). In Lagrange variables, the solid is characterized by the labeled positional correlation functions

$$S_{ij}(\tau) =< r_i(t) * r_j(t+\tau) >= R_i * R_j + < u_i(t) * u_j(t+\tau) > \tag{4.160}$$

where $S_{ij}(\tau)$ are the correlation functions between the particles labeled by their equilibrium positions; the $u_i(t)$-deviation of ith particle from its equilibrium position.

Such a presentation is meaningless for a liquid because of the active permutation of the particles. It is the main reason for the introduction of Euler's variables instead of Lagrange's which are adequate for solids.

Naturally, this means that the processes in which permutation of the particles is manifested (such as diffusion) are not taken into account.

So, if one deals with a crystalline solid, there are no problems in the development of the lattice theory of elasticity with further transition to a corresponding continuum theory in the long wavelength and small-frequency approximation. Such a procedure is a direct generalization of that for a one-dimensional oscillatory chain which was demonstrated above. In addition, if it is necessary, the alternative long wavelength and high-frequency as well as short wavelength continuum approximations may be taken into account.

In the case of such disordered media as polymer networks the situation is quite different. The reason is that contrary to the crystalline solid it is impossible to introduce a rigid microscopic reference frame at the atomic level because the polymer network as well as gel is not solid but liquid at this level. The possibility to justify our intuitive perception of rubber as solid is provided by existence of the nets leading to the formation not only of the volume but also of the shear rigidity. It is rather clear in common theory of rubber elasticity, but such a theory in its continuum limit is strongly different from the classical theory of elasticity. However, as it is shown in (Alexander, 1998), the theory of elasticity of polymer networks can be combined with the theory of elasticity if one considers the initial state as a stressed one. Initial stresses, which are usually small enough in crystals or polycrystals, have to be taken into account if dealing with such complicated solids as polymer networks. Namely, initial stresses (e.g. a negative pressure) provide a mechanism of shear stiffness formation strongly different from the usual one caused by angle deformation. The negative pressure providing shear stiffness is only a "network" part of the full pressure. The latter also includes a "liquid" part, so that their sum is equal to zero (Alexander, 1998).

The statement about the role of initial stresses is universal for all disordered solids, although the manifestations of their influence are rather different. If we deal, e.g., with glasses which are intuitively grasped as solids, the effect of internal stresses can be traced in the solidification process. It is a rather complicated relaxation process in a quenched liquid leading to formation of "composite" media consisting of compressed (more dense) regions and tensed (less dense, "soft") regions. From an energetic viewpoint the relaxation process may be considered as wandering at the surface of constant energy in a state space, between saddles to a local minimum. The presence of these "soft" domains allows describing the universal low temperature anomalies in the physical properties of glasses (Anderson et al., 1972).

4.6 Models of Non-Fourier Heat Conduction

Some systems and processes are described by continuous models so complicated that even if direct numeric simulation of the microscopic dynamics is possible, it is still difficult to validate the macroscopic model. The reason is that it is not clear which results of the simulation can be compared with the predictions of the model. The following example of non-Fourier heat conduction models demonstrates how the modal reduction of the continuous equations can allow a direct comparison with the numeric simulation data.

It is well-known that the Fourier law of heat conduction implies an infinite speed of signal propagation and is thus inconsistent with causality (Cattaneo, 1958; Vernotte, 1958; Chandrasekharaiah, 1986, 1998; Christov and Jordan, 2005). Numerous modifications were suggested to recover the hyperbolic character of the heat transport equation (Chandrasekharaiah, 1986, 1998). Perhaps, the most well-known is the lowest-order approximation known as the Cattaneo-Vernotte (CV) law (Cattaneo, 1958; Vernotte, 1958). In its one-dimensional version it is written as

$$\left(1 + \tau \frac{\partial}{\partial t}\right) \vec{q} = -\kappa \nabla T \tag{4.161}$$

where κ is the standard heat conduction coefficient and τ is the characteristic relaxation time of the system. The latter can be of macroscopic order (Christov and Jordan, 2005). Importance of the hyperbolic heat conduction models for description of a nanoscale heat transfer has been recognized in (Heino, 2007; Shiomi and Maruyama, 2006).

Only a few papers dealt with the numeric verification of such laws from the first principles (Volz et al., 1996). As it is well-known now from numerous numeric simulations and few analytic results, the relationship between the microscopic structure and applicability of the Fourier law is highly nontrivial and depends both on size and number of dimensions of the model in question (Lepri et al., 2003). To the best of our knowledge, no similar analysis has been performed to test the hyperbolic models of heat conduction.

Here we will concentrate on the study of size and temperature effects on the non-stationary heat conduction in two simple one-dimensional models with a conserved momentum – Fermi-Pasta-Ulam (FPU) chain and a chain of rotators (CR). From the perspective of stationary heat conduction, these two systems are known to belong to different universality classes. Namely, in the FPU chain the heat conduction coefficient diverges with the size of the system (Lepri et al., 1997), whereas in the CR model it converges on a finite value (Gendelman and Savin, 2000; Giardina, 2000; Savin and Gendelman, 2001). So, it is interesting to check whether and how this difference between the models reveals itself in the problem of non-stationary heat conduction.

In order to investigate this process, one should choose the parameters to measure. This question is not easy, since the simplest CV law already has two independent coefficients, whereas more elaborate approximations can include even more constants. Besides, we would not want to restrict ourselves to a particular approximate equation. Instead, it seems reasonable to look for some quantity which will characterize the process of non-stationary conduction and can be measured from the simulations without relying on a particular approximate equation. For this sake, we choose the characteristic *length* which characterizes the scale at which the non-stationarity effects are significant.

In order to explain the appearance of this scale, let us refer to a 1D version of the CV equation for temperature:

$$\tau \frac{\partial^2 T}{\partial t^2} + \frac{\partial T}{\partial t} = \alpha \frac{\partial^2 T}{\partial x^2} \tag{4.162}$$

where α is the temperature conduction coefficient.

Let us consider the problem of non-stationary heat conduction in a one-dimensional specimen with periodic boundary conditions $T(L, t) = T(0, t)$, where T(x,t) is the temperature distribution, L is the length of the specimen, t>0. If this is the case, one can expand the temperature distribution to Fourier series:

$$T(x, t) = \sum_{n=-\infty}^{\infty} a_n(t) \exp\left(\frac{2\pi inx}{L}\right) \tag{4.163}$$

with $a_n(t) = a^*_{-n}(t)$, since T(x,t) is a real function.

Substituting (4.163) with (4.162), one obtains the equations for the time evolution of the modal amplitudes:

$$\tau \ddot{a}_n + \dot{a}_n + \frac{4\pi^2 n^2 \alpha}{L^2} a_n = 0 \tag{4.164}$$

The solutions of Eq. (4.164) are written as:

$$\begin{aligned} a_n(t) &= C_{1n} \exp(\lambda_1 t) + C_{2n}(\lambda_2 t) \\ \lambda_{1,2} &= \frac{1}{2\tau}\left(-1 \pm \sqrt{1 - \frac{16\pi^2 n^2 \alpha\tau}{L^2}}\right) \end{aligned} \tag{4.165}$$

where C_{1n} and C_{2n} are constants determined by the initial distribution.

From (4.165) it immediately follows that for sufficiently short modes the temperature profile will relax in an oscillatory manner:

$$
\begin{aligned}
& n > \frac{L}{4\pi\sqrt{\alpha\tau}} \\
& a_n(t) \sim \exp\left(-\frac{t}{2\tau}\right)\exp(i\omega_n t),\ \omega_n = \frac{1}{2\tau}\sqrt{\frac{16\pi^2 n^2 \alpha\tau}{L^2} - 1}
\end{aligned}
\tag{4.166}
$$

If the specimen is rather long ($L >> 4\pi\sqrt{\alpha\tau}$), then for small wavenumbers (acoustic modes):

$$
\begin{aligned}
\lambda_1 &\approx -\frac{1}{\tau} \\
\lambda_2 &\approx -\frac{4\pi^2 n^2 \alpha}{L^2}
\end{aligned}
\tag{4.167}
$$

The first eigenvalue describes a fast initial transient relaxation, and the second one corresponds to a stationary slow diffusion and, quite naturally, does not depend on τ. So, we can conclude that there exists a critical length of the mode

$$
l^* = 4\pi\sqrt{\alpha\tau} \tag{4.168}
$$

which separates between two different types of the relaxation: oscillatory and diffusive. The oscillatory behavior is naturally related to the hyperbolicity of the system. Existence of this critical scale characterizes the deviance of the system from the proper Fourier law and does not depend on a particular choice of the model.

In order to measure l^* without relying on any particular empiric equation of the non-stationary heat conduction, the numeric experiment should be designed in order to simulate the relaxation of thermal profile to its equilibrium value for different spatial modes of the initial temperature distribution. For this sake, we simulate the chain of particles with conserved momentum with the Hamiltonian

$$
H = \sum_n \frac{1}{2}\dot{u}_n^2 + V(u_{n+1} - u_n) \tag{4.169}
$$

for n=1...N and with periodic boundary conditions. In order to obtain the initial nonequilibrium temperature distribution, all particles in the chain were embedded in the Langevin thermostat. For this sake, the following system of equations was simulated:

$$
\begin{aligned}
& \ddot{u}_n = V'(u_{n+1} - u_n) - V'(u_n - u_{n-1}) - \gamma_n \dot{u}_n + \xi_n \\
& n = 1 \ldots N
\end{aligned}
\tag{4.170}
$$

where γ_n is the relaxation coefficient of the nth particle and the white noise ξ_n is normalized by the following conditions:

$$
\langle \xi_n \rangle = 0,\ \langle \xi_n(t_1)\xi_k(t_2) \rangle = 2\sqrt{\gamma_n \gamma_k} T_n \delta_{nk}(t_1 - t_2) \tag{4.171}
$$

where T_n is the prescribed temperature of the nth particle. The numeric integration has been performed for γ_n=0.1 for every n and within the time interval t=250. After that, the Langevin thermostat was disabled and relaxation of the system to a stationary temperature profile was studied for various initial distributions T_n for two particular choices of the nearest – neighbor interaction described above (FPU and chain of rotators):

$$V_1(x) = \frac{1}{2}x^2 + \frac{1}{4}x^4 \tag{4.172}$$

$$V_2(x) = 1 - \cos x \tag{4.173}$$

In order to study the relaxation of different spatial modes of the initial temperature distribution, its profile has been prescribed as

$$T_n = T_0 + A\cos\frac{2\pi(n-1)}{Z} \tag{4.174}$$

where T_0 is he average temperature, A – amplitude of the perturbation, Z – the length of the mode (number of particles). The overall length of the chain L has to be a multiple of Z in order to ensure the periodic boundary conditions. The results were averaged over 10^6 realizations of the initial profile in order to reduce the effect of fluctuations.

A typical result of the simulation is presented in Fig. 4.7a, b. The chain of rotators of the same length N = 1,024 and the same modal wavelength Z = 64 demonstrates a qualitatively different relaxation behavior for different temperatures – the oscillatory one for lower temperature and the smooth decay for higher temperature. This observation suggests that the critical wavelength, if it exists, should decrease with the temperature increase. However, its existence should be checked for constant temperature and varying wavelength.

Such simulations are presented in Fig. 4.8 (for the chain of rotators) and Fig. 4.9 (for the FPU chain). In both models, one observes oscillatory decay for the short wavelengths and smooth exponential decay for relatively long waves. It means that for both models there exists some critical wavelength l^* which separates two types of the decay and thus the effect of the non-stationary heat conduction is revealed.

The results presented in Fig. 4.9 allow one to conclude that the critical wavelength for the FPU chain for a given temperature may be estimated as $512<l^*<1024$. The interpretation of Fig. 4.8 is not that straightforward. It is clear that $32<l^*<128$, but for Z = 64 the result is not clear. Within the accuracy of the simulation, it seems that only a finite number of the oscillations is observed. Such behavior is not consistent with the CV equation, since expressions (4.166 and 4.167) suggest either an infinite number of the oscillations (4.166), or single crossing of the

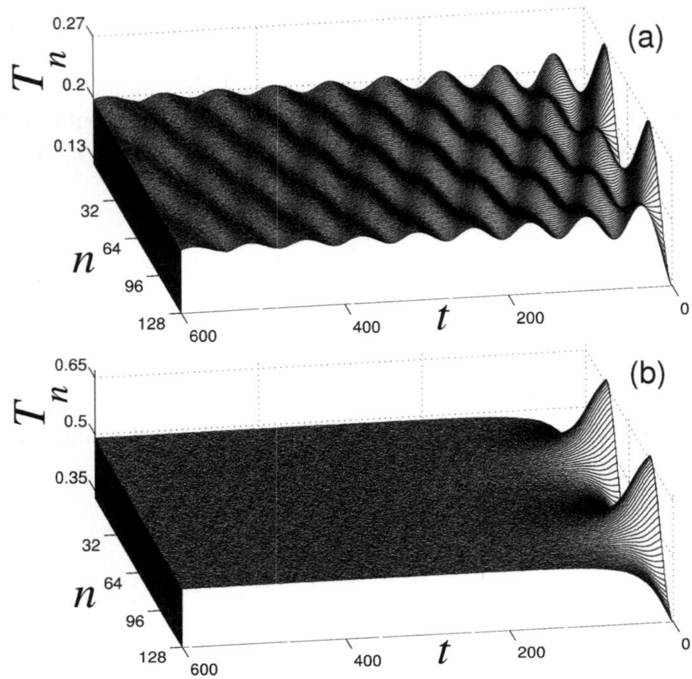

Fig. 4.7 Relaxation of the initial periodic thermal profile in the chain of rotators, Z=64, L=1024, (**a**) T_0=0.2, A=0.05 (oscillatory decay) and (**b**) T_0=0.5, A=0.15 (smooth decay of the initial thermal profile)

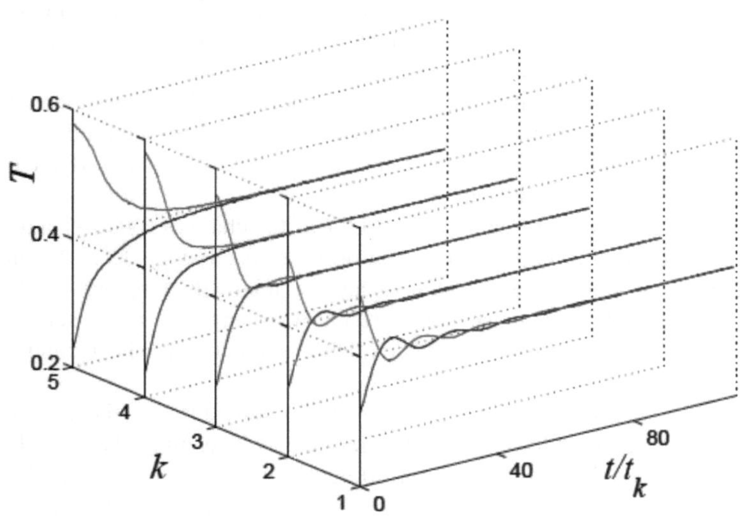

Fig. 4.8 Evolution of the relaxation profile in the chain of rotators with change of the mode length Z. Time dependence of the mode maximum T(1+Z/2) (*lines starting from above, red online*) and minimum T(1) (*lines starting from below, blue online*) are depicted with average temperature T_0=0.4 and $Z=2^{k+3}$, k=1,. . .,5, scaling time $t_k=2^{k-1}$. For all simulations the length of chain is L=1,024

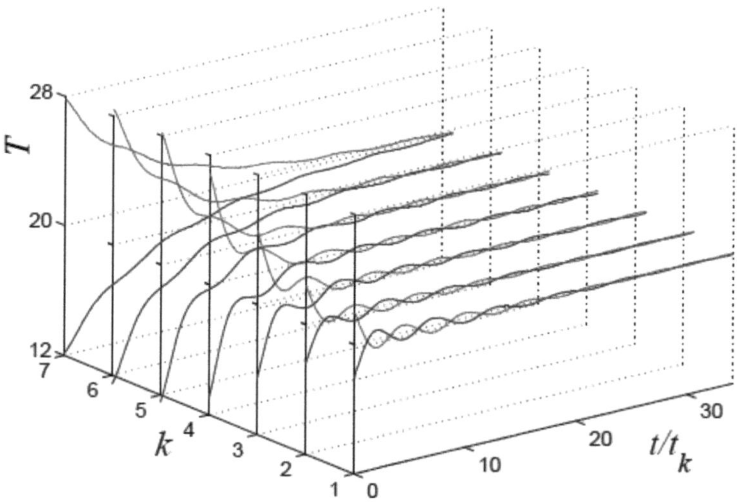

Fig. 4.9 Evolution of the relaxation profile in the FPU chain with change of the mode length Z. Time dependence of the mode maximum T(1+Z/2) (*lines starting from above, red online*) and minimum T(1) (*lines starting from below, blue online*) are depicted with average temperature $T_0=20$ and $Z=2^{k+3}$, k=1,. . .,7, scaling time $t_k=2^{k+1}$, length L=1,024

average – temperature level, or no such crossing at all (4.167). A possible explanation is that if the modal wavelength is close enough to the critical, the low-order CV model is not sufficient and more nonlocal effects should be taken into account.

The latter observation has motivated us to inspect another prediction of the CV model – the independence of the amplitude decrement of the relaxation profile on the wavelength in the oscillatory regime (4.166). The results of simulation are presented in Fig. 4.10 (chain of rotators) and Fig. 4.11 (FPU). One can see that for the chain of rotators the above prediction more or less corresponds to the simulation results. For the FPU chain the decrement is strongly dependent on the wavelength, and at odds with the CV model.

So, the effects of the non-stationary heat conduction can be easily revealed in simple one-dimensional models of dielectrics. There exists a critical modal wavelength l* which separates between oscillating and diffusive relaxation of the temperature field; existence of such a critical scale is inconsistent with the Fourier law. If the size of the system is close to this critical scale, more exact models should be used for the computation of the non-stationary heat flow. In both models studied the critical size decreases with the temperature increase. As for the CV model itself, it seems in the chain of rotators it is inconsistent with the simulations in the vicinity of the critical wavelength. In the FPU chain, this model is also wrong in the oscillatory regime. One can speculate that this difference between the two models is related to their difference with respect to the stationary heat conduction – saturating versus size dependent (Lepri et al., 2003).

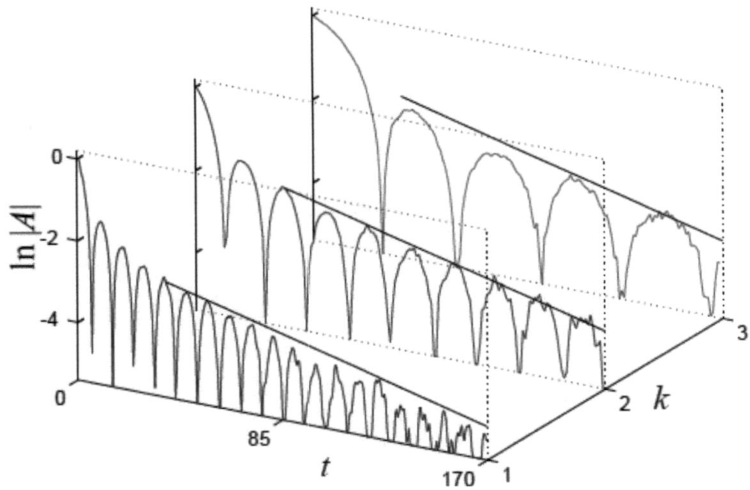

Fig. 4.10 Exponential decay of the normalized oscillation amplitude in the chain of rotators $A(t)=(T(1+Z/2)(t)-T_0)/A)$ for the average temperature $T_0=0.3$, initial amplitude $A=0.05$ and different periods of the thermal profile $Z=2^{k+3}$, k=1, 2, 3. The straight lines illustrate the decay of the maximum envelope according to $A=\exp(-\lambda t)$ with universal value $\lambda=0.015$ for all three simulations

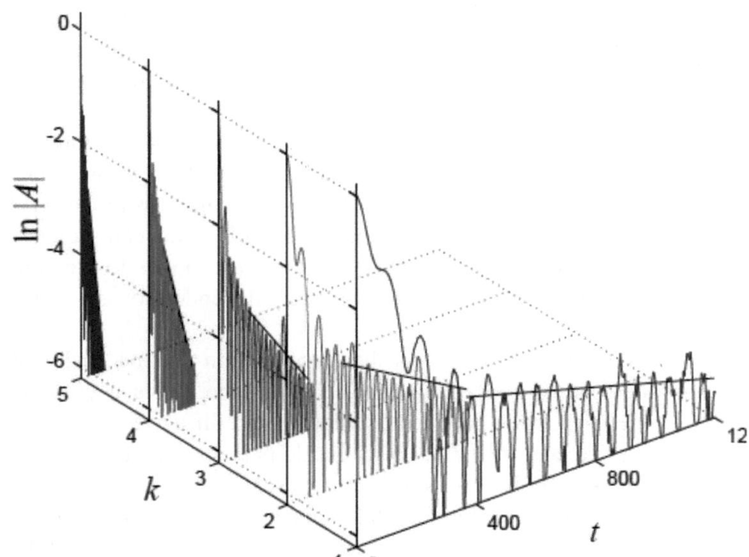

Fig. 4.11 Exponential decay of the normalized oscillation amplitude in the FPU chain $A(t)=(T(1+Z/2)(t)-T_0)/A$ for the average temperature $T_0=10$, initial amplitude $A=0.05$ and different periods of the thermal profile $Z=2^{k+3}$, k=1,. . .,5. The straight lines illustrate the decay of the maximum envelope according to $A=\exp(-\lambda t)$ with values $\lambda=0.0015$\$, 0.003, 0.008, 0.024 and 0.06 for k=1, 2, 3, 4 and 5

4.7 Concluding Remarks

Different parts of this chapter comprise a series of problems from simple enough (classical and Timoshenko's beams) to those of moderate complexity, and finally to very complicated contact problems. We would like to demonstrate that even in relatively simple cases, the genuine understanding requires a deeper analysis leading to the construction of tractable models. This way, one can understand the nature of transversal vibrations in the elastic beam. Contrary to longitudinal and torsional vibrations, the transversal vibrations contain both a "periodic" component (the only present in the infinite beam and under simple support conditions) and two boundary layers which can be approximately separated in accordance with Bolotin's idea.

As for the Timoshenko beam, it becomes clear that this model contains several limiting cases and cannot be considered simply as a generalization of the classical beam theory. The latter is approximately valid for thin beams only; however, even in this case, one can obtain the high frequency (shear type) limit, which does not exist in the classical theory. So, four different tractable models "pop up" and make the Timoshenko model completely understandable.

While considering the elastic planar dynamical problem, essentially more complex than the dynamics of one-dimensional systems, the tractable models describe different types of dynamic behavior close to pure dilatation or pure shear. Such an approach allows clarifying the physical nature of corresponding mathematical simplifications. Besides, close connection is established between the dynamics of the classical beam and Raleigh waves which are commonly considered as completely different phenomena. At last, the substantiation of one-dimensional tractable models is obtained.

In the case of the contact problem, we deal with a situation in which an analytical solution is impossible if one directly treats the initial formulation of the boundary problem. We show that, although the preliminary derivation of the tractable models is rather complicated, the final boundary problems have a clear physical meaning and admit an efficient analytical study.

The last section of this chapter is devoted to a very significant problem of another kind involving the relationship between the dynamical (continuum) and the thermodynamical description of the energy transfer in an oscillatory chain. It is shown that dynamical analysis substantiates the modification of the common equation of heat conduction if fast transient processes are involved.

References

Alexander, S.: Amorphous solids: their structure, lattice dynamics and elasticity. Phys. Rep. **296**, 65–263 (1998)

Anderson, P.W.: Basic Notions of Condensed Matter Physics. Benjamin/Cummings, Menlo Park, CA (1984)

Anderson, P.W., Halperin, B.I., Varma, C.M.: Anomalous low-temperature thermal properties of glasses and spin glasses. Philos. Mag. 25, 1 (1972)

Andrianov, I.V., Awrejcewicz, J., Manevitch, L.I.: Asymptotical Mechanics of Thin-Walled Structures. Springer, New York, NY (2004)

Arnold, V.I.: Mathematical Models of Classical Mechanics. Springer, New York, NY (1980)

Bolotin, V.V.: Asymptotic method of study of the eigenvalue problems for rectangular domains. In: The Problems of Continuum Mechanics, pp. 60–72. Gostehizdat Moscow (1961)

Born, M., Huang, H.: Dynamical Theory of Crystal Lattices. Oxford University Press, Oxford (1954)

Born, M., von Karman, Th.: Uber Schwingungen in Raumgittern. Physik. Z. **13**, 297–309 (1912)

Cattaneo, C.: Sur une forme de l'équation de la chaleur éliminant le paradoxe d'une propagation instantanée. C. R. Acad. Sci. **247**, 431 (1958)

Chandrasekharaiah, D.S.: Thermoelasticity with second sound: a review. Appl. Mech. Rev. **39**, 355–377 (1986)

Chandrasekharaiah, D.S.: Hyperbolic thermoelasticity: a review of recent literature. Appl. Mech. Rev. **51**, 705–730 (1998)

Cherepanov, G.P.: Mechanics of Brittle Fracture [in Russian]. Nauka, Moscow (1974)

Christov, C.I., Jordan, P.M.: Heat conduction paradox involving second-sound propagation in moving media. Phys. Rev. Lett. **94**, 154301 (2005)

Downs, B.: Transverse vibration of a uniform, simply supported Timoshenko beam without transverse deflection. Trans. ASME, Ser. E, J. Appl. Mech. **43**, 671–674 (1976)

Erdelyi, A.: Asymptotic Expansions. Dover, New York, NY (1956)

Flory, P.J.: Principles of Polymer Science. Cornell University Press, Ithaca, NY (1953)

Galin, L.A.: Contact Problems of Theory of Elasticity [in Russian]. Gostechizdat, Moscow (1953)

Gendelman, O.V., Savin, A.V.: Normal heat conductivity of the one-dimensional lattice with periodic potential of nearest-neighbor interaction. Phys. Rev. Lett. **84**, 2381–2384 (2000)

Giardina, C., Livi, R., Politi, A., Vassalli, M.: Finite thermal conductivity in 1D lattices. Phys. Rev. Lett. **84**, 2144 (2000)

Guntze, R.: Ermittlung transversaler Ergenfreguenzen dunner Rechteckscheiben und Vergleich mit der Nahezung on Timoshenko. Acta Mechanica. **7** (1969)

Heino, P.: Thermal conduction simulations in the nanoscale. J. Comput. Theor. Nanosci. **4**, 896–927 (2007)

Huang, T.C.: The effect of rotatory inertia and of shear deformation on the frequency and normal mode equations of uniform beams with simple end conditions. Trans. ASME, Ser. E, J. Appl. Mech. **83**, 579–584 (1961)

Korenev, B.G.: Introduction to the Theory of Bessel Functions [in Russian]. Nauka, Moscow (1971)

Korenev, B.G.: Constructions on elastic foundations. In: Building Mechanics in the USSR 1917–1967 [in Russian]. Stroijizdat, Moscow (1969)

Lekhnitsky, S.G.: Theory of Elasticity of Anisotropic Body [in Russian]. GITTL, Moscow (1960)

Lepri, S., Livi, R., Politi, A.: Heat conduction in chains of nonlinear oscillators. Phys. Rev. Lett. **78**, 1896–1899 (1997)

Lepri, S., Livi, R., Politi, A.: Thermal conduction in classical low-dimensional lattices. Phys. Rep. **377**, 1–80 (2003)

Manevich, L.I., Vorob'eva, N.I.: On approximate equations of the axisymmetric problem of elasticity theory for a transversely-isotropic foundation. Prikl. Mekh. **8**, 33–38 (1972)

Manevitch, L.I., Pavlenko, A.V., Koblik, S.G.: Asymptotic Methods in a Theory of Elasticity of Orthotropic Bodies [in Russian]. Vyscha Shkola, Kiev (1982)

Muravsky, G.B.: About model of elastic foundation. Build. Mech. **6** (1967) [in Russian]

Nowacki, W.: Theory of Elasticity [in Russian]. Mir, Moscow (1975)

Prescott, J.: Elastic waves and vibrations of thin rods. Phil. Mag. **33**, 225 (1942)

Savin A.V., Gendelman, O.V.: Heat conductivity in the chain of rotators. Fiz. Tverd. Tela. (Leningrad) **43**(341) (2001). [Sov. Phys. Solid State **43**, 355–364 (2001)]

Sedov, L.I.: Two-Dimensional Problems in Hydrodynamics and Aerodynamics [in Russian]. Nauka, Moscow (1980)

Shiomi, J., Maruyama, S.: Non-Fourier heat conduction in a single-walled carbon nanotube: classical molecular dynamics simulations. Phys. Rev B. **73**, 205420 (2006)

Timoshenko, S.P.: On the transversal vibrations of bars of uniform cross-sections. Phil. Mag. **43**, 253 (1922)

Timoshenko, S.P.: Engineering Mechanics. McGraw-Hill, New York, NY (1956)

Vernotte, P.: Les paradoxes de la theorie continue de l'equation de la chaleur. C. R. Acad. Sci. **246**, 3154 (1958)

Vlasov, V.Z., Leontyev, N.N.: Beams, Plates and Shells on Elastic Foundation [in Russian]. Fizmatgiz, Moscow (1960)

Volz, S., Saulnier, J.B., Lallemand, M. Transient Fourier-law deviation by molecular dynamics in solid argon. Phys. Rev. B. **54**, 340–347 (1996)

Vorobiova, N.I., Koblik, S.G., Manevitch, L.I.: Axisymmetric contact problem taking account of adhesion and slip. J. Appl. Math. Mech. **43**, 590–598 (1979)

Afterword

Difficulties mastered are opportunities won.
Winston Churchill

There is no reason to repeat in the afterword what this book is about – the authors made their best efforts to say that in the body of the monograph. There is also no reason to explain once more why the tractable models are desirable, interesting, or useful. We hope that the reader is more or less convinced at this point. So, perhaps, here is a good opportunity to say a few words about some important issues which were almost completely omitted in the book – intentionally or not.

It seems that, after all, the most important and controversial issue of this sort is a problem of mathematical rigor. As it was mentioned above a few times, there are two basic ways to get the tractable model – to make simplifying physical assumptions from the beginning and to get the model immediately or to simplify an already existing mathematical model by means of some asymptotic procedure. The former way poses no problem for mathematical rigor – the "mathematics" starts from already tractable equations. In this case, the lack of mathematical rigor is buried in physical assumptions.

In the case where the asymptotic reduction appears, one could be interested in the accuracy of the approximation obtained. For instance, one can enquire whether the series obtained will converge and how many terms should be kept in order to obtain the required accuracy. If the series is asymptotic, one can enquire how many terms should be kept in the expansion in order to achieve optimal accuracy. In the book, these enquiries remain almost completely unanswered, except the problem of asymptotic decomposition of the anisotropic elastic problem in Chap. 4, and few exact solutions presented in other places.

This lack of answer is, of course, not occasional. For nonlinear problems, explicit computation of high-order corrections is extremely complicated, if at all possible; clarification of the asymptotic nature of the expansion obtained may be even more difficult. In nonlinear dynamics there are only a handful of nontrivial problems where the convergence of the perturbation series can be rigorously proven. Even if this is the case, the estimations of the convergence radius are sometimes extremely pessimistic. For instance, in some particular examples based on the Kolmogorov-Arnold-Moser theorem, the convergence is proven for $\varepsilon \sim 10^{-50}$ – hardly of any

L.I. Manevitch, O.V. Gendelman, *Tractable Models of Solid Mechanics*, Foundations of Engineering Mechanics, DOI 10.1007/978-3-642-15372-3, physical significance. Still, even in these marginal examples, it turns out that the asymptotic expansions yield reasonable results – even if the small parameter is of order unity. The only way to establish that is to compare the approximate formula with the results of some experiment – physical and/or numeric. From a mathematical viewpoint, such comparison, of course, proves nothing: the "coincidence" with the numeric simulation can easily blow up if the higher-order terms are taken into account.

In addition, there exists a well-known principle in experimental physics – the accuracy of computations should not exceed the accuracy of measurements, in order not to create a false impression of high accuracy. There exist some problems in physics, where exceptional experimental accuracy can be achieved – for instance, a measurement of the anomalous magnetic moment of an electron. The relative error of these measurements can be substantially less than 10^{-10}; needless to say, similar accuracy is required from the theory which describes this phenomenon. That is why the laborious calculations of high-order approximations in quantum electrodynamics are absolutely necessary. Unfortunately, in mechanics such a level of experimental accuracy is far beyond the achievable limits. That is why sometimes the computation of high – order corrections may be even harmful – it can cause the false impression of an improved overall accuracy, whereas the accuracy of the underlying physical assumptions or available experimental techniques does not allow such improvement at all. To conclude, the authors believe that the tractable model, which is based on clear physical assumptions, allows computations in explicit or at least comprehensible form and demonstrates reasonable agreement with available experimental data, has the right to exist in science despite the lack of mathematical rigor – especially if there is no other alternative.

According to popular saying, as there are no grains without straw, there is no book without mistakes. The authors will be very grateful to any reader who will let them know about the mistakes, misprints or who would like to share with them any other comments. Please send such correspondence to Professor Oleg Gendelman, Faculty of Mechanical Engineering, Technion – Israel Institute of Technology, Haifa, 32000, Israel, e-mail: ovgend@tx.technion.ac.il

Index

MIX
Papier aus verantwortungsvollen Quellen
Paper from responsible sources
FSC® C105338

If you have any concerns about our products,
you can contact us on
ProductSafety@springernature.com

In case Publisher is established outside the EU,
the EU authorized representative is:
Springer Nature Customer Service Center GmbH
Europaplatz 3, 69115 Heidelberg, Germany

Printed by Libri Plureos GmbH
in Hamburg, Germany